MODERN
MICROPROCESSORS
THIRD EDITION

V. KORNEEV
A. KISELEV

CHARLES RIVER MEDIA, INC.
Hingham, Massachusetts

Editor: David Pallai
Cover Design: The Printed Image

CHARLES RIVER MEDIA, INC.
10 Downer Avenue
Hingham, Massachusetts 02043
781-740-0400
781-740-8816 (FAX)
info@charlesriver.com
www.charlesriver.com

This book is printed on acid-free paper.

V. Korneev and A. Kiselev. *Modern Microprocessors Third Edition.*
ISBN: 1-58450-368-8

Library of Congress Cataloging-in-Publication Data
Korneev, V. V. (Viktor Vladimirovich)
  [Sovremennye mikroprotsessory. English]
  Modern microprocessors / V. Korneev and A. Kiselev.— 3rd ed.
      p. cm.
  Includes bibliographical references and index.
  ISBN 1-58450-368-8 (pbk. : alk. paper)
  1. Microprocessors. I. Kiselev, A. (Andrew), 1965- II. Title.
  TK7895.M5K67 2004
  621.39'16—dc22
                          2004012322

Printed in the United States of America
04 7 6 5 4 3 2 First Edition

Table of Contents

Introduction

The invention of the first microprocessor by Intel® in 1971 ushered in the era of widespread computerization. "Thanks to microprocessors, computers have become a universal, commonly available commodity," said Ted Hoff, one of the inventors of the first microprocessor. He, along with his colleagues Federico Faggin and Stan Mazor, has been inducted into the United States National Inventors Hall of Fame, and the invention itself has been declared one of the greatest achievements of the twentieth century.

In just over a quarter century, microprocessors have truly come a long way. The first Intel chip, 4004, operated at 750 KHz, had 2,300 transistors, and cost about $200. Its performance was 60 thousand operations per second. In 2000, it was stated that microprocessor Alpha from DEC® company held the record figures of 700 MHz, 15.2 million transistors, 2 billion operations per second, and a price of $300. To date, processor clock frequency has exceeded 2 GHz, the number of transistors has gone over 50 million, and the peak performance is more than 7 billion operations per second.

Comparing these figures of microprocessor industry advancements confirms the assessment attributed to the Intel founder and the chairman of its board of directors Gordon Moore: "If the car industry had been progressing at the rate the semiconductor industry has, today a Rolls-Royce would cost three dollars, could get half-a-million miles per gallon of gas, and would be cheaper to throw away than to pay for parking."

In a society in which information processing is becoming the main labor activity, such intensive technological development has been caused by the increased demand on the new work tools: computers. Computerization is one of the main directions of scientific and technical progress and is its concentrated expression. The quantity and quality of computers that a country builds and its computerization level in the most diverse areas is one of the main criteria of its economic and military potential.

Every year a group of U.S. experts compiles a list of critical technologies that encompasses practically all directions of manufacturing, research, and design that affect the military and economic status of the country. Microelectronic technologies traditionally take the first place in this list.

Microprocessors, the most complex microelectronic devices, incorporate the most advanced achievements of scientific and technological thought. In the environment of aggressive business competition and huge capital investments inherent to this industry, producing each new microprocessor model is in one way or another associated with the next scientific, engineering, or technological breakthrough; within the individual model family, however, the designers try to use the standardized, worked-out solutions and preserve software compatibility among models to the extent possible.

Studying such an intensively developing and science-intensive subject area as microelectronics in general and microprocessor equipment in particular is quite an interesting and difficult task, requiring one to constantly perfect and replenish his knowledge and be familiar with the developments in the related scientific and technological areas. To solve application problems effectively, any modern professional computer technology specialist must have a clear idea of the current state and the future development prospects of the hardware base.

The task of elucidating the state of a dynamically changing subject area is a difficult one, inevitably resulting in the famous Achilles and Tortoise paradox. Understanding all the difficulty of the task, the authors, nevertheless, attempt to lay out the main ideas, concepts, and directions that define the development of the modern microelectronic technologies to the fullest extent possible. However, the main goal of the book is to get readers interested in the given area and induce them to further their self-education.

The book comprises five chapters. Chapter 1 expounds the concepts; the other chapters provide explanatory reference.

Chapter 1 provides general understanding about the main directions in which microprocessor architecture, structure, and functional organization are developing. The current state and development prospects of the modern computing technology elementary base—memory chips, programmable logic integrated circuits, and universal and signal microprocessors—are considered. Architectural and structural methods of increasing integrated circuit operating speed, as well as the practiced approaches to measuring processor performance, are analyzed. The main concepts of constructing superscalar microprocessors, long-instruction word processors, and promising multithread microprocessors are also set forth.

Chapter 2 deals with the CISC and RISC architecture microprocessors and gives general characteristic of the modern universal microprocessors. Several microprocessors from manufacturers from different countries are considered.

In the other chapters, problem-oriented microprocessors that are specialized for signal processing, multimedia applications, and the creation of massively parallel computer systems and neural processing are considered.

Chapter 3 presents main families of signal microprocessors from Motorola, Analog Devices, and Texas Instruments as well as new signal-processing devices—media and communications microprocessors. Specifics of digital signal processing (DSP) and hardware-support means of implementing DSP algorithms in signal and media microprocessors are considered. Comparative analyses are made of the effectiveness of signal microprocessor architectures when performing typical signal-processing tasks.

Chapter 4 describes transputer-type microprocessors oriented toward use in massively parallel computer systems. Their architectural and structural specifics are considered in detail.

Chapter 5 is devoted to organization of neural network computations and implementation of neuroprocessors.

The readers should answer the review questions at the end of each chapter to ensure quality assimilation of the material.

The book is intended for undergraduate and graduate students studying informatics and computer science, as well as for all those who hold interest in modern computer technology.

1 Modern Processor Architecture

1.1 THE PROCESSOR AS A DIGITAL INFORMATION-PROCESSING DEVICE

1.1.1 Digital Information Processing

In a book devoted to exploring microprocessors, it is logical to expect answers to the questions what is a processor, what kind of processors are there, and where among them microprocessors belong. The answer to the first question is that a processor is a digital information-processing device. Digital or discrete information-processing devices manipulate *data elements* that have a finite set of discrete values. Each data element is defined by at least two attributes: a name that sets this element apart from the rest, and a value. For example, a data element representing a bit of information can take on only one of two values, namely 0 or 1.

Data elements are stored in *memory elements*. To store a data element that can take on 2^n distinct values, an ordered set of n memory elements can be used, $n \in \{1, 2, ...\}$, each of which can store one bit of information. Thus, for $n = 3$, the set of distinct values consists of the following eight members: 000, 001, 010, 011, 100, 101, 110, and 111. To store the value 011, for example, the first memory element holds 0, the second holds 1, and the third also holds 1.

All digital devices are built using logic elements, called *gates*, and memory elements. Gates implement logic functions, the first of which that must be mentioned

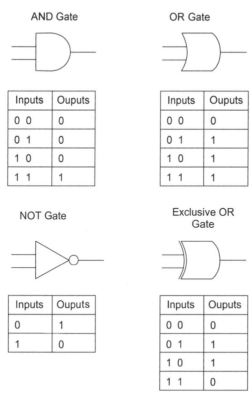

FIGURE 1.1 Graphical gate symbols and their truth tables.

are two-variable functions AND, OR, and exclusive OR and a one-variable function NOT. Symbolic graphical representation of the gates and the functions they implement are shown in Figure 1.1.

Gate inputs and outputs assume logic values 0 or 1, which can be represented, for example, by voltage levels when implemented electronically or by pressure levels when implemented using pneumatic media. Which voltage levels represent logic 1 and 0 is decided by convention.

Gates are used to build digital device blocks, such as triggers, for example, which are one of the memory types that holds one information bit. One possible trigger schematic diagram is shown in Figure 1.2, *a*.

When values 1 and 0 are applied to the input 1 and input 0, respectively, the trigger's output 1 assumes value 1, which sets the trigger's state to logic 1. The trigger will remain in this state until it is switched into logic 0, by placing logic 0 and 1 on its inputs 1 and 0, respectively. In the 0 state, the trigger's output 1 assumes value 0. The trigger's property to preserve its state is achieved by the feedback links. Figure 1.2, *b* shows a graphic symbol for a trigger.

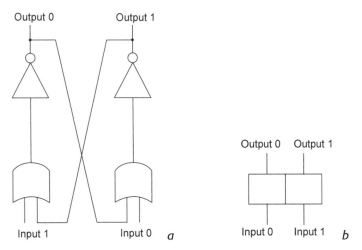

FIGURE 1.2 Trigger: *a*—schematic diagram, *b*—graphic symbol.

For the trigger state changes to take place after signals have been placed on its inputs, clock signals are used. *Clock signal* is a periodic pulse sequence. For the sake of simplicity, let's assume that a pulse is represented by logic 1, with the gap between pulses corresponding to logic 0. Moreover, the time during which a pulse is maintained equals $\tau/2$, where τ is the duration of one *clock cycle*: A single complete traversal of the pulse, from the rising edge, through the time when the value of the clock is 1, through the falling edge, through the time when the value is 0, until the start of the next rising edge. The reciprocal of the cycle time, the $1/\tau$ value, is called *clock frequency*. It is also sometimes called *clock rate* or *clock speed*.

Figure 1.3 shows a trigger whose output state changes only when the clock signal is at logic 1.

It is intuitively clear that the higher the clock frequency, the faster a digital device is. However, the highest clock frequency a gate can stably operate at is determined by the physical characteristics of the gate's primary building blocks. When this clock frequency is exceeded, the clock pulses are distorted to an unacceptable degree and levels 0 and 1 cannot be reliably differentiated.

Next in importance after the trigger logic block is the *register*, which in the simplest case is an ordered collection of triggers. Registers are used to store data elements. For example, a register consisting of one trigger (a degenerative case) stores one bit; a register consisting of eight triggers stores eight bits, or one byte; a register consisting of 16 triggers stores two bytes, or a word; and, accordingly, a register consisting of 32 triggers stores a double word. The concepts of a bit, a byte, a word, and a double word have become commonly accepted in the computer industry.

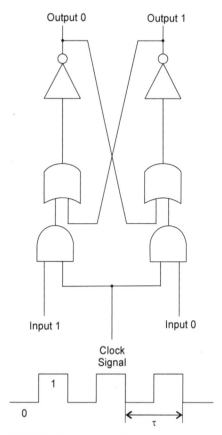

FIGURE 1.3 A clocked trigger.

Besides triggers, memory elements based on other physical principles are used to store data: capacitor-charge conservation, magnetic domain-orientation conservation, and other physical effects. Memory elements are grouped into *cells*, each of which stores one data element—a byte, a word, or a double word, for example. An ordered sequence of memory cells forms a *memory module*. Which type of data elements a memory cell stores (a bit, a byte, a word, and so on) is decided when the memory is designed. The important thing is that each memory cell has a unique address, which is used to access individual cells during memory operations. The value that a memory cell with the address A holds is usually represented as [A]. Two memory operations have been defined:

■ Writing a data element to a memory cell, with the data source and the address of the memory cell to which the data is to be written specified in the write operation instruction.

■ Reading a data element from a memory cell, with the address of the memory cell and the register into which the read-out data is to be placed specified in the read operation instruction

An important memory module characteristic is its *size*, which is defined as the maximum number of data elements it can store.

Actually, to understand how a processor is organized, it is also necessary to examine how data is exchanged between registers. Figure 1.4 shows a *register file* consisting of $n + 1$ registers (each register is represented by one trigger) and a multiplexer. The *multiplexer* is a block that transfers a value stored in the trigger of one arbitrarily selected register to the corresponding trigger of another register or to the corresponding triggers of several registers in this file.

To transfer data between the registers, a logic 1 must be placed on one input of the AND gate. The other input of the AND gate is connected to the output of the register trigger, whose state is being transferred. At the same time, logical 0s must be placed on the corresponding inputs of the AND gates that are connected to the outputs of the triggers of the rest of the registers. The output of the OR gate, to which the outputs of the previously mentioned AND gates are connected, is connected to one of the inputs of each of the AND gates. The outputs of the AND gates are connected to the corresponding inputs of all triggers of the file's registers. When a logic 1 is placed on the other input of the AND gate connected to a trigger's input, this trigger switches into the same state as the trigger whose state is being transferred.

Consequently, data is transferred between registers by placing control signals on the multiplexer's control inputs to select the register whose contents are to be transferred and placing corresponding control signals to set the triggers of the registers to which the data is to be sent. A desired data transfer from register R_i to register R_j can be written as $R_j \leftarrow R_i$, $i \neq j$, $i \in \{0, 1, ..., n\}$, $j \in \{0, 1, ..., n\}$. This notation can be called an interregister data *transfer instruction*.

It must be noted that a wired OR can be used in the multiplexer instead of an OR gate. In this case, the outputs of all AND gates connected to the triggers' outputs are connected together to a bus. Accordingly, inputs of the gates used to set the triggers of the register file are also connected to this bus. Figure 1.5 shows a schematic for implementing interregister transfers using a wired OR.

Using some initial number of primary modules built from AND, NOT, OR, and XOR gates, more complex modules can be built, from which still more complex modules can be constructed. Any digital data-processing device can be constructed in this way [1]. For example, Figure 1.6, *a* shows a schematic for an adder for two bits of binary numbers stored in corresponding registers. This circuit adds two i^{th} order bits and a carry bit from the $i - 1$ position, producing an ith order sum and a carry bit to the $i + 1$ position. This type of addition circuit is called a *full adder*. A flowchart of adding two binary numbers is shown in Figure 1.6, *a*.

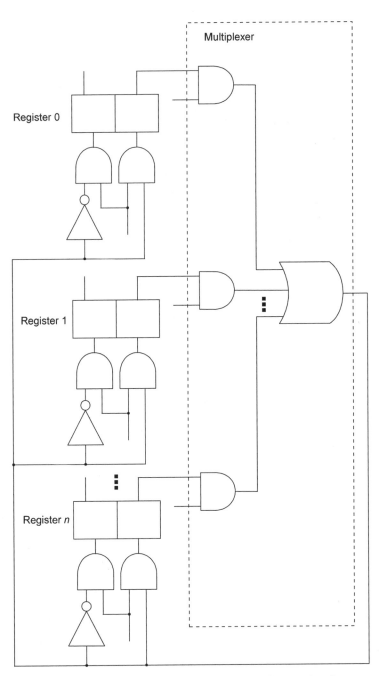

FIGURE 1.4 Schematics of a multiplexer for transferring data between registers.

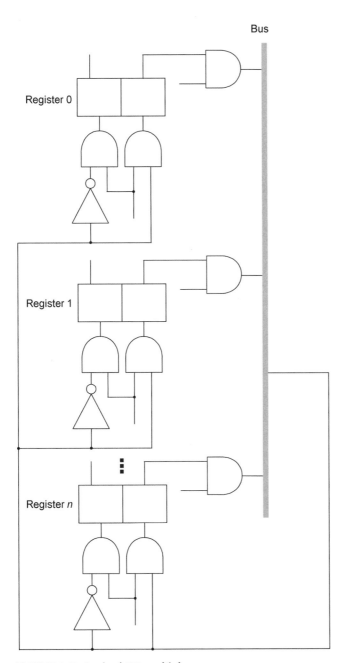

FIGURE 1.5 A wired OR multiplexer.

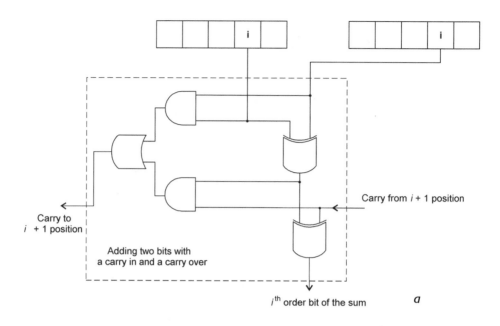

Carry to $i + 1$ position

Carry from $i + 1$ position

Adding two bits with a carry in and a carry over

i^{th} order bit of the sum

a

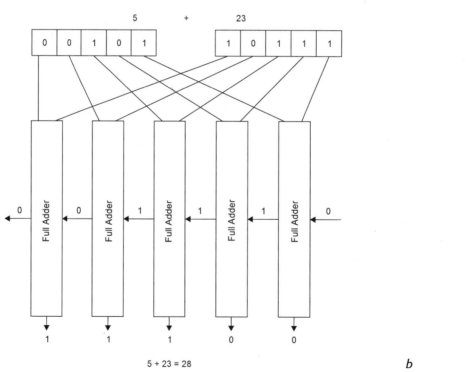

5 + 23 = 28

b

FIGURE 1.6 *a*—full adder circuit diagram, *b*—flowchart of adding two 5-bit numbers.

1.1.2 Implementing Computations by Hardware

Before moving on to describing processors themselves, let's consider how a digital data-processing device is constructed and how calculations are performed in this device so that we can subsequently understand why processors are needed, as well as how data can be digitally processed without resorting to processors.

To provide a tangible example, let's consider how a concrete computational process is implemented. Let's say that a filter $y_i = \sum_{j=1}^{4} w_j \times x_{i-j+1}$ that takes a specified infinite x_{-3}, x_{-2}, x_{-1}, x_0, x_1, ... sequence as the input and produces a y_0, y_1, ... sequence as its output needs to be constructed. Figure 1.7 shows one possible way to construct a digital device to perform the needed computation [2].

Registers, interregister data-transfer circuits, functional units such as adders and multipliers, and circuits for transferring data between functional units are used to construct the filter device. Details on how to construct functional units from gates can be found in Kartsev [1]. To understand how processors are constructed, it is important to realize that functional units use as operands values from the registers and place operation results into registers.

The device shown in Figure 1.7 functions as follows. In a clock cycle, the x_i value is placed on the input of the R_1 register and is stored in it. In the following cycle, the result of multiplying x_i by w_1 is stored in the R_2 register, while the x_i is transferred from the R_1 register to the R_3 register. In the same cycle, x_{i+1} value is placed into the R_1 register. It is now easy to see that the y_i value will emerge on the output of the device six cycles after the x_i value was placed into the R_1 register.

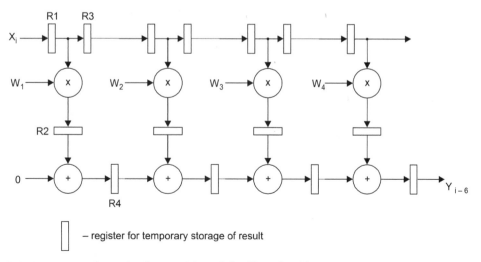

- register for temporary storage of result

FIGURE 1.7 Hardware implementation of the filter algorithm.

In the given device, multipliers and adders receive their data from the registers that were previously connected to them according to the algorithm being implemented. To implement a different algorithm, a different set of functional devices, registers, and, to connect them, multiplexers are needed.

Because all data in the device depicted in Figure 1.7 is transferred between the registers and functional units at clock signals, increasing the clock frequency increases the frequency with which results are output.

This method of performing computations is called *hardware-* or *circuit-implemented*, as opposed to the software-implemented computations. When computations are performed by hardware, all multiplexer (bus) control signals are fixed, that is, either specified combinations of constant values (0 or 1) or outputs of specific circuit blocks are placed on the corresponding control inputs.

A question arises whether the given hardware implementation of computation is the only one possible. The answer to this question is of course not. A circuit to implement hardware computation of the needed algorithm can be constructed using one multiplier and one adder, for example. The great number of ways to implement the same algorithm by hardware creates a problem of choosing one among them. When making a choice of the hardware circuit to implement an algorithm, you must have a criterion that the suitable circuit must meet. This criterion can be minimizing computation time or minimizing the number of circuit parts. In this respect, a hardware implementation of an algorithm can be considered successful if it provides the extreme value of the applied criterion.

The hardware implementation of the algorithm $y_i = \sum_{j=1}^{4} w_j \times x_{i-j+1}$ shown in

Figure 1.7 demonstrates two methods used to decrease computation time: parallelism and pipelined processing.

The concept of the *parallelism* consists of performing data manipulations in several functional units simultaneously.

Pipelining is a special case of parallelism where the manipulations necessary to obtain an individual result from the homogeneous population of the results—this being y_i, $i = 0, 1, \ldots \ldots, N$ in our example—are broken down into a sequence of algorithm steps of equal duration. Each individual step of the algorithm is executed by a separate unit, and the aggregate of the data manipulations performed by all the units composes the whole algorithm for the specified transformation. Results produced by each unit in the pipeline are latched in a register, and the next unit uses this previously stored result, which allows the execution of the current step for y_i to be combined in time with the execution of the preceding step for y_{i+1}. In the given example, the pipeline devices that execute the algorithm steps perform operations of multiplying by the w_1, w_2, w_3, w_4 coefficients and addition operations. Each step is executed in parallel with the rest of the steps, and when the pipeline is fully loaded, four values of the y_i, $i = 0, 1, ..., N$ output sequence are calculated simultaneously. An advantage of the pipeline method is that after all pipeline units have been loaded, the next result is produced in a time period one step long. This is when it takes the combined time of all pipeline steps to obtain this result.

1.3.3 Implementing Computations by Software

The hardware method to perform computations requires that a custom-made circuit suitable for the specific algorithm be built from registers and functional units. That's why implementing another algorithm in effect requires building a new computational device. This is a tedious task when it must be done in applied computations.

To be able to use the same hardware circuit to implement different algorithms, computations are done by software. When computations are done by software, first a collection of registers and functional units is set up, with multiplexers added to exchange data between them. A set of instructions (manipulations) to express an algorithm for a required computation is also defined. When executed, each instruction produces a collection of control signals, which are input into the multiplexers, resulting in data transfer between the registers and possible data conversion in the functional units. The actual effect of an instruction is determined by onto which multiplexers the control signals it produces are placed and what these signals are.

Consequently, when computations are performed by executing different instruction sequences (individual instructions executed in variable order), different algorithms can be implemented using the same arrangement of registers, functional units, and multiplexers. It must be pointed out, of course, that the initial arrangement order as well as the set of the utilized instructions can be more suitable to perform one type of algorithms and hinder effective execution of other types. However, this is the price extracted for using the same device to implement different algorithms.

It is important to understand that instructions are input into the described digital data-processing device externally, from another device that must be specially constructed for each individual algorithm. Consequently, the whole device consists of two parts: one used to store and process data and which does not change for any algorithm, and one that provides instruction sequences required to perform computations and which does change for each algorithm. You should not think that such digital data-processing device construction is purely hypothetical. This is exactly how the English digital device Colossus, used to decipher German communications during the WWII, was constructed. Alan Turing, the inventor of the Turing machine, participated in building this device. A Turing machine uses a tape with cells to store data and has a computational unit external to the tape that issues an instruction sequence that modifies the tape's cells.

Let's try to picture a hypothetical digital data-processing device suitable, among other tasks, for implementing the previously described filter. The machine will consist of a register file, a memory unit, and a functional unit capable of executing a certain instruction set, including addition and multiplication. Let's equip this device with a multiplexer to transfer data between the registers and between the registers and the functional unit, storing the result in one of the registers, as well as to transfer data between the memory and the registers.

The instruction set necessary to perform the $y_i = \sum_{j=1}^{4} w_j \times x_{i-j+1}$ computations

consists of the following instructions:

$\mathbf{R_1} \leftarrow \mathbf{R_2}$—Contents of R_2 register are sent to R_1 register.

$\mathbf{R_1} \leftarrow \mathbf{R_2}$ op $\mathbf{R_3}$—Contents of R_2 and R_3 registers are modified by an op operation and the results stored in R_1 register.

$\mathbf{R_1} \leftarrow \mathbf{[R_2]}$—Contents of the memory cell whose address is stored in R_2 register are copied to R_1 register.

$\mathbf{[R_1]} \leftarrow \mathbf{R_2}$—Contents of R_2 register are written to the memory cell whose address is stored in R_1 register.

To perform computations by software, the locations of the processed data and the computation results must be explicitly defined. The initial values of the employed registers also need to be set, as well as the values of the memory cells if data is to be read from them.

To perform needed calculations, we assume that values x_i, $i = 0, 1, ..., n$ are stored in the memory cells with addresses $A_{\text{srt}(x)}$, $A_{\text{srt}(x)+1}$, ..., $A_{\text{srt}(x)+n}$, whereas the calculated values y_i, $i = 0, 1, ..., n$ are written to the memory cells with addresses $A_{\text{srt}(y)}$, $A_{\text{srt}(y)+1}$, ..., $A_{\text{srt}(y)+n}$. Additionally, to obtain the value y_0, the values of $x_0, x_{-1}, x_{-2}, x_{-3}$ must be defined. If these values are not defined (initialized with 0s, for example), then you must realize that the first correctly calculated value in the y_i, $i = 0, 1, ..., n$ sequence will be y_3.

The software implementation of the computations suggested in [2] requires that an x_i, $i = 0, 1, ..., n$ sequence be stored into the memory, and the A_x, A_y, W_1, W_2, W_3, W_4, E_3, E_2, E_1 registers be set to the following initial values:

$$A_x := A_{\text{srt}(x)} - 1;$$

$$A_y := A_{\text{srt}(y)} - 1;$$

$$W_1 := w_0;$$
$$W_2 := w_1;$$
$$W_3 := w_2;$$
$$W_4 := w_3;$$
$$E_3 := x_{-3};$$
$$E_2 := x_{-2};$$
$$E_1 := x_{-1}.$$

The actual calculations execute the computation-cycle instruction sequence (n+1) times. Consequently, a control action must be generated to first transfer the contents of the E_3 register to the E_4 register, then to transfer the contents of the E_2 register to the E_3 register, and so on. After an $[A_y] \leftarrow C_1$ instruction is executed, an $E_4 \leftarrow E_3$ instruction has to be executed again.

The body of the cycle follows:

$E_4 \leftarrow E_3$—transferring contents of the x_{i-3} element or value 0

$E_3 \leftarrow E_2$—transferring contents of the x_{i-2} element or value 0

$E_2 \leftarrow E_1$—transferring contents of the x_{i-1} element or value 0

$$A_x \leftarrow A_x + 1$$

$E_1 \leftarrow [Ax]$—transferring contents of the x_i element

$$C_1 \leftarrow W_1 \times E_1$$
$$C_2 \leftarrow W_2 \times E_2$$
$$C_1 \leftarrow C_1 + C_2$$
$$C_2 \leftarrow W_3 \times E_3$$
$$C_1 \leftarrow C_1 + C_2$$
$$C_2 \leftarrow W_4 \times E_4$$
$$C_1 \leftarrow C_1 + C_2$$
$$A_y \leftarrow A_y + 1$$
$$[A_y] \leftarrow C_1$$

Operations are executed on the content of 12 registers: E_1, E_2, E_3, E_4, W_1, W_2, W_3, W_4, C_1, C_2, A_x, and A_y. It must be noted that mnemonics can be used to name registers when writing programs; this makes programs easier to understand.

1.1.4 Structure and Functional Organization of Processors

Software-implemented computations require the issuing of a specific instruction sequence to perform a necessary computation. Naturally, it is easier to construct a device that issues an instruction sequence that the algorithm being used requires than to build a computing unit from individual electronics components, including registers, memory, and functional units; however, even this is quite a laborious process.

Actually, the processor concept is an offshoot of an idea of John von Neumann, the creator of the first stored-program computer, EDVAC. The idea involved storing the instruction sequence, together with data, necessary to perform a computation in memory. An algorithm to execute this stored instruction sequence in proper order was implemented in hardware. This hardware-implemented algorithm was later called a *processor*.

The essence of this algorithm comes down to a cyclic execution of the steps shown in Figure 1.8.

Let's consider the steps comprising the functional algorithm of a processor in more detail.

In the first step, called "fetching the next instruction from the memory," an instruction is fetched from a memory cell whose address is contained in the special register—instruction counter, called Program Counter (PC). This instruction is then placed into the instruction register.

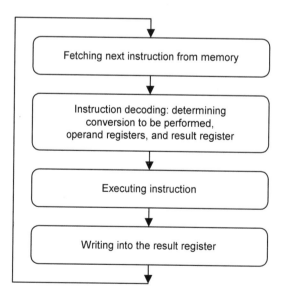

FIGURE 1.8 Functional algorithm of processor.

Because the contents of the instruction counter determines which instruction is to be executed next, let's look how values of this register are generated.

It is assumed that the program instructions are contained in a memory cell range with consecutive addresses, such as A_{srt}, A_{srt+1}, A_{srt+2}, ..., where A_{srt} is the starting address of the range. This consecutive placement of the instructions is intuitively understood as the order in which the instructions were written by the programmer when expressing the algorithm of the problem's solution using the given processor instruction set. Therefore, as a rule, to determine the address of the instruction to be executed next, the value of the program counter simply needs to be counted up by one, hence its name.

To change the instruction execution sequence, when the instruction to be executed next is not the one whose address is the incremented by one address of the instruction being currently executed but an instruction from a memory cell with an A_{jmp} address, special jump instructions are issued. Jump instructions change the contents of the program counter.

Consequently, the address of the next instruction is either determined by incrementing the program counter by one or generated by an executed jump instruction.

In the instruction-decoding step, the instruction placed into the instruction register is decoded to determine what functional transformation this instruction specifies, as well as what registers and memory cells are to be used as the data source and the destination for the results of the instruction execution. It is actually during the instruction-decoding step that preparations for placing control signals onto the multiplexers to transfer data between the registers, functional units, and memory cells are carried out.

The functions of the instruction execution and storing the result steps are understood from their names.

Let's consider a hypothetical processor DLX [4] as an example.

The processor instruction set consists of three types of instructions:

- Read/Write
- Conversions in the arithmetic/logical unit (ALU)
- Jumps

Formats of these instruction types are shown in Figure 1.9, *a*, *b*, and *c*, respectively.

Depending on the value of the 6-bit operation code, read/write instructions do the following:

- Move data from the source register rs1 to the destination register rd
- Move a 16-bit constant from the immediate operand field to the destination register rd
- Move data from registers to the memory and from the memory to registers

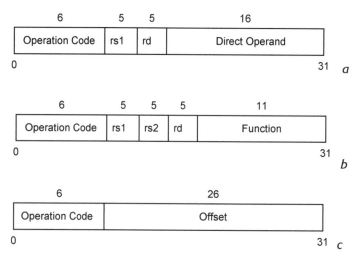

FIGURE 1.9 Instruction formats of DLX processor: *a*—read/write, *b*—ALU conversions, *c*—jumps.

The ALU conversion instructions perform the operation specified in the *operation code* and *function* fields on both or one of the operands stored in the *rs1* and *rs2*, subsequently storing the result of this operation in the *rd* register.

Jump instructions can be unconditional and conditional, that is, executing a jump only when some condition specified in the operation code evaluates to *true*. To execute a jump, the contents of the 26-bit *offset* field of the jump instruction are added to the current value of the instruction counter. The new value of the instruction counter determines the address of the instruction to be executed after the jump instruction. Depending on whether it is positive or negative, the constant in the instruction's *offset* field can either increase or decrease the value of the instruction counter.

From the number of bits in the instruction fields used to specify the operand sources registers *rs1* and *rs2* and the result register *rd*, it becomes clear that the instruction system requires using 32 registers: R_0, R_1..., R_{31}. Two of these registers are special purpose, the rest are universal and can be used to either read or write data. The R_0 register has a constant value of 0. The R_{31} register is used to store the old program counter value when a jump instruction is executed, which modifies the value of the instruction counter to PC+offset.

The specialization of the R_{31} register allows a subprogram mechanism to be added. A jump instruction transferring the control to some code fragment comprising a subprogram can be executed from any point in the program. The subprogram performs some conversion frequently required by the user, calculating a sin(x) value, for example. Each subprogram ends with a jump instruction in which the offset field is formed using the value from the R_{31} register; upon completion of the subprogram, this makes it possible to automatically return from the subprogram to the instruction that follows the jump instruction that had induced the jump to the subprogram.

The structure of the DLX processor is shown in Figure 1.10.

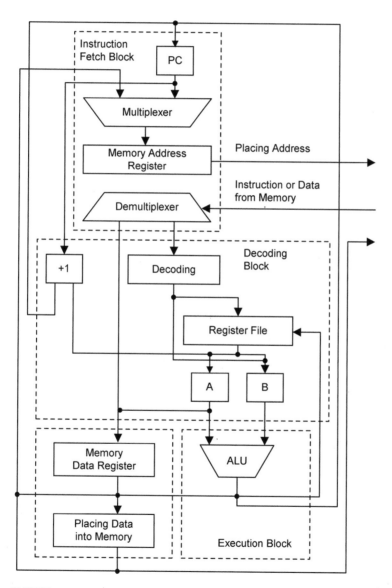

FIGURE 1.10 DLX processor structure.

All processor registers are 32 bits wide. The data transfer paths inside the processor shown in the block diagram also are 32 bits wide. Only data transfer paths are shown in the block diagram; the control signal transfer paths are not shown. Let's consider how the processor blocks the implementation by hardware of the processor algorithm, shown in Figure 1.8.

The multiplexer of the instruction fetch block provides the memory either with an instruction address supplied by the PC or data address that is generated in the ALU when executing a move instruction specifying the exchange between registers and a memory cell.

The address is applied to the memory over the address bus, and in a read access an instruction or data from the memory cell with the specified address are returned over the data bus to the processor; in a write access, data that had been placed on the data bus is written into the memory cell with this address.

In a case when an instruction is received from the memory, the demultiplexer of the instruction fetch block transfers it to the decoding block. When data is received from the memory in the process of executing a move instruction, the ALU writes this data into the register specified in the instruction.

The value of the program counter can either be incremented by one by a special block that functions in parallel with the ALU or a jump address generated by the ALU is written into the program counter when a jump instruction is executed.

The decoding block identifies the register file registers that are indicated in the instruction. The contents of the operand registers *rs1* and *rs2* are latched in the A and B registers, respectively. If a directly specified operand is used, it is also latched into the B register.

After the operation is executed in the ALU, the result can also be stored in one of the registers of the register file.

In addition to data from the register file, the value PC+1 can be written into the A and B registers when a jump instruction is executed.

Having defined the structure and functional organization of the processor, let's consider how a computational program for it (see Listing 1.1) can be specified and how it will differ from the program presented in Section 1.1.3.

LISTING 1.1

```
M: Ax ← Asrt(x) - 1;

Ay ← Asrt(y) - 1;

W1 ← w0;

W2 ← w1;
```

$W_3 \leftarrow w_2;$

$W_4 \leftarrow w_3;$

$E_3 \leftarrow x_{-3};$

$E_2 \leftarrow x_{-2};$

$E_1 \leftarrow x_{-1};$

$i \leftarrow 0$ // Initialization of the i register, in which the

        // number of the performed iterations will be stored.

M1: $E_4 \leftarrow E_3$ // Transferring contents of the x_{i-3} element.

$E_3 \leftarrow E_2$ // Transferring contents of the x_{i-2} element.

$E_2 \leftarrow E_1$ // Transferring contents of the x_{i-1} element.

$Ax \leftarrow A_x + 1$

$E_1 \leftarrow [A_x]$ // Transferring contents of the x_i element.

$C_1 \leftarrow W_1 * E_1$

$C_2 \leftarrow W_2 * E_2$

$C_1 \leftarrow C_1 + C_2$

$C_2 \leftarrow W_3 * E_3$

$C_1 \leftarrow C_1 + C_2$

$C_2 \leftarrow W_4 * E_4$

$C_1 \leftarrow C_1 + c_2$

$A_y \leftarrow A_y + 1$

  $[A_y] \leftarrow C_1$

$i \leftarrow i + 1;$

```
jump (i ≤ n) M1 // Jump to the instruction with the M1 address

                // if i<=n; go to the next instruction if i>n.

stop // Terminate program
```

In essence, the differences are few, but they are all fundamental. The i register, which stores the number of times the program body has been executed, has been introduced. After each execution, the value in the i register is incremented by one. The next conditional branch instruction examines the $i \leq n$ condition for being true. If the condition is true, control branches to the instruction with the M1 address. If the condition is false (that is, $i > n$), no further repetitions of the cycle body are carried out. The next instruction, stop, completes the computations.

The calculations are launched by loading the M address of the first instruction of the program into the program counter.

1.1.5 Processor Architecture

The processor architecture is the processor's structure as seen by a programmer. In general, processor architecture can be said to be its instruction set [1]. In turn, the structure defines the components that make up a processor and how these components are interconnected. Finally, the functional organization defines how the components interact when computations are carried out.

There are different types of instructions: register ← register, register; register←register, memory; register←memory, memory; memory←memory, memory; memory←register, memory; register←memory; memory←register, and so on. The instruction type is defined by source of the operands and the destination of the result. Thus, in an instruction of the type register←register, register, the operand sources are registers, and a register is used to store the result.

Instructions are divided into two classes: scalar and vector. An instruction is of the *scalar* type when its operands and the result are numbers (scalars).

A *vector* instruction is one whose operands, and possibly the result, are vectors (arrays of numbers). Elements in a vector are modified by one vector instruction. An example of a vector instruction is an instruction that multiplies two next elements of two arrays, then adds the obtained product to the contents of a specified register, and finally modifies memory addresses to access the next two elements of the array. This operation sequence is repeated the number of times specified in the instruction body.

The introduction of vector instructions is due to the desire to speed up processing of data arrays by eliminating time expenditures on fetching and decoding processing instructions that are the same for all components of the input arrays.

However, using vector instructions requires a programmer to prepare a vectorized program code, which, generally speaking, is equivalent to designing parallel programs.

A reasonable question arises as to whether it is possible to develop the optimal processor architecture and whether it will ever be developed. Historically, processor architectures have been developed as results of thought-out choices, based on the potential task sets to be executed on these processors and the upward compatibility of the already existing software for them. The problem of processor architecture standardization will be examined later in this book. Although it has been a subject of numerous studies, it looks like it will be a long time before a universal processor architecture will be created. However, the hardware resources in possession of the processor developers allow them to combine in one processor all known architectural techniques for increasing performance, only taking heed of their mutual compatibility.

1.1.6 Interrupts

In Section 1.1.4, two of the main computer components were considered: the processor and the memory. A third component is the input/output devices, also called peripheral devices, through which information—programs and data— are entered into computers. Peripheral devices come in a great variety, and discussing their particularities is not among this book's objectives. However, all peripheral devices interact with the processor and the memory in a common way, due to the necessity to coordinate actions of the processor and peripheral devices with respect to the memory.

Actually, the I/O can be implemented by software, without having to expand the already introduced concepts. Software input and output are allocated two registers each; one of them is used to store a data element, the other to store the ready status bit. The processor and a peripheral execute the following protocol.

When the peripheral device is ready (or, in the trade parlance, wants) to send data to the memory, it analyzes the status of the ready bit of the processor's input register. We will assume that when this bit is 1, then the processor has not finished reading the previous data element the peripheral had supplied, and the peripheral must wait until the processor finishes reading this data element and sets the ready bit of the input register to 0. When the peripheral detects that the ready bit of the input register is 0, it can begin writing next data element to the input register.

Likewise, a peripheral examines the status of the ready bit of the output register when outputting data. If this bit is 1, then the peripheral can take the data element from the output register and, having done this, must set the ready bit to 0. Accordingly, prior to placing data into the output register, the processor must examine the value of the output register ready bit and place data into the output register only when this value equals 0. When the value of the output register ready bit is 1, the processor must wait until the peripheral releases the output register.

Software I/O requires that the processor execute a program containing conditional jump instructions that examine the ready bit status of the input or output registers with a predetermined cyclicity. When such an instruction detects that the value of, for example, the output register ready bit is 1, it executes a jump to itself. It will continue doing this until the peripheral removes data from the output register and sets the register bit to 0. Having detected this in a next execution loop, the jump instruction transfers program control to the instruction that follows it and is the first instruction of the routine that continues output of data elements.

It is certainly possible to conceive of a case of a harmonious interaction between the processor and a peripheral device, when the program parts responsible for examining the ready bits and inputting or outputting data are located in the body of the program in such a way that the peripheral device manages to read out or write in the next data element in time for the next inspection. But such coordination of the operating speeds of the processor and the peripheral devices is very resource-intensive, and the program must be rewritten when the characteristics of the peripheral devices or the processor change. However, a more important circumstance hindering the achievement of such harmony is that the execution moments of the conditional jump instructions and the reaction times of the peripheral devices cannot be predicted. Consequently, implementing output by software entails time losses due to having to execute conditional jump instructions, which serve only to synchronize the processor and the I/O devices.

To avoid execution of these extra conditional jump instructions, the concept of interrupts was introduced. This concept's essence consists in that in the next-instruction fetching step, prior to fetching the next instruction the processor examines the values of the I/O registers value bits. If it discovers that the next data element must be input or output, an interrupt is triggered: the current PC value is saved in some register and the address of the first instruction of the interrupt-processing program—an interrupt handler—that performs input or output of data is loaded into the PC. The last instruction in the interrupt handler restores the saved PC value to the PC, thus returning the processor to execution of the interrupted program.

The interrupt handler and the program it interrupts use the same register set of the processor; therefore, when an interrupt is triggered, in addition to the PC, the contents of the registers must also be saved. This can be done either as a part of the processor algorithm with the corresponding hardware expenditures or as a part of the interrupt handler. In the latter case, the interrupt handler must temporarily save the contents of the registers in the memory and restore them after it finishes processing the interrupt.

It is obvious that the memory cells in which an interrupt handler temporarily stores the PC and the registers must not be used by other programs to store their data; otherwise, this data will be lost when an interrupt is processed. Usually this memory range is organized as a stack, and a special register—Stack Pointer (SP)—is assigned to point to the first available memory cell in this range. When data is pushed onto the stack, the SP is incremented by the value equal to the number of the cells written to; accordingly, when data is popped off the stack, the SP is decremented by the number of the read cells.

The execution of one program by the processor is called a computational process flow. The saved values of the PC and registers form the process environment. Placing process environment onto the stack allows multilevel interrupts to be organized, when interrupt handlers can be interrupted.

1.1.7 Hierarchic Memory Organization

The processor as it has been presented so far is logically fully operational and even is the ideal processor with which every programmer would like to work. However, in reality, processors have much more complex architecture due to the individual features of the elementary components from which they are built. Processor performance is increased by raising the clock rate that the logic and memory elements operate at, as well as increasing the number of these elements, which allows processors using parallelism and pipelining to be constructed. Each elementary base-development level has limitations on the upper value of the clock rate and the number of elements arising from the physical characteristics of the electric impulse propagation, reliability of the elements, and the processor cost.

Limitations are also imposed by the gap between the speeds of logic elements and memory elements. This gap has a tendency to increase as the integration level (number of elements in a unit of volume) and the speed of the basic elements increase. Therefore, the architecture of any type of processor must be based on the principles that, for the assumptions made regarding the organization of computations, make it possible to overcome the negative effects that a slow, as compared with the processor, memory has on the performance.

Memory can be implemented using basic elements based on different physical information storage principles: logic circuits (triggers), capacitors with current switches, magnetic domains, and so on. These can be organized into registers, static memory chips, dynamic memory chips, ferro-magnetic memory, disk memory, and other memory block types. The speed and capacity characteristic for each memory type are determined by the physical and cost limitations. Moreover, for any memory type, the access time increases, as its capacity gets larger; also, faster memories are more expensive per each data storage unit.

The ideal memory must provide the processor with instructions and data without causing the processor to idle while awaiting the data it has requested from the memory. Moreover, the memory must be large. In modern computer systems, the access time is reduced by employing hierarchic memory organization.

Memory access time depends on its type and capacity. Under these circumstances, a two-level memory can be built. The first level, or the main memory, will consist of the larger capacity but slower memory type, whereas the second-level buffer memory will consist of smaller capacity but faster access memory type. Data to be processed is first sent from the main memory to the buffer memory, from where it is fetched for processing with the results sent back to the main memory afterward. If the computation can be organized in such a way that a next portion of data can be delivered from the main memory to the buffer memory before the processing of the previous portion of data has been completed, then for all practical reasons data will be processed at the speed of the buffer memory. It is obvious that computations cannot always be organized like this, and the processor can sometimes idle while awaiting data from the main memory. However, in a number of cases, combining the transfer of data from the main memory to the buffer memory with the processor's using data from the buffer memory gives an effect of lowering the overall memory access time. In other words, the processor sees the memory as having the capacity of the main memory and the speed of the buffer memory.

This effect of lowering memory access time is achieved by exploiting a universal property of all programs known as *locality of reference*. There are two types of locality: *temporal* and *spatial*, each of which is further divided into *data* and *instructions* locality. The general principle of temporal locality is that once a location is referenced, there is a good chance that it will be referenced again in the near future. An example of temporal instruction locality is loop instructions: once the loop is entered, all the instructions in the loop will be referenced again before the loop terminates. Some examples of data exhibiting temporal locality are counters and look-up tables. The general principle of spatial locality is that when an instruction or data is accessed, it is very likely its neighbors will be accessed soon, too. A sequential instruction stream is an example of spatial instruction locality. Arrays or strings demonstrate spatial data locality.

Subsequently, implementing the main memory of the two-level memory as a two-level memory itself, a three-level hierarchic memory can be built; continuing this process, a hierarchic memory with any number of levels can be built.

Three types of memory operation mechanisms are based on using the locality of reference principle: cache memory, page or segment memory, and interleaved memory.

1.1.8 Cache

Locally processed data can arise during computations and not be concentrated in one area when statically located in the main memory. Because of this, buffer memory can be organized as associate, or content-addressable, memory. Here data together with its main memory addresses is stored in the buffer memory. Associative memory allows selecting data elements whose positional values match those in the selection template, which is also called a key [1]. For example, if the key has a 1 in the third bit position and a 0 in the fifth, then all associative memory cells that have 1s in their third bit positions and 0s in their fifth positions will be selected.

Accessing associative memory using the address field as a key allows data to be selected regardless where it is located in the associative memory. Storing addresses along with the data allows data to be uniquely identified by its addresses when transferring data between different memory levels and to be easily found on any level. This type of buffer memory is called *cache*. Cache makes it possible to flexibly coordinate the data structures needed for dynamic computations with the static data structures of the main memory.

Cache is a collection of fixed-size addressable cache blocks, or lines, into which contents of consecutive main memory cells are copied. Cache lines are typically 16, 64, 128, or 256 bytes long.

Because cache contains copies of the contents of the main memory cells, memory `read` and `write` operations are processed differently on the cache level. `read` operations present no problems. If the required data is available in the cache, it is delivered to the processor. The case when the cache has the required data is called a cache hit. The case when the cache does not have the required data is called a cache miss. In case of a miss, the required data is copied from the main memory into the cache.

However, when a memory `write` is performed, provisions must be made to ensure cache data coherence, that is, correspondence of cache data to the main memory data. This is achieved by modifying those areas in the main memory whose corresponding areas in the cache have been modified. Operations to maintain cache coherence are carried out in parallel with the mainstream calculations. Several methods of maintaining cache coherence are available, as are several cache operating modes.

The simplest to implement, but not the most efficient in terms of speed, is by modifying the main memory right after data in the cache is changed. With this method, the processor has to idle while the main memory is being updated. A valid copy of the cache is maintained in the main memory, and nothing needs to be done when cache lines are replaced. This type of cache is called *write-through*.

In another method, if the memory cell to which a `write` needs to be done is reflected in the cache, only its corresponding cache cell is modified. If the memory cell is not cached, then the data is written directly to the main memory. Cache changes are reflected upon the main memory only when replacing cache lines. This type of cache is called *write-back*.

There also is a buffered write-through type, where changes to the main memory are buffered and do not slow down the processor for the main memory write. The data is then written to the main memory as the cache controller gets access to the main memory.

There are three most-often-used methods of implementing a cache, which differ by the amount of hardware needed for their implementation. These are direct-mapped cache, set associative cache, and fully associative cache.

In the direct-mapped cache, each main memory cell address is represented as consisting of three fields, with the upper bits of the cell address representing the tag, middle bits representing the cache-line index, and the lower bits representing the byte offset in the line. For example, in a 16-bit address, the high five bits represent the tag, the next seven bits represent the cache-line index, and the last four bits represent the byte offset within the cache line. In this case, a cache line contains values of a sequence of 16 main memory cells in the increasing order of their addresses.

A direct-mapped cache consists of a set of cache lines, each of which contains a tag component and cache memory elements whose addresses are indicated by offsets relative to the start of the cache line. In the previous example, the cache contains a total of 2^7, or 128, cache lines each holding 2^4, or 16, bytes.

A univocal correspondence between individual main memory cells and where in the cache these cells can be located is established; namely, a memory cell is always located in the cache line specified by the index bits of the address at the position specified by the offset bits of the address.

Only two things can happen when the processor accesses the memory: the contents of the addressed cell are either cached or they are not. The first case is called a cache hit; the other is called a cache miss.

Whether a memory cell reference has produced a cache hit or miss is determined by examining the tag of the cache line whose index matches the line-index bits of the cell address. If its tag value matches the value of the tag bits of the cell address, then a copy of the addressed memory cell is in the cache line in the position determined by the offset bits of the address.

If the tag value of the corresponding cache line does not match the value of the tag bits of the memory cell address, a memory block with the tag bits specified in the address needs to be transferred into the cache. In case of a write-through cache, the required cache line is simply swapped in from the corresponding memory block. When a buffered or write-back cache is used, all previously modified cache lines need to be posted to the main memory prior to swapping the required cache line in from the main memory. Otherwise, the results of the previous write instructions will be lost.

The presence of the required memory block in the cache can be determined by a single comparison of the specified line's tag with the memory address tag bits. Memory blocks are brought into cache lines whose locations are fixed. These factors make for a rather small number of hardware parts necessary to implement this type of cache.

However, this implementation has obvious shortcomings. If a program alternatively references memory cells whose addresses have the same address index bits but different tag bits, each memory access results in a replacement of the cache line and the data brought in from the main memory.

In the associative cache, two parts of the memory address are used to map a particular memory cell to its location in cache. The higher address bits represent the tag, and the lower bits represent the offset.

A needed cache line is found by matching its tag with the tag bits of the address. An associative cache can have any number of lines, the only limitation being the number of possible tag values. Therefore, when locating a needed cache line, the tag values of all cache lines need to be compared with the tag bits of the address. If the search is done consecutively, line by line, the comparison process takes an unacceptably long time. Therefore, the comparisons are done simultaneously on all cache lines using the associative memory principles, hence the name of this type of cache.

When the needed memory block is not in the cache, one cache line must be replaced with the needed memory block. Various algorithms are used to determine which line needs to be replaced, or evicted. Among them are the cyclic, the least-used line, the least recently used line algorithms, and others.

Advantages of associative cache are obvious. A shortcoming is the difficulty organizing associative search, which requires a rather large number of hardware parts, consequently limiting the cache capacity.

A set-associative cache combines the two described approaches: the cache is comprised of a set of associative cache blocks. Unlike the direct-mapped cache, the middle bits of the address specify not the line index but the number of one of the associative blocks. During data search, tags are compared only for the block set whose number matches the middle address bits. Based on the number n of the associative blocks, a set-associative cache is called n-entry. Usually, the n is selected to be 4 or greater, within reasonable limits.

The typical modern memory hierarchy for one-processor architectures is shown in Table 1.1.

TABLE 1.1 Typical Memory Hierarchy

Level	Access Time	Typical Size
Registers	1 processor cycle	64–256 bytes
L1 Cache	1–2 processor cycles	32 KB
L2 Cache	3–5 processor cycles	256 KB
Main Memory	12–55 processor cycles	up to 4 GB

1.1.9 Page Memory

Another method after cache to reduce memory access time, taking advantage of the locality of reference principle, is page or segment memory organization [5]. This method makes virtual memory of much greater size than the physical memory available to a program. The difference between page and segment memory is that pages have fixed size, whereas segments can be of any size, the latter circumstance requiring segment length to be taken into account when working with them.

If a program were using physical memory of the size equal to the size of the virtual memory, the access time would be much greater than the access time of the physical memory being actually used (not to mention the questions of such large memory being possible to implement and the costs if it were).

Memory is divided into pages whose size usually is 2^m bytes, typically $6 < m \leq 12$. The majority of the pages or segments are stored in the external, most often disk, memory. Only the pages containing the data currently being processed are stored in the main memory. It is the exact localization of data that makes it possible to reduce computation time, despite the time losses caused by transferring pages between the main memory and disk storage.

Processor instructions use virtual addresses made up of two fields: the page number of the virtual page field and the offset of the cell address within the page field.

Each process being executed by the processor is associated with a page-mapping table. This table has a number of lines equal to the maximum number of memory pages allowed for use by the program.

Along with such registers as PC and SP introduced earlier into the process environment, specially for the purposes of organizing paged virtual memory a page table pointer (PTP) register has been added. The user structure and the page table are created by the processor's operating system when a program is launched for execution.

Each row of the page table corresponds to one virtual page and can be selected using the virtual page number (VPN) field of the virtual address.

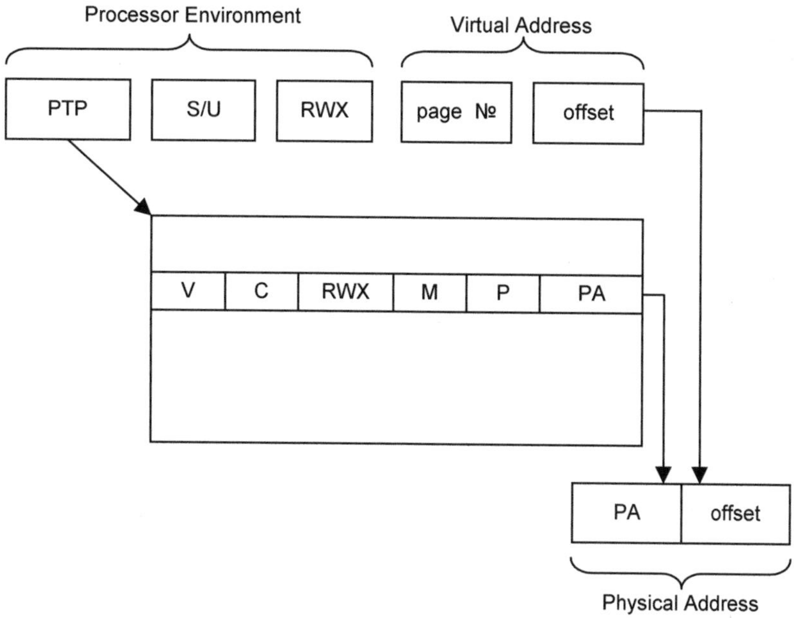

FIGURE 1.11 Mapping virtual addresses onto physical addresses.

A table row consists of the following fields:

V—mapping bit
C—page modification bit
RWX—page access rights
M—page location bit
P—page-caching enable bit
PA—physical page address

The mapping bit V determines whether or not a virtual page with the indicated VPN has been mapped onto the main memory. For example, if V=0, then the page has been mapped onto the main memory. When V=1, the virtual page is not in the main memory, and accessing memory location with this virtual address activates the virtual memory-to-main memory-mapping mechanism.

The page modification bit C is used to increase the efficiency of page handling. When a page is loaded from the disk into the main memory this bit is set to 0. If the contents of the page have been changed, bit C is set to 1. When a page is unloaded from the main memory to the disk, the value of this bit is examined. If it equals 0, meaning no changes have been made to the page, there is no need to do the actual disk writing of the page contents.

Page access rights determine what operations can be performed on this page. For example, if:

R=1—page contents can be read

R=0—page contents cannot be read

W=1—page can be written to

W=0—page cannot be written to

X=1—page can be both read from and written to

A virtual memory reference is considered valid if the RWX field of the processor environment matches the RWX field of the page being referenced.

The M bit indicates whether a page is located in the main memory or on disk. For example, M=1 means that the page is located in the main memory, and the physical address (PA) field specifies the upper part of the main memory physical address. M=0 means that the page is located on disk, and the PA field specifies the place on the disk where the page is located. When this page is referenced, it will be brought into the main memory, the PA field updated to reflect its main memory location, and the M bit set to 1.

Finally, the P bit is employed to increase efficiency of the hierarchic memory operation. For example, a user, knowing its program, can disable caching a memory block because he needs to read or write only one memory cell in the block. In this case caching the whole block is definitely an unnecessary operation.

To conclude the review of the page memory organization, it must be noted that the mechanism, although logically operational, is rather inefficient because it requires two memory references when using a virtual address. The first reference is needed to access the page table and the other to access the data itself.

To speed up translation of the virtual addresses into physical addresses, a special translation look-aside buffer (TLB) is used. This buffer contains N rows, each of which consists of a VPN and a number of the corresponding physical page. An N number of virtual pages that were used for last memory references are mapped onto the buffer. In this way, localized memory references are created conditions for rapid virtual-address-into-physical-address translations, without time expenditures on selecting pages from the page table. It is understood that the N cannot be too large because the TLB must be fast and, as a rule, is built using a limited amount of hardware.

Therefore, when the necessary virtual address is not found in the TLB or a corresponding physical address is not mapped for it, the previously described virtual-into-physical address mapping mechanism is launched. The obtained physical address along with a corresponding virtual address is then written into the TLB. If the buffer happens to be full at this time, one of its virtual-address-into-physical translation entries is replaced with the new data. The replacement policy can be implemented using various algorithms, for example, the least used or a randomly selected entry can be evicted.

It must be noted that although the considered method of accelerating different translation types by using an associative buffer to store translation tables requires substantial time expenditures, it is widely used in processor architectures.

Ideally, the task of localizing data of programs written in high-level programming languages is carried out by compilers taking into account the particularities of the hardware mechanisms that transfer data between different levels of memory hierarchy or perform page mapping. However, not all users are satisfied with the efficiency provided by compilers; therefore, processor and operating system instructions are devised that provide control over data transfers between different memory hierarchy levels and main memory page placement.

1.1.10 Interleaved Memory

Memory interleaving also takes advantage of one of the variations of the locality of reference. In this type of locality, the data being accessed has a fixed address order; for example, a next address is calculated as a sum of the previous address and a certain fixed offset d, which usually equals 1. When the order of memory accesses is fixed, the memory can be arranged as a sequence of n blocks, where n is called the interleave factor. Block k ($k = 1, \ldots, n-1$) is used to store the data needed for the $k + i^*n$ ($i = 0, 1, \ldots$) memory access. As any n successive memory accesses are addressed to different blocks, they can be executed in parallel. Subsequently, the resulting effect is decreased access time to a large-size memory.

In this case, memory can be accessed in parallel at n different addresses that lie in different blocks. Therefore, interleaved memory speeds up instruction fetching, excepting branch instructions, of course. Similarly, by placing sequential elements of data arrays in different memory blocks, their processing can be accelerated.

Hierarchic and interleaved memory can be used side by side.

1.1.11 Microprocessors

The complexity of a processor—which in its simplest version is just a hardware implementation of the algorithm that controls execution of instructions that in turn control data transfer between registers, memory, and the functional units—grows quite rapidly. This complexity is a consequence of the architectural and structural solutions devised and implemented by the computer designers to increase efficiency of processors. The originality of these solutions lies in that they make it possible to close the gap between the operating speeds of memory and logic elements and that they rationally distribute a limited amount of hardware between processor blocks to maximize the performance.

Until the 1990s, processors were built from discrete elementary components (vacuum tubes, transistors, resistors, and capacitors connected by conductors to form the necessary circuit) or small- and/or medium-scale integration circuits (elements containing hardware equivalent necessary for building several triggers). Difficulty building a processor out of tens of thousands of discrete elements and the requirements for the reliability of such a construction and its electric power consumption and cooling imposed rather stringent upper limits on the advances in the processor development.

Vector pipelined processors, whose production is associated with the construction of CRAY-1 [6], crowned the development of processors built from discrete components. The main architectural principle of these processors consists in that if it is not possible to speed up execution of all instructions, then a group of instructions allowing execution speed acceleration is selected. If a substantial part of computations can be placed on these instructions, the resulting effect will be the overall computation acceleration. The instructions selected for this group were vector instructions operating on vector registers. A vector register is a high-speed buffer that is loaded before a vector instruction or a sequence of chained vector instructions is executed. Contents of vector registers are operands for vector instructions. For processors built from discrete components, the pipelined vector architecture is optimal in terms of the performance-to-hardware amount ratio.

In the 1970s, along with the processors built from small- and medium-scale integrated circuits, development of large-scale integrated circuits gave rise to microprocessors with a small number of limited width (4–8 bits) registers.

At first, these microprocessors were in the micro category in terms of both the memory size and the performance. However, by the mid-90s, increasing integration levels made it possible to gradually implement on one crystal microprocessors that contained all processor-performance-increasing methods that have been developed within the framework of the traditional one-processor approaches. Single-crystal microprocessors substantially surpassed specially designed custom multicrystal processors in terms of the efficiency-for-dollar ratio. The particularities of the VLSI circuit, including microprocessors, design, and production, make them economically justifiable only when mass-produced (in production runs of millions of pieces). This has practically eliminated possibilities of building specialized crystals (in hundreds or thousands of pieces) required for specific multicrystal processor projects. Moreover, a custom-designed multicrystal processor with high efficiency characteristics at the time it was designed would take several years to build. By the time it was built, it would be significantly inferior to the universal microprocessors commercially available at that time.

Mass-produced high-performance microprocessors served as a basis for creating personal computers, workstations, and parallel supercomputers, which revolutionized society's access to information. This development impressed the contemporaries to such a degree that the term *killer microprocessors* [7] was coined, which reflected the very fact that microprocessors started to be used to build computers and computer systems of all types.

Therefore, nowadays the terms *microprocessor* and *processor* are interchangeable. It would be even more correct to say that microprocessors are the only type of processors available. This, in fact, is the answer to the second question posed at the beginning of the chapter.

1.2 MICROELECTRONICS MANUFACTURE OUTLOOKS

1.2.1 Very Large-Scale Integration Circuits

Characteristic features of the contemporary elementary base—very large-scale integration (VLSI) circuits—are a great number of transistors on a die and a relatively small number of outputs on chip. Therefore, VLSI chips are equivalent to logically complete devices.

Currently, mostly complementary metal-oxide semiconductor (CMOS) VLSI chips are used. In the CMOS technology, electronic circuits are built from rectangular or polygonal conductance semiconductor areas connected by metal or polysilicon conductors or isolated by dielectric. Silicon dioxide (SiO_2) film is used as dielectric. Figure 1.12 shows a schematic diagram, cross section, and topology of an inverter [8] implementing a NOT gate.

In the inverter's schematics shown in Figure 1.12, *a* consists of two MOS transistors with *p*- and *n*-type channels whose drains are connected to the output. The use of two types of MOS transistors is what constitutes the CMOS technology complementary situation. When a ground-level signal is applied to the circuit's input, the lower transistor will close while the upper will open; this action will cause a power-supply-level signal to appear at the circuit's output. Accordingly, when a power-supply-level signal is applied to the circuit's input, a ground-level signal will appear on its output, because the upper transistor will be closed and the lower one open. Thus, the output signal level is always inverted with respect to the input signal level, in compliance with the NOT gate logic.

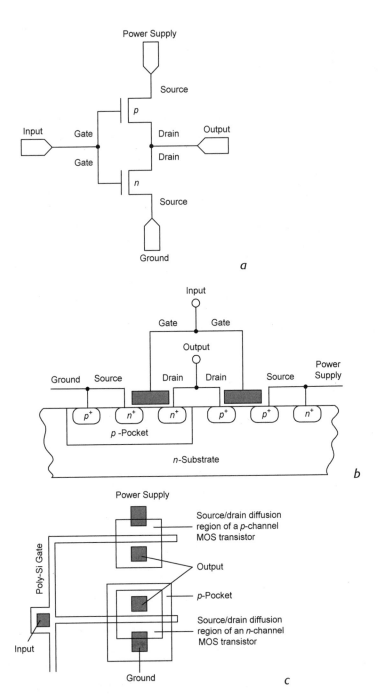

FIGURE 1.12 Logic schematic of a NOT gate and its integrated circuit implementation: a—inverter's schematics, b—a cross section of the silicon wafer, c—inverter's topology.

Figure 1.12, *b* shows a schematic cross section of a silicon wafer fragment containing *p*- and *n*-type conductance regions and polysilicon transistor gates. The corresponding topology of the inverter is shown in Figure 1.12, *c*.

Accordingly, transistors are formed as a multiplayer structure of the source and drain regions, separated by dielectric from the gate region. Metal conductors interconnect individual gate circuits when building digital device blocks and the device itself. The conductors are routed as intended for this purpose and separated by dielectric layers.

The general process of constructing a microcircuit follows: A cylindrical silicon monocrystal is drawn from a melted silicon bath. This cylinder is sliced into wafers about 0.6mm thick and 200mm in diameter. (A transition to using 300-mm-diameter wafers is in process.) The surface of this relatively thin wafer is polished and then covered with a thin film of silicon dioxide (SiO_2) in a pure oxygen environment under high temperature and pressure.

The following technology is employed to form *p*- and *n*-type conductance semiconductor regions: The wafer is coated with a thin film of photoresist. (Photoresist is a material resistant to etching agents but loses this property when exposed to light of certain wavelengths). During the design of an integrated circuit, templates are made for each layer of the circuit. The templates have openings shaped like the regions that need to be formed on their corresponding layers. A specific template is used at each stage of the integrated-circuit-manufacturing process. A template is placed over the photoresist-coated wafer after which the wafer is exposed to light. The photoresist in those areas of the wafer that were covered by the template does not change, but in the areas under the openings in the template, which are under the influence of light, it becomes susceptible to the action of the etching agent and dissolves during etching. Dry etching is used in modern technological process. Here the wafer is etched by ionized gas, which forms volatile compounds when entering into chemical reaction with the silicon dioxide film. Thus, dry etching removes the silicon dioxide film in those areas where the photoresist was exposed to light and leaves it unchanged in the areas where it was not.

Next, the remaining photoresist is removed from the wafer, which is now ready for the next technological operation—diffusing into the silicon substrate donor or acceptor dopants. In the diffusion process, the dopant atoms are uniformly inserted into the silicon lattice, forming semiconducting regions of the required conductance type.

Similarly, if a region of one conductance type needs to be created within a region of the other conductance type, the previously described process is applied.

After the semiconductor structures of transistors have been formed, they need to be interconnected to form a circuit. Only a planar conductor layer can be built, and the conductors in it cannot intersect. Therefore, to build a complex circuit, several conductor layers are built, with the conductors in them connected using interlayer junctions. Modern microprocessors require up to six or more conductor layers to build a high-speed circuit and to minimize its power consumption. The importance of being able to use a larger number of conductor layers when building a complex circuit is demonstrated by the following example: the North-wood Pentium 4 processor has 55 million transistors located on the 145 mm^2 area and six conductor layers. In comparison, the Athlon™ XP processor has 54.3 million transistors on an only 101 mm^2 area but with nine conductor layers.

There are two technologies, either one or both of which can be used to form conductor layers. In one of them, each conductor layer is formed as follows: the silicon wafer is first coated by thin oxide film; then the oxide film is etched off those places where electric contact is needed, and the wafer is coated by a thin film of metal, for example, aluminum. Then, this metal layer is coated by photoresist, covered by a template protecting the areas where conductors are to be, and the wafer exposed to light. The metal covered by the exposed photoresist is removed by etching and what is left forms the needed connector layer. Using this technology has one shortcoming that prevents the increase of the number of connector layers. Formation of the next conductor layer makes the wafer surface uneven, with the created metal conductors sticking out above the oxide film. It is possible to achieve a certain degree of flatness by building up an oxide film of appropriate thickness with the subsequent polishing, but this process is quite labor intensive. However, if the wafer surface is not sufficiently flat, when the photoresist is exposed through a template, the projected image will be distorted, causing changes in the parameters of the connections being built and even their breaking.

The other technological process is used to form copper connections. In general, it amounts to the following: as usual, a wafer is coated with photoresist over which a template with openings where the conductors are to be is imposed. The wafer is then exposed and etched, obtaining grooves in the oxide film where the photoresist was exposed. Afterward, a coat of copper is deposited on the wafer using the electron-beam deposition method. The wafer is then polished to the oxide film, which results in copper left only in the previously etched grooves. This technological process allows the wafer surface to be polished flat after each conductor layer is formed, which in turn allows a greater number of connector layers to be obtained than by using the first method.

Of course, the technological process of producing integrated circuits is substantially more complex than described in the previous examples. However, its essence consists exactly in a layered mass formation of conducting, semiconducting, and dielectric regions in each of the layers.

Reducing the size of the transistor parts results in improvement of all integrated circuit parameters: the operating speed increases, power consumption per transistor decreases, and reliability gets higher. The technological limits on the crystal transistor size due to the physical limitations are about 0.02 micron. Current 0.09-micron technology used in processor manufacture will be replaced with the 0.065-micron technology in 2005; 0.045-micron technology will be in use by 2007, and in 2009, the 0.032-micron technology level will be reached, yielding 10^9–10^{10} transistors on a crystal.

In their *Technology Roadmap for Semiconductors*, 2001 edition, several semiconductor manufacturers presented their prognosis for future developments in microprocessor production technology. According to this report, they will be in the direction of increasing the transistors-per-crystal density, using new types of three-dimensional transistors with three gates, increasing the number of the conductor layers, and simultaneously increasing the clock rate and lowering power supply voltage, energy consumption per transistor, and heat dissipation.

In addition to the obstacles to the further miniaturization stemming from physical limitations, there also are financial ones. For each successive integrated-circuit generation, the cost of technology doubles and the lead times increase. For example, it took six months to put the Pentium® processor into production; this increased to nine months for the subsequent Pentium Pro. In many aspects, even now the level of mass-production technology is determined by economic considerations. Increasing the number of conductor layers raises the production reject ratio exponentially; increasing the microprocessor crystal area also increases the percentage of defective crystals.

Different functions performed by and application areas of VLSI circuits have caused their specialization. They can be rather loosely broken down into the following classes:

- VLSI circuits implemented in hardware data-processing algorithms: universal and signal microprocessors and microcontrollers, including interface circuits for building multiprocessor systems
- Dynamic and static memory microchips
- Programmable logic-integrated circuits (PLIC)

1.2.2 Microprocessors and Microcontrollers

Universal microprocessors are used in all types of computing devices: personal computers, workstations, and recently in massively parallel supercomputers. Moreover, universal microprocessors are also used in telecommunications equipment and automatic control equipment and are embedded in industrial automation machinery. These processors' main feature is their advanced facilities for performing efficient floating-point operations with 32- and 64-bit, and longer, operands. Lately, these processors have been being equipped with functional units for processing multimedia information.

Digital signal processors are aimed at real-time processing of digital streams obtained by digitizing analog signals. Because of this, they process mostly low-bit-count integers. However, modern signal processors can perform floating-point operations with 32–40-bit operands. Moreover, a class of media processors has emerged that are complete audio- and video-information processing systems.

Control systems' embedded processors, including those in domestic equipment, are the most specialized and have the greatest variety of functions. The total number of the types of chips with different instruction sets exceeds 500, and they all hold their market share steadily, owing to the existence of equipment in which they are used.

Before starting a detailed discussion of various types of microprocessors, the organization of memory and variable logic-integrated circuits will be briefly considered. This knowledge is necessary to better understand the specifics of microprocessors and their use in building computers.

1.2.3 Memory Microchips

MAIN TYPES OF MEMORY

Increased processor speeds elevate the bandwidth requirements for the other computer subsystems, most important, for the memory subsystem bandwidth. Designers use the fastest and, accordingly, the most expensive memory in the most important system components. Widely used is the hierarchic memory architecture with higher-speed cache at its apex. Intensively used portions of the instruction code and the processed data are placed in cache to speed up access to them. In actual systems, from one to four cache levels are used. The first-level cache L1 is, as a rule, located on the same crystal as the microprocessor and works at its clock rate; the second-level cache, L2 and lower, are usually located off the microprocessor crystal and are built using high-speed memory microchips. A tendency toward placing the L2 cache also on the processor crystal has been observed.

Currently, the most widely used are volatile-static and dynamic-type memory microchips, which lose the stored information when the power is removed. Static-memory chips use triggers to store data bits, requiring 4–6 transistors to store one bit. Dynamic memory uses capacitors to store bits. Logic 1 and 0 are represented by charged and discharged states; the state represents which logic level is a matter of convention. Because only one transistor is needed to control capacitor charging and discharging, the amount of hardware necessary to store one bit is four times less than in static memory. However, unlike static memory, dynamic memory needs to be refreshed to prevent losing information because of the capacitor discharge caused by the leakage current. Refreshing is performed by reading the capacitor charge state and refreshing this state by a subsequent write. Refreshing lengthens the memory access cycle, lowering its bandwidth.

DYNAMIC MEMORY

To date, relatively low-speed Dynamic Random Access Memory (DRAM) is often used as the main memory at the bottom of the memory hierarchy. Memory cells in chips have array organization and are addressed by specifying row and column numbers. To access a memory cell, the cell address needs to be placed onto the chip's address inputs followed by first Row Address Strobe (RAS) and then Column Address Strobe (CAS).

Along with the regular RAM, fast page mode DRAM (FPM DRAM) used to be widely used. This type of memory provides faster access to often-used data element sequences by placing these elements in one row of the memory cells array.

A DRAM read access cycle is started with activating the array row by issuing the row address and a RAS signal. Then, the memory cell containing the needed data is activated by issuing the column address and CAS signal. After validity of each element has been verified, the data is sent to the processor. The column is then deactivated and preparations for the next cycle are performed. This step causes the processor to wait for the completion of the memory cycle because during column deactivation no other operations are performed. The output data buffer is disabled until either the next cycle is imitated by issuing the address of the next column in the currently selected row followed by a CAS signal or next data row is requested.

In case of FPM DRAM, the next column is activated by a CAS signal on the assumption that the next requested data element is located in the next memory cell of the same row. Activating the next column produces improved results only when memory cells in a given row are read sequentially.

Memory-block transfer operations are described by an expression showing the number of cycles needed to read a four-item data block. For DRAM, this expression is 5-5-5-5, meaning that five cycles are needed to read each data item. For FPM DRAM, the expression is 5-3-3-3, meaning that five cycles are needed to read the first data element; only three cycles are needed to read each of the remaining three data items, because they are assumed to be all in the same row, and there is no need to issue the row address to read them.

In a system operating at a high clock rate, extended data output (EDO) DRAM is used to ensure reliable processor and memory interaction.

In many aspects, EDO DRAM operation is similar to FPM DRAM: first the row is activated, and then the column. But after the data item is found, instead of deactivating the column and disabling the output buffer (as is done in FPM DRAM), EDO DRAM stores the output data in supplementary output registers until the next column is accessed or next cycle started. This action extends the time interval during which output data is available, hence, the name extended data output. By keeping the output buffer enabled, EDO DRAM eliminates the wait state, and burst transfers are executed faster. Access time of this type of memory is about 30 ns in the page mode. Block operations are described as 5-2-2-2 cycles.

Burst EDO (BEDO) DRAM transfers data in bursts. Data access is pipelined. Page access cycle is broken down into two stages. In read operations, data is placed into the output register in the first stage; in the second stage, logical levels corresponding to the contents of the output register are formed on the data bus. Pipelining lowers the number of wait cycles even more, to 5-1-1-1.

The main shortcoming of EDO and BEDO DRAM is that these types of memory microchip are intended to work at frequencies up to 66 MHz, whereas the system bus operating frequency is substantially higher these days (75 MHz, 83 MHz, 100 MHz, 200 MHz, 266 MHz, and getting still higher).

The considered dynamic memory types are asynchronous, because the time the memory data appears on the computer system bus is not synchronized with the clock fronts. Data arriving at the bus right after the clock front will be sensed only at the next clock pulse. Synchronous DRAM (SDRAM) is faster than asynchronous. It has access times of 7–10 ns and is built using Bi-CMOS technology.

The main feature of SDRAM is that all its operations are synchronized with the processor clock signals. This simplifies building control interfaces and reduces the column access time. SDRAM has an internal burst counter that can be used to increment the column address in the burst access mode. This allows a new memory access to be initiated before completing the previous one.

Inside the chip, the memory is interleaved: it has two blocks, and consecutive accesses are addressed to different blocks. The block mode access time for SDRAM is expressed the same as for BEDO: 5-1-1-1; however, unlike the latter, PC100 and PC133 SDRAM modules can work at 100 MHz and 133 MHz, respectively.

SDRAM has an 8-byte data bus; this makes PC100 bandwidth 800 MBps and 1.06 GBps for PC133.

A further development of SDRAM is SDRAM-II, developed by Samsung®, the leader in the memory microchip production industry. Another name for this type of memory is double data rate SDRAM (DDR SDRAM). Two read or write operations are executed in one clock cycle: at the rising and falling clock edges. DDR SDRAM bandwidth is 1.6 GBps at 100 MHz system bus clock rate. This memory is also designated as DDR200, reflecting the fact that its effective operating frequency is 200 MHz. Similarly, DDR266, DDR333, and DDR400 memories are used for 133 MHZ, 166 MHz, and 200 MHz system bus clock frequencies. To show what bandwidth in MBps these memory modules have, they are also designated as PC1600 (DDR200), PC2100 (DDR266), PC2700 (DDR333), and

PC 3200 (DDR400). For example, using 0.11-micron technology, Micron produces DDR400 512 MB memory chips, from which standard 512 MB DIMM can be built.

Plans exist to build in the future DDR-II memory that in burst mode will allow four accesses per clock cycle. Accordingly, this type of memory is also called quadra data rate (QDR).

To date, dynamic memories based on the data-channel concept—Rambus® DRAM (RDRAM) and SyncLink DRAM (SLDRAM)—are considered to be the most promising.

Rambus technology, named after the company that developed it, is based on a high-speed interface that provides data transfer rates up to 600 MBps via the Rambus channel—a serial data bus. The effective bandwidth of a memory module reaches into 480 MBps, which is ten times the effective bandwidth of memory modules based on EDO DRAM chips. Access time to a row of memory cells is an equivalent of 2 ns per byte; the latency (access time to the first byte in the data array) is 23 ns.

Even faster is a new version of RDRAM, direct RDRAM. This type of memory provides data transfer rates of 1.6 GBps over one channel and up to 6.4 GBps over four channels.

RDRAM-based memory subsystem comprises the following components: a controller, a channel, a dual in-line memory module (DIMM) built using RDRAM chips, and a terminator. The controller is built using a specialized Rambus ASIC Cell (RAC) chip, which generates memory subsystem control signals at up to 400 MHz frequency. Up to four independently functioning channels can be connected to one RAC controller. The channel provides an electric link between the controller and the memory chips. It also contains a control block for the memory chips connected to the channel (up to 32 RDRAM microchips) and 30 high-speed data lines transferring data at both clock edges. To suppress signal reflections, a terminator is installed at the end of the channel. Channel lines are divided into a 16- or 18-bit data bus, a 5-bit row address bus, and a 3-bit column address bus. Using separate buses for the row and column addresses increases the memory subsystem speed.

Data and control signals are transmitted over the channel in 8-bit bursts under the Direct Rambus™ protocol. For transferring large data blocks, Rambus memory is the most advantageous choice in terms of the efficiency/cost ratio. This explains the wide use of RDRAM in game stations (Nintendo®) and graphic workstations (Silicon Graphics®).

Rambus and Intel have the rights to the RDRAM production technology, and other manufacturers have to license it from them. With the aim of creating an open-memory standard equaling the RDRAM technical characteristics, IBM®,

Apple®, Motorola®, and Micron Technology® formed an association named SyncLinc Consortium.

The SLDRAM they developed is similar to the RDRAM organization. A SLDRAM subsystem is made up of a controller; an instruction; an address; and a data-transfer channel, memory microchips, or modules (SL modules); and a terminator. Instructions, addresses, and control signals are transmitted over a one-way 10-bit bus named CommandLink; data is transferred over a two-way 18-bit data bus named DataLink. Data and instructions are transferred in 4- or 8-bit bursts.

Up to eight memory devices (SLDRAM chips or SL modules containing several chips) can be connected to the memory subsystem controller. When the system is powered up, each memory chip is assigned an individual number. This allows its speed and location in the memory subsystem to be taken into account when determining temporal delays during signal transfers. The values of a chip's read and write delay times are written into its control registers, making it possible to achieve simultaneous response from all memory chips. Taking speeds of the memory subsystem's individual chips into account allows their different modifications to be used, differing in both speed and capacity.

Depending on the speed of the memory modules, the channel can operate at different speeds. To date, SLDRAM is available working at both 200 MHz clock edges. Production of 400 MHz and higher SLDRAM memory in the near future is planned.

Using fast and expensive dynamic memory as the system main memory is not always economically practical. More often, inexpensive, slow main memory is used in combination with a small, fast static-memory cache.

STATIC MEMORY

Static random-access memory (SRAM) has 15–20 ns access time and as a rule is used to build cache.

The simplest caches operate in asynchronous mode. Here, the processor sends an address to the cache; the cache finds the specified location and sends the data that was asked for to the processor. At the beginning of each access, as a rule, an extra cycle is used to examine tags. For asynchronous static memory, block-data `read` access time is expressed as 3-2-2-2, and as 4-3-3-3 for `write` operations.

Incoming addresses are buffered in synchronous cache. During the first cycle, SRAM stores the requested address in a register. During the second cycle, data is extracted and sent. Because the data address is stored in a register, synchronous SRAM can obtain the next address while the processor is receiving data from the previous request. Synchronous SRAM can combine sequential data elements into bursts without having to receive and decode additional addresses. Access time for this type of memory is reduced by 15 to 20 percent compared to asynchronous, down to about 10 ns.

Pipelined data-burst exchange mode is used to lower the execution time of block `read` and `write` operations. Memory that supports such mode is called pipelined burst SRAM. The pipelining consists in adding an output buffer into which data read from memory cells are placed. Serial `read` accesses to the memory are executed faster, without delays for accessing memory grid to get the next data item. Pipelined burst SRAM `read` and `write` operations are expressed as 3-1-1-1.

Sometimes developers try to solve the problem of increasing main memory speed by building cache into dynamic memory microchips. This type of memory from Mitsubishi Electronics® is called cached DRAM (CDRAM). Each 4- or 16-bit microchip of this memory is equipped with 16 KB of fast-cache memory. The exchange of data between the dynamic and static memory is done in 128-bit words. Enhanced DRAM (EDRAM) from Ramtron® has an 8 KB cache for each 4 Mb of dynamic memory. The exchange in this case is done in 2,048-bit words. Because of their high speeds, CDRAM and EDRAM are usually used in systems without L2 cache.

VIDEO MEMORY CHIPS

Along with regular dynamic memory microchips (FPM DRAM, EDO DRAM, and SDRAM), specialized memory types oriented for supporting bit-stream operations are used in computer video subsystems. Currently, the main specialized memory types are synchronous graphics RAM (SGRAM), multibank DRAM (MDRAM), video RAM (VRAM), window RAM (WRAM), and 3D-RAM.

Operating principles of SGRAM are similar to those for SDRAM. At the same time, SGRAM supports several additional functions specific to the application of this type of RAM. Special block write modes—Block Write and Write-per-Bit— were added to optimize large bit-block transfer operations, which are characteristic of video adapter operations. The Block Write mode allows the value stored in the Color register to be written into eight memory cells simultaneously. The Write-per-Bit mode is used to fill up memory cells with data with individual cell-masking capability. Compared with DRAM, this memory type allows the execution speed of video operations to be increased fivefold.

Developed by MoSys®, MDRAM is mostly used in the video cards from Tseng Labs. The specific feature of this memory is its block organization. Memory consists of many independently functioning 32-bit-wide banks holding 32 KB each and connected to the internal multi-bit bus. Owing to the block organization, microchips of practically any size can be built. Thus, if the standard types of video memory come in 1 MB, 2 MB, 4 MB, 8 MB, and so on, MDRAM can be 2.5 MB, 3.5 MB, and so on. This feature makes it possible to lower video card prices while at the same time providing the required color rendering and resolution. For example, only 2.25 MB of SGRAM are needed to implement 1024 × 768 True Color video mode, whereas 4 MB of regular video memory is required to implement the same mode.

VRAM has been developed to provide the 170 MBps bandwidth that is needed to support the 1024 × 768 × 16 M video mode at 75 KHz frame rate. For this purpose, VRAM is equipped with an additional port. Data from VRAM is sent in 4-KB blocks over the internal bus to the special built-in serial access memory (SAM); SAM converts data into a serial bit stream, which is then sent to digital-to-analog converter (DAC) to form the video signal. Between data writing operations from the memory core into SAM, VRAM can service processor requests without delays.

Developed by Samsung, WRAM is a less-expensive alternative to VRAM. Like VRAM it also has two ports, which allow it to exchange data simultaneously with the processor and the DAC. The WRAM's simplified internal architecture made it possible to lower its price by 20 percent, as compared with VRAM, and at the same time raised the exchange operation speed by 50 percent. Memory has hardware support of the screen-scroll function, filling a portion of the screen with a two-color pattern and masking individual memory areas. WRAM can operate at 50 MHz providing 960 MBps bandwidth.

3D-RAM was developed by Mitsubishi Electronics for use in video cards in tandem with a 3D graphics accelerator. This memory contains a CDRAM array and an ALU block, which allows some image manipulation operations to be performed.

1.2.4 Programmable Logic Integrated Circuits

There has always existed the need to rapidly build a prototype of the developed device to ascertain its operability and eliminate potential mistakes made during its development. Moreover, due to various technical and economic factors a need may arise to perform certain computations by hardware. This need may be caused, for example, by the requirement for higher computation efficiency that cannot be achieved by software. This situation may happen in a case when an instruction must be fetched before actual computation can be carried out with the processor lacking the proper arrangement of registers, functional units, and data paths to transfer data among them to realize the given algorithm by hardware efficiently. Another example may be modifying a hardware implementation of an algorithm, depending on the type of the processed data, to reduce the number of gates as compared with the situation where for each type of processed data an individual hardware implementation of the processing algorithm is constructed.

To provide a solution to the previously described problems, programmable logic-integrated circuits (PLIC) have been developed. PLIC are LSI matrix circuits that can be programmed to form in one chip an electronic circuit equivalent to a circuit containing from ten to several hundred standard logic-integrated circuits [10]. Compared to other microelectronic technologies, including the master-slice wafer and application-specific integrated circuits, the PLIC technology provides a record-short engineering design cycle (from several hours to several days), minimum design costs, and maximum equipment modification flexibility.

Logic, I/O, and switching blocks are configured by loading the PLIC with the bit sequence obtained during the development of the circuit. Depending on the PLIC family, logic, I/O, and switching blocks have various degrees of complexity and different functional capabilities.

Currently several main PLIC producing companies on the world market produce reprogrammable erasable programmable logic devices (EPLD) and reconfigurable field programmable gate arrays (FPGA).

The EPLD PLIC use ultraviolet erasable read-mostly memory in which to store the configuration; the FPGA PLIC use static memory for this purpose.

A FPGA circuit is a two-dimensional array of logic cells interconnected by logic keys. The FPGA static memory is filled with a specific bit sequence and affects the logic cells and, connecting them, logic keys, producing the required electric circuits—registers, counters, logic circuits, and so on—interconnected as required. FPGA chips also have an input for writing the bit stream into the static memory, as well as I/O elements for linking with other microchips.

During development of PLIC-based systems, all design stages are completed by the designers in one workplace using CAD systems. Each company PLIC manufacturer develops and manufactures a custom CAD system that provides implementation of all design stages for each programmable logic type.

Similarly to writing programs for universal processors, some instrumental means are necessary for programming PLIC: programming languages, compilers, optimizers, an so forth. The end result of using these tools is a bit stream that fills the static memory of FPGA chips. CAD systems allow designers to use standard library components to create logic circuits that implement specific algorithms, to perform modeling and analyzing of their functional and timing characteristics, to optimize the developed circuits with respect to the equipment and the task execution time, and to translate the developed circuits into bit streams that define processor's logic. XILINK, for example, supplies complete mathematic applications for designing and using FPGA and EPLD devices. Most popular CAD systems, such as ViewLogic® and Mentor Graphics®, running on Sun™ and HP® PCs and workstations, respectively, have schematics editors and modeling systems for designing PLICs. These CAD systems use hardware-engineering programming languages such as VHDL and Verilog to program PLIC. Efforts are also being made to use for this purpose data-flow computing languages as well as the traditional programming languages such as C, for example.

There is a radical point of view that PLIC will eventually replace regular microprocessors because the user will be able to implement the system he needs using software, taking into account all the specifics of his application. However, PLICs are more likely to follow another direction in the VLSI circuit development, which is the integration of PLIC and processors. Here, the PLIC is built using either off-the-shelf blocks developed by the manufacturer or blocks that are intellectual property of other firms (IP blocks). IP blocks can be soft, that is, their topology not fixed, or hard, that is, their topology fixed. This makes it possible to achieve high efficiency of the hardware implementation and greater protection of intellectual property rights.

Consequently, a reconfigurable processor can be built using one or several FPGA microchips. This processor will have the advantages of a hard-logic-specialized processor, but it can be programmed like a universal processor to solve any type of problem.

A distinguishing PLIC architecture feature of the XILINX XC2000, XC3200, and XC400 family FPGA is its field of configurable logic blocks and I/O blocks interconnected by switch blocks.

One of the FPGA basic elements, a configurable logic block (CLB), can perform any logic function by means of the look-up table (LUT) into which a specific bit sequence is loaded (see Figure 1.13). The function that a CLB performs can be changed an unlimited number of times by loading the LUT with a different bit sequence.

CLBs of different FPGA chips differ in the number of LUTs, number of LUT inputs (from two to four), which allow functions of the corresponding number of variables to be implemented, as well as the number of registers per block.

I/O blocks, like CLBs, can be programmed to make any type of electrical connection of the circuit implemented inside a programmable logic chip to the external world via a corresponding contact on the chip.

Currently the following microchips are produced by XILINX:

XC7200 and XC7300 series of the EPLD type—Contain from 18 to 144 multi-input macrocells, each of which is a programmable one-bit ALU with a built-in trigger latch. The cells are interconnected by a matrix switch. These microchips can be used to implement nonconventional ALUs, decoders, counters, and so on.

FPGA type XC2000 and XC3000 series—Contain from 2,000 to 9,000 equivalent gates and up to 320 CLBs.

FPGA type series XC3000A, XC3100A, XC4000, and XC5000—Contain from 2,000 to 25,000 gates. Have up to 1,024 CLBs implementing 5-, 9-, or 20-variable functions with the execution time up to 2 ns. Series XC4000 chips can implement up to 2,560 triggers and have up to 32-Kb built-in RAM

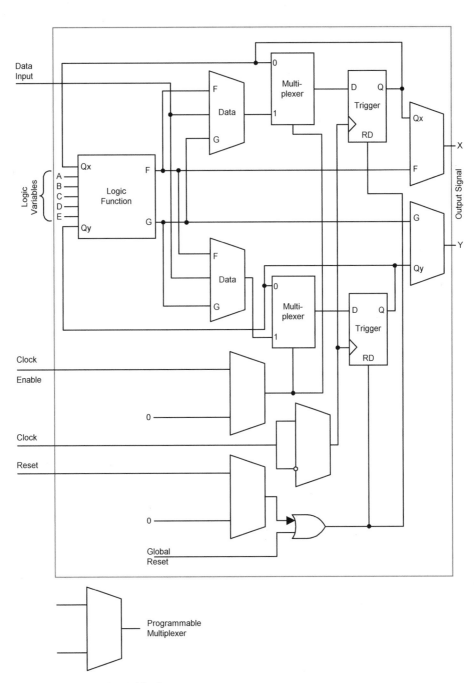

FIGURE 1.13 PLIC logic block.

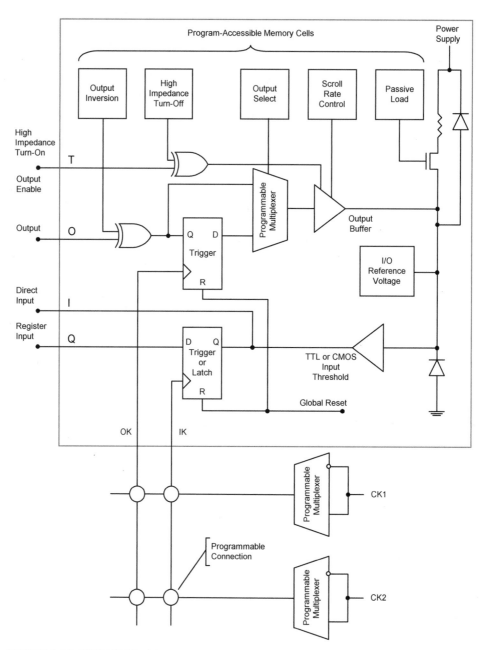

FIGURE 1.14 PLIC I/O block.

FPGA type XC4000E series—Developed on the basis of the XC4000 series using improved 0.5-micron technology with three metallization (conductor) layers. Provides a trigger-switching rate 1.5 times higher than the XC4000 and 60 percent more efficient arithmetic operation execution with a lower relative cost. A new function of configuring the built-in RAM makes it possible to implement synchronous/asynchronous, one-/two-port exchange modes.

FPGA type XC6200 series—Designed specially to construct coprocessors. It has a built-in 8/16/32-bit programmable interface (FastMAP) that performs direct exchange with the bus of the main processor. The increased configuring speed (1,000 times higher compared with the previous series) allows partial reconfiguration of chips while executing the current task. A built-in RAM (36–256 KB) can be accessed via the FPGA logic, FastMAP interface, or using both methods. This series had an innovative architecture and was popular in academic research but found no commercial application. It is no longer available.

XC8100 series—One-time programmable FPGA. It is executed on the base of the MicroVia antifuse CMOS technology with three metallization layers. The technology makes for high logic-component density and low power consumption.

XC9500 series are CPLD type—Contain an in-system chip reprogrammable up to 10,000 times. Contain from 800 to 6,400 work cells or from 36 to 288 macrocells analogous to the XC7300 series.

XC4000XLA and XC4000XV series of FPGA type—Executed using 0.35-micron and 0.25-micron CMOS technologies, respectively, and have five metallization layers. Maximum number of gates is 500,000.

Virtex-E and Virtex-EM series are FPGA—Executed using 0.18-micron CMOS technology and have six metallization layers and over 40,000 CLBs. I/O blocks are PCI bus compatible (32/64 bits, 33/66 MHz).

Virtex II series are FPGA type—Executed using 0.15/0.12-micron technologies. Have eight metallization layers and up to 10 million gates. CLBs have eight LUTs for specifying functions and an 18-bit fast multiplier. Have a provision for using custom and IP blocks.

All XILINX series chips have the highest degree of protection against copying.

1.2.5 Development of Computer System Microelectronic Components

To design a high-performance computer system, it is important to integrate as many data-processing and storing functions on the chip, as well as the interface with user and other computer systems, as possible.

Desire to integrate function sets is motivated by several factors. First, with single-chip implementation, bandwidths of the interfaces between data-processing and storing subsystems are not limited by the number of pins on the chip's casing and can be made as large as needed. Second, construction of the motherboard on which the chip is mounted becomes simpler as fewer components need to be mounted on it; this increases computer system reliability and performance and reduces its cost. Third, the number of output pins on the chip casing can be reduced, because the interface with the other components of the computer system (such as with the information-displaying devices) is minimized. Finally, resources of millions of transistors that can be placed on the chip are utilized. These transistors can be used to build a specialized system consisting of a collection of task-oriented blocks as well as to build a parallel system from a set of identical processors.

Along with the task of constructing single-chip systems, there is the problem of constructing fast interfaces between chips in multichip systems, for example, between processor and memory chips. In Pentium Pro, this problem is solved by placing two chips—the actual microprocessor and L2 cache—into one casing. Another possible solution to this problem is multichip micro-assemblies in which packageless VLSI circuits are mounted on the silicon substrate and are connected by interchip conductor layer, perhaps by even several of them.

1.3 ARCHITECTURAL FEATURES OF MODERN MICROPROCESSORS

1.3.1 Classification of Microprocessor Architectures

When trying to set forth the collection of concepts and technical solutions comprising the basis of the modern microprocessor architecture, inevitably a problem arises associated with so rapid and far from linear developments in this area of knowledge and technologies. At each stage of the development, a particular subset of interrelated architectural concepts is dominant due to the elementary basis capabilities and the preferences of a relatively small number of microprocessor manufacturing companies. Further, these concepts can be intercombined giving rise to another architecture. Therefore, in presenting architectures, a certain historic order of presenting their concepts as they appeared and were employed in constructing microprocessors must be followed.

At first, processors had instruction sets formulated by their designers based on the analysis of that problem area for the solving of which the computer system was being designed. The optimization criterion for a proposed instruction set was the minimal size of the program aimed at solving the necessary problems. In the optimization process, instructions were introduced that used registers and memory cells as operands and employed complex address-generation schemes that used index registers for performing in one instruction both the operand processing and address modification for executing the next instruction. Analysis of the program code generated by high-level compilers showed that only a limited set of simple instructions of the *register ← register, register, memory ← register* and *register ← memory* formats was used. Compilers cannot use complex instructions efficiently. This observation has contributed to the formation of the concept of the reduced instruction set computer (abbreviated RISC) [4].

Another circumstance that actually led to creation of RISC processors was the development of the Cray-type pipeline processor architecture [6]. These processors use separate instruction sets for memory operations and for processing information in the processor registers. Each of these instructions is broken down into a few stages of equal execution time (instruction fetching, decoding, executing, and storing the result); this makes it possible to build an effective processor pipeline capable of executing the next instruction every cycle.

However, pipelined instruction execution caused problems associated with data and control dependencies between instructions that are loaded for execution into the pipeline sequentially. For example, if the next instruction uses results of the previous instruction, it cannot be executed for several cycles that are necessary to obtain this result. Similar problems arise with execution of conditional-branch instructions, when the data determining the branch direction are not ready by the time the conditional-branch instruction has been decoded.

These problems are solved in two ways. In the first, the compiler defines the order instructions are launched for execution in the pipeline and inserts *nop* (no operation) instructions when the next instruction cannot be launched. In the other method, special hardware in the processor monitors instruction interdependencies and eliminates conflicts.

Processors can be divided into two classes: reduced instruction set computer (RISC) and complex instruction set computer (CISC).

The former have processing instructions of the type *register ← register, register* and load and store instructions of the *memory ← register* and *register ← memory*, respectively. Functional modifications can be performed only on the content of registers with the result also placed into a register.

As a rule, instructions in CISC processors come in various formats and require a variable number of memory cells for their storage. Because of this, instruction type can be determined only in the process of its decoding when being executed. This fact makes the processor control unit more complex and is an obstacle to raising the clock frequency to the level of the RISC processors built on the same elementary base.

It is obvious that RISC processors are more efficient in those application areas in which structural methods of reducing memory-access time can be used productively. If a program generates memory-access addresses at random and each data item is used by only one instruction, then, in effect, processor efficiency is determined by main memory-access time. In this case, using reduced instruction set only lowers the efficiency, because it necessitates exchanging operands between memory and registers instead of memory transfers of the *memory* ← *memory, memory* type. The programmer must take into account the necessity to localize the processed data so that when data blocks are cached, all data in them will be processed. If a program is written is such a way so that the data is located in random order and only a small part of data in the transferred blocks is processed, the execution speed will drop manifold, down to the operating speed of the main memory. As an example, Table 1.2 lists performance benchmark results of an Alpha 21066 233 MHz processor running Hadamard transform with $n=8-20$.

TABLE 1.2 Alpha 21066 Microprocessor Performance Executing Hadamard Transform

n	Performance in Conventional Algorithmic Operations, MIPS
8	150
10	133
11	73
≥ 12	20

The given example demonstrates that although data is located in the on-chip cache, the performance is high. However, as soon as the data volume exceeds the cache size and memory accesses are made to addresses distributed uniformly over the memory range, the performance drops by more than seven times.

Microprocessor development is accompanied by the constant desire to maintain upward software compatibility and to raise performance by perfecting the architecture and increasing the clock frequency. The goals of maintaining upward software compatibility and raising performance, generally speaking, contradict each other. Thus, for example, x86 processors, which up to Pentium Pro belong to the CISC class, have lower clock frequencies as compared with RISC microprocessors built under the same technological standards. In some applications, these x86 microprocessors demonstrate much lower performance than the RISC processors implemented using the same elementary base. However, upward software compatibility of the x86 processors of different generations assured them a stable, dominant market position.

High-language sequential program source code is compiled into machine code that reflects the program's static structure, that is, the ordered instruction set in the computer's memory. The execution process of a program with specific input data sets can be represented by the program's dynamic structure, that is, by a set of instruction sequences in the order of their execution.

The parallelism degree of a program can be raised by changing its relevant static or dynamic structure. Because the static structure of a program has a univocal correspondence to its source code (assuming the same compiler is always used), changing it comes down to changing the source code, which in general is not always possible. The dynamic structure, on the contrary, can be changed with the static structure remaining unchanged. The main goal of this change must be increasing the degree of the parallel instruction execution.

The extent to which the dynamic structure of a program can be permitted to be changed is determined by the relations existing within the set of instructions: control and data dependencies. The superscalar processor architecture is described using the execution window model. When a processor is executing a program, it is as if moving the execution window over the program's static structure, and in this way restricting the set of instructions that are checked for data and control interdependencies. Instructions in the window can be executed in parallel only if they are not interdependent.

For elimination of dependencies caused by branch instructions, a prediction method allowing extraction and conditional execution of the instructions of the predicted branch is employed. If it turns out later that the prediction was right, the results of the conditionally executed instructions are accepted. If the prediction was wrong, the processor state is restored to what it had been at the moment the decision to make the branch was made.

Instructions placed into the execution window can be data dependent. These dependencies are caused by using the same memory resources (registers, memory cells) in different instructions. Therefore, for the program to execute correctly, these resources must be used in the order specified in the program.

All types of data dependencies can be classified by the association type: RAR—read after read, WAR—write after read, WAW—write after write, and RAW—read after write.

An example of various instruction data dependencies is shown in Figure 1.15.

Some data dependencies can be eliminated. RAR actually corresponds to the absence of dependencies, because in this case, the order in which instructions are executed does not matter. Only RAW is a real dependence, because in this case it is necessary to read not the old data but the previously written new ones.

Extraneous data dependencies are created as a result of WAR and WAW. The WAR dependence consists in that an instruction must write a new value into a memory cell or a register, which then must be read. Extra dependencies are caused

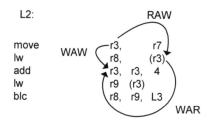

FIGURE 1.15 Instruction data dependencies.

by various factors: unoptimized program code, insufficient number of registers, attempts to economize on memory, or loops in programs. It is important to note that write operations can be performed into any available resource, not only into the one stated in the program.

After extraneous control and data dependencies have been eliminated, instructions can be executed in parallel. The task of generating the parallel instruction execution schedule is given to the microprocessor hardware. This schedule takes into account the existing instruction interdependencies and the available functional processor modules.

Modern microprocessors widely use the principle of pipelined execution of elementary operations. Pipelining internal processes allows results to be obtained after each processor cycle.

STRUCTURAL PARALLELISM OF PROCESSORS WITH DECOUPLED ARCHITECTURE

The striving to use inherent natural parallelism in calculating integer address expressions and in the actual floating-point data processing has led to the development of decoupled processor architectures [11].

Roughly, a decoupled architecture processor, as shown in Figure 1.16, consists of two linked subprocessors, each of which is controlled by its own instruction stream.

By convention, these processors are called address A-processor and executive E-processor. A- and E-processors each have their own register sets A0, A1, and so on, and X0, X1, and so on, respectively, and instruction sets. The A-processor performs all address calculations and generates read and write memory accesses. It is a regular integer processor and as such can also perform any integer operations not related to address calculations. The E-processor performs floating-point calculations.

Data extracted from the memory either is placed into the FIFO AA queue and used by the A-processor or is placed into the FIFO AE queue to be sent to the E-processor. When the E-processor needs data from the memory, it takes it from the AE queue. If the AE queue is empty, the E-processor stalls until data arrives,

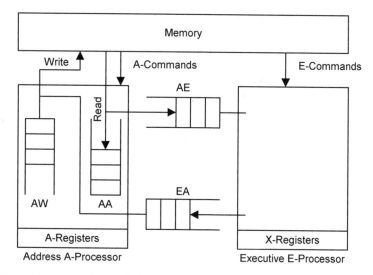

FIGURE 1.16 A decoupled architecture processor.

which solves the problem of synchronizing operations of the A- and E-processors. When the E-processor produces data that must be sent to the memory, it places it into the FIFO EA queue.

When data is written to the memory, the A-processor sends the address to the FIFO AW memory write queue immediately after it calculates it, without waiting for the data to arrive into the EA queue. The A-processor matches the address/data pairs by selecting the first elements in the EA and AW queues and transferring these pairs to the memory. Naturally, if one or both of these queues are empty, the memory transfer is postponed.

When data is read from the memory, the A-processor sends the address to the memory, indicating the AA or AE queue into which data must be read from the memory.

Decoupled architecture makes it possible to achieve scalar processing efficiency characteristic of vector processors. It achieves this by prefetching data from the memory and automatically launching several sequential loops of the cycle in the A-processor. The task of splitting the original program into programs for the A- and E-processors is resolved on the compiler level or by a special splitting block.

INSTRUCTION PREFETCH AND BRANCH PREDICTION

The main principle defining development of the superscalar microprocessors [12] consists in building microprocessors containing the greatest number of functional devices possible while preserving traditional sequential programs. This means that compilers and the processor hardware load the parallel functional units of micro-processor by themselves, without programmer's involvement.

A typical architecture of a superscalar microprocessor is shown in Figure 1.17.

The main blocks of a superscalar microprocessor are instruction fetch and branch prediction, instruction decoding, instruction interdependence analysis, renaming and dispatching, register, fixed-point processing unit, memory control, and executed-instruction ordering blocks.

Superscalar processing requires extracting several instructions from the memory in one cycle to load parallel functional units. Because of this necessity, increased requirements are placed on the microprocessor/memory interface. Modern processors address this issue by using separate hierarchic instruction and data caches.

Branch instructions create serious problems for effective loading of functional units. When the PC needs to be modified, at least one clock cycle is needed to de-code the branch instruction, modify the PC, and fetch the instruction from the new PC address. These delays cause unproductive cycles in the processor pipelines.

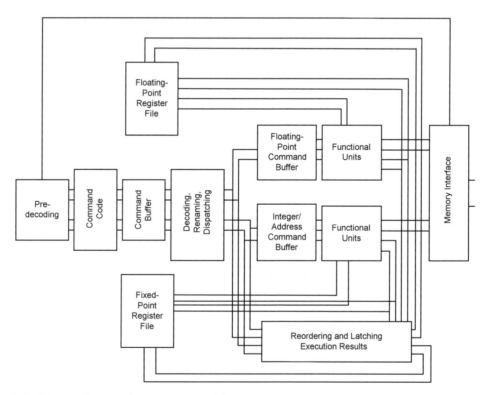

FIGURE 1.17 Superscalar processor architecture.

Postponed branches can be employed to prevent the arising of empty pipe-line cycles. Here one or more instructions following the branch instruction are executed unconditionally. More sophisticated solutions employ the following methods:

- Branch predicting
- Changed-order instruction execution
- Conditional (predicate) execution

When a branch is predicted, the instructions in the predicted branch are fetched and executed without waiting for the evaluation of the branch condition determining the actual branch. In case of a wrong prediction, all results of the instructions executed because of this prediction must be discarded. The number of the processor cycles lost in this case depends on the number of functional units and the depth of their pipelines. In modern processors the number of cycles lost because of a wrong prediction can reach several hundreds. Moreover, the amount of hardware used to clean up the consequences of wrong prediction is sufficiently large and can negatively affect attempts at increasing processor clock frequency.

The goal of changing instruction-execution order is to load idling functional units by snatching out of the instruction stream those instructions for which there are ready operands and functional units available to execute them. However, in this case, instruction interdependencies in terms of registers and other used resources must be examined. The absence of dependence must be determined rapidly; otherwise, changing instruction-execution order will make no sense.

To optimize executable program code, trace cache can be used [13]. This type of cache maps instructions selected from the microprocessor's instruction cache onto the physical contiguous instruction-memory area. The instruction stream that fills trace cache is optimized to increase the efficiency of the already executed instruction trace when it is selected again. A trace is optimized against the background of its execution by the processor.

To reduce the number of lost processor cycles caused by cache misses when executing branch instructions, caches are equipped with branch-prediction capabilities, whose main mission is to increase the probability of the required instruction being in the cache.

Execution of conditional branches consists of the following stages:

- Decoding a conditional branch instruction
- Evaluating the branch condition
- Calculating branch address
- Transferring control if a branch is made

Special efficiency raising methods are used at each stage.

Stage 1—Either additional bits in the instruction field or instruction pre-decoding during fetching from the instruction cache is employed to speed up instruction decoding.

Stage 2—Often, when an instruction has already been fetched from the cache, the branch condition has not yet been evaluated. In order not to hold back the instruction stream, the branch is predicted using one of the several available methods. Some branch-prediction methods use static information either from the program's binary code or specially generated by the compiler. For example, certain operation codes generate branches more often than other codes, or branching is more probable (with loops), or the compiler can set a flag indicating the branch direction. Static information obtained during program tracing can also be used.

Other branch predictors use information that is dynamically generated during program execution. This is usually information about the history of a particular branch, which is stored in the branch table or in the branch-prediction table. A branch-prediction table is organized similarly to cache, using the associative principle; its elements are accessible at the address of the instruction whose branch is being predicted.

In some implementations, an element in a branch-prediction table is a counter whose value is incremented when a correct prediction is made and decremented when the prediction is wrong. This way the counter value determines the prevailing direction of branches.

At the moment when the actual value of the branch condition has been evaluated, the branch history is updated. If a prediction was wrong, correct instructions must be fetched. Results of the conditionally executed instructions must be discarded.

Stage 3—To determine a branch address, usually an integer-addition operation needs to be performed that adds the offset specified in the branch instruction to the current PC. Although this operation does not require extra processor cycles to access the registers, address calculation can be sped up by using a buffer containing branch addresses used earlier.

CONDITIONAL INSTRUCTION EXECUTION IN VLIW PROCESSORS

Very Long Instruction Word (VLIW) is an alternative to the superscalar processing. This method consists in specifying a collection of concurrently executable instructions in an instruction word. Such programs are prepared by compiler.

Within the IA-64 architecture framework being developed by Intel and HP, long instructions are organized based on instruction bundles that are constructed from three instructions and a special template field.

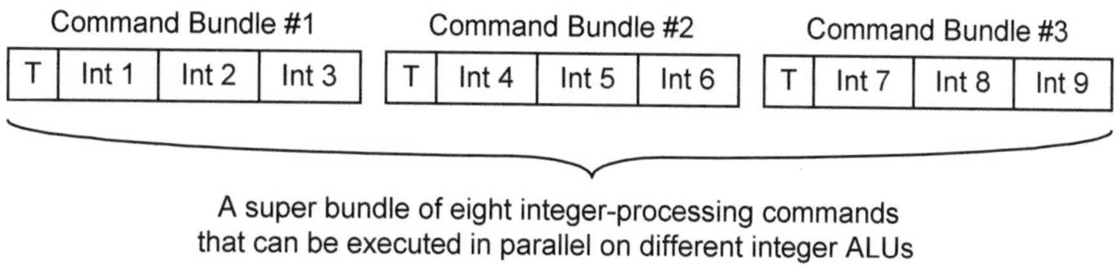

A super bundle of eight integer-processing commands
that can be executed in parallel on different integer ALUs

FIGURE 1.18 An instruction super bundle.

The possible three-instruction bundles are as follows:

i1 || i2 || i3 —Instructions i1, i2, and i3 are all executed concurrently.

i1 & i2 || i3—First the i1 instruction is executed, then the i2 and i3 instructions are executed concurrently.

i1 || i2 & i3—First the i1 and i2 instructions are executed concurrently, then the i3 instruction.

i1 & i2 & i3—Instructions i1, i2, and i3 are all executed sequentially.

The template field is used to control the instruction execution and organize super bundles from several adjacent bundles. The template indicates which of the instructions in the bundle or in the neighboring bundles can be used in parallel on different functional devices. For example, Figure 1.18 shows a super bundle built of eight integer instructions.

The linking template is created when the program is compiled and all information pertaining to the control of the concurrent instruction execution is placed into it. Depending on the number of the needed functional devices, all eight instructions can be executed all concurrently in one cycle, in two cycles with four instructions per cycle, or some other way.

The IA-64 instruction format follows: an instruction code, three 7-bit operand fields, one destination and two sources (only registers can serve as operands), special fields for floating- and fixed-point arithmetic operations, and a special 6-bit predicate field.

Efficient functional unit loading in LIW/VLIW processors is obtained by using predicated instruction execution and instruction prefetch.

The predicated execution mechanism is based on adding a special predicate field to instructions. Predicated execution eliminates the need to use conditional branch instructions. Branch instructions and two alternative branches, one of which is executed depending on what the predicate of the branch instruction evaluates to, are no longer used. Instead, the same predicate as in the branch instruction is evaluated by a separate special instruction, and its value is saved in the special predicate register. Instructions of one of the alternative branches use the U value of the calculated predicate, whereas the instructions of the other alternative branch use the *not U* value of the predicate. Instructions of both branches are executed, but only the results of that instruction branch whose predicate evaluates to *true* will be accepted. This changing of branch instructions to predicated execution is called *if-conversion* [14] and is carried out by compilers. On the left side in Listing 1.2, a fragment of a program using branch instructions is shown; on the right side, a corresponding code fragment using predicated execution and predicate evaluation instructions is shown.

A corresponding flowchart of the code fragment is shown in Figure 1.19.

LISTING 1.2

```
if (a ← 10)              pred_lt p1(-U), p2(U), a, 10

c = c+1                  add c, c, 1 (p2)

else if (b > 20)         pred_gt p3(-U), p4(U), b, 20 (p1)

d = d+1                  add d, d, 1 (p4)

else e = e+1             add e, e, 1 (p3)
```

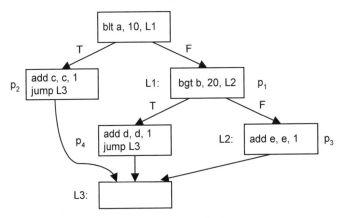

FIGURE 1.19 Flowchart of Listing 1.2 code fragment.

Using predicated instructions simplifies the task of loading a processor's functional units, shifting the job of generating conditional instructions to the program compilation stage. However, to efficiently support predicated instruction execution, the following mechanisms are needed:

- A special instruction field to indicate the predicate operand
- A predicate register file
- The discarding of results of the instructions whose predicate operand evaluates to false
- The segregation of the collection of conditional instructions

One way of preventing an instruction word from getting longer is to limit the instruction types that can be conditionally executed to only those instructions whose formats have fields for the predicate. For example, only MOV move instructions can be conditionally executed. For this purpose, the special conditional move format CMOV is introduced.

Conditional instructions are used to a various degree in such microprocessors as Alpha, Advanced RISC Machines ARM, Philips TriMedia, MIPS R x000, Sun SPAR, TMS 320 C6xx, and in the x86 architecture. Intel introduced the CMOV instruction into its Pentium Pro and Pentium II microprocessors in 1995. These instructions were introduced in the Sun SPARC, DEC Alpha, and R x000 microprocessors in 1991, 1992, and 1995, respectively [15].

There are various restrictions on using conditional instructions in modern processors. Thus, some instructions perform only conditional moves. Another approach is implemented in the ARM microprocessor: all instructions are conditional and can use 16 predicates; however, nested predicates of conditional instructions cannot be used.

Both specially designated registers and general-purpose registers can be used as predicate registers. In the latter case, more often than not a register shortage arises. Moreover, the values of the predicate registers must be read the same as other operands, which increases the number of register file ports. Therefore, introduction of a separate predicate file is fully justified.

Of course, the preferred use of the conditional instructions is to parallel short alternative branches.

Traditional branch instructions can be used to execute cycles with a great number of repetitions (for example, in the order of a thousand) organized using branch instructions with the known-in-advance handoff to the initial instruction. Moreover, branch instructions are needed to perform seldom-used program-code-block processing, for example, exceptions.

Introduction of new instructions requires hardware to specify predicates and work with them. However, there always is a relative excess of processing devices that cannot be loaded with direct calculations. Therefore, these processing devices can be used to calculate predicates toward the overall increase of the processor efficiency.

Using conditional instructions introduces elements of associative data processing to microprocessors. This, in general terms, can have a substantial effect on the program design and implementation style when developing the predicate work logic.

The framework developed by the Moscow SPARC Technology Center NArch architecture [16–18] proposes [18] using hardware-executable conditional instructions. It stipulates using a special predicate register file and a special functional device, allowing up to three predicates to be calculated in one cycle. Together with the predicates calculated by the processor's arithmetic units, the total number of the predicates calculated in one cycle can reach up to six.

There are primary predicates and secondary, which are calculated using primary predicates. Examples of primary predicates [18] are given in Listing 1.3:

LISTING 1.3

```
int a, b;

float c, d;

if (a==b)                           0: cmp_eq %r1, %r2, %p1

.......... . .                      .............................. . .

if (c<=d)                           n: fcmp_le%r3, %r4, %p2
```

Variables a, b, c, and d are initially placed in register %r1, %r2, %r3, %r4, respectively. Predicate values are written into registers %p1 and %p2 of the predicate file.

The possible calculations of compound predicates with receiving three predicate values in one cycle are demonstrated in Listing 1.4.

Listing 1.4

```
int a, b, c, d;

............... .

if (a==b && c<d | |                 0: cmp_eq%r1, %r2, %p1

a !=b && c<=d)                      cmp_l %r3, %r4, %p2
```

```
                                      cmp_le  %r3, %r4, %p3

                              1: land      %p1, %p2, %p4

                                 land      !%p1, %p2, %p5

                                 land      !%p4,!%p5, %p6
```

Character ! in Listing 1.4 denotes logic NOT; logic OR operation is implemented by performing logic AND operation (LAND) over inverted operands in a special functional unit. In this way, it takes two cycles to calculate a compound predicate.

The nature of code compiled in NArch architecture is shown in Listing 1.5.

Listing 1.5

```
int a, b, c, d, x;

.................

if (a==b && c==d)            0: cmp_eq  %r1, %r2, %p1

x++1;                           cmp_eq  %r3, %r4, %p2

                             1: land      %p1, %p2, %p3

                             2: nop

                             3: add       %r5, 1, %r5 ? %p3
```

Character ? precedes the predicate of a conditionally executed instruction.

A series of SPEC-92 tests that are characterized by multiple, short linear code sequences and branch instructions with short alternate branches were conducted. It is claimed in [18] that in six of such tests the efficiency can be raised on average by 23 percent, owing to the use of NArch architecture with the conditional instructions fully supported by hardware.

INSTRUCTION DECODING, RESOURCE RENAMING, AND DISPATCHING

Regardless of where the fetched instructions are to be executed, be it either a superscalar microprocessor or a VLIW microprocessor, they are next decoded and resources for their execution allocated. At this stage the essential data dependencies between instructions (RAW) are determined, the nonessential dependencies (WAW, WAR) are negotiated, and the instructions are distributed among the instruction buffers of the functional units.

Instruction decoding produces one or several ordered triples, each of which contains:

- The operation being executed
- Pointers to the operands
- Pointer to where the result is to be stored

The mechanism of dynamic mapping of the logic resources defined by the program text onto the microprocessor physical resources is used for negotiating extraneous WAR and WAW dependencies caused by the scarcity of logic resources (memory cells, registers). Using this approach, one logic resource can be associated with several values in different physical resources, each of whose values corresponds to the logic value at one of the moments in the sequential program execution.

When an instruction produces a new value for a logic register, the physical resource into which this value is placed is given a name. The subsequent instructions that use this value are given the name of the physical resource. This procedure is called *register renaming*. There are two main register-renaming methods.

In the first method, the physical register file is larger than the logic. When a renaming needs to be done, one register is taken from the list of the available physical registers and a logic name is associated with it. If there are no free physical registers available, instruction dispatching is halted until such become available.

This renaming mechanism is implemented as follows: suppose an instruction sub r3, r3, 5 needs to be executed (subtract constant 5 from the value in register r3 and place the result into register r3). Logic register names start with a lowercase letter; physical register names start with an uppercase letter. Suppose, also, that at the moment when the instruction is executed, physical register R1 corresponds to logic register r3. Let register R2 be the first in the list of the available free registers. Therefore, register r3 is replaced with R2 in the result field of the sub r3, r3, 5 instruction. The executable instruction is now sub R2, R1, 5. Any instruction following sub that uses its result must use R2 for an operand.

The remaining issue is that of returning physical registers to the available free registers list after data was read from them for the last time. One way is to associate a counter with each physical register. The counter is incremented by 1 every time an operand is renamed in an instruction that uses this physical register. Accordingly, when an operand is used, the value of the counter is decremented by 1. When the value of the counter reaches 0, the physical resource must be placed into the available free registers list.

Another renaming method uses the same number of logic and physical registers and maintains their univocal correspondence. There also is a buffer with an entry for each instruction initiated for execution. This buffer is called *reorder buffer* because it is also used to set the instruction order during interrupts. This buffer can be considered to be a FIFO queue implemented as a circular buffer with the *start* and *end* indicators.

Instructions are placed at the end of the buffer. When an instruction is completed, its result is placed into the queue entry previously assigned to it, regardless of where in the queue this entry may be. If an instruction is executed by the time the instruction reaches the buffer's start, its result is then placed into the register file, and the instruction itself is removed from the buffer. An instruction placed into the buffer and not executed by the time it reaches the buffer's start because there has been no operand value remains in the buffer until it receives this value. Several instructions at once can be fetched from the queue or placed into it, but the FIFO discipline is always maintained.

The logic register value can be placed either into a physical register or into the reorder buffer. At the moment an instruction is decoded, its result value is associated with the corresponding result position of the ordered instruction triple in the entry of the reorder buffer in which the instruction is located. Also, an entry is made in the value correlation table that shows that a result value has been found in the corresponding buffer entry. The source and instruction result fields are used to access the table's entries. The table shows that the needed value is located in the corresponding register or that it can be found in the reorder buffer. When the reorder buffer is filled, instruction dispatching is suspended.

A renaming operation execution will be considered using the `sub r3, r3, 5` instruction as an example. Let's assume that the `r3` value is or will be in entry 6 of the reorder buffer. Register `r3`, being the source, is substituted with the corresponding result field of the buffer's entry 6. The `sub` instruction is placed at the end of the reorder buffer, into entry 7, for example. This number is then written into the table to be used by the instructions, the users of the result. It must be noted that when operands become ready, the reorder buffer introduces a flow-oriented computation model.

Regardless of the renaming method used, extraneous data dependencies are eliminated in superscalar processors.

The issue of conflicts when accessing a shared resource—memory cells—is in essence the same as when accessing registers.

To calculate a memory address, at least one addition needs to be performed. After calculating the address it may have to be converted into a physical address; this task is carried out by the translation look-aside buffer.

INSTRUCTION EXECUTION

Ordered triple instructions consist of the operation code, physical source operands, and physical destination operands. After an ordered instruction triple is formed and placed into the buffer, a dynamic examination of the operands being ready for the instruction's execution is conducted.

Ideally, an instruction is ready for execution as soon as its source operands are ready. However, several restrictions are associated with the availability of physical resources, such as execution units, switches, and ports of register files (or of the reorder buffer). Various methods are employed to organize the execution window: one-queue, two-queue, or a reservation station method.

If there is only one queue, there is no need to rename the registers, because the availability of the operand values can be marked by a reservation bit associated with each register. A register is reserved when the instruction that modifies it is launched for execution. A register is released when the instruction execution is completed. If no resources have been reserved for the instruction, its execution is suspended.

In the multiple-queue method, a separate queue is organized for one-type instructions, for example, a floating-point instruction queue or a memory-operation instruction queue.

The third method involves using a reservation station. This station comprises a collection of elements, each of which has positions for placing operation codes, the name of the first operand, the first operand itself, the availability indicator for the first operand, the same elements for the second operand as for the first one, and the result register name. When an instruction completes execution and produces a result, the result's name is compared to the operands' names in the reservation station.

If an instruction is waiting for this result in the reservation station, the data is written into the corresponding position and its availability indicator is set. When all instruction operands are available, they are launched for execution. The reservation station keeps track of the operands' availability. When during dispatch an instruction is placed into the reservation station, all available operands from the register file are copied into this instruction's field. When all operands are ready, the instruction is executed. Sometimes not the operands themselves but the pointers to their locations in the register file or in the reorder buffer are stored in the reservation station. The reservation station is a type of processor that controls data flow.

INSTRUCTION TERMINATION

The concluding stage of instruction execution is changing the processor status according to the executed instruction. The purpose of this stage is to preserve the sequential program execution model when individual instructions are in reality executed in parallel and when branch instructions are executed predicatively.

Two main methods are employed to change the processor status. They both are based on using two states: the state that has been changed as the result of the operation, and the state that needs to be restored.

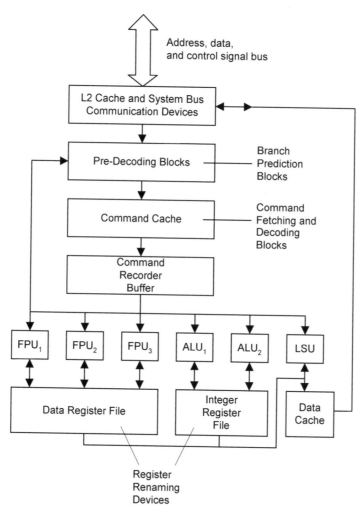

FIGURE 1.20 Superscalar processor architecture.

Using the first method, the processor's state is saved in a set of control points or in the transaction look-aside buffer, which are later used to restore the processor's state.

The other method involves examining the logic (architectural) and physical processor state. The physical state is changed immediately upon completion of a `next` instruction. The architectural state is changed when the result of the conditionally executed instructions become known. This method is implemented using a reorder buffer: results from the buffer are sent to the architecture register file and to the memory.

For each instruction the reorder buffer maintains corresponding values of the program counter and of other registers that are needed to service the interrupts properly.

Figure 1.20 shows main components of a superscalar microprocessor. The main functional units are ALU and FPU, loading/storing unit, register files, and separate instruction and data caches. The auxiliary units provide dynamic scheduling of the computations. There also are an L2 cache connection unit, instruction-reordering block, and a pre-decoding block.

DEVELOPMENT DIRECTION OF PROCESSORS WITH INSTRUCTION-LEVEL PARALLELISM

As has already been said, superscalar processors attempt to implement parallel execution of instructions within the sequential program model framework. After the sequential instruction flow is extracted, only the necessary data dependencies between the instructions are established. Sufficient information about the instruction order in the source program is preserved to maintain its order when interrupts are generated.

A typical superscalar processor fetches instructions and examines them as it executes them. The purpose of this examination is to detect and process branch instructions and to identify the instruction type to further dispatch it to the appropriate execution block or to a memory buffer. Some operations to relax data dependencies are also performed, register renaming, for example. In a VLIW processor, functions executed dynamically in a superscalar processor are implemented statically by the compiler.

At least two circumstances limit the superscalar processor architecture efficiency. First are some limitations on the degree of the instruction-level parallelism, even if the most advanced superscalar computation techniques are used. Second, the complexity of a superscalar processor increases at the same rate, and even faster, as the number of concurrently executed instructions.

The nature of these limitations is that resource demands increase sharply when conditionally executing nested branches, which limits the number of instructions that can be executed. Moreover, the size of the execution window (the number of active instructions that can be executed concurrently) limits a program's parallelism, because the instructions located outside the window are not considered for parallel execution. Most likely, the upper parallel limit for superscalar processing is seven or eight instructions launched simultaneously for execution in each cycle.

An alternative to superscalar processing is using VLIW. The advantages of VLIW follow. First, the compiler can examine the instruction interdependence and select concurrently executable instructions more efficiently than can be done by the hardware of a superscalar processor circumscribed by the size of the execution window.

Second, a VLIW processor has a less complex control unit and can potentially operate at higher clock frequencies. However, a serious factor limits efficiency of VLIW processors. This factor is the branch instructions, which depend on the data whose values become known only during the process of computational dynamics. The execution window of a VLIW processor cannot be very large due to the compiler's lack of information on dynamically generated dependencies. This shortcoming prevents instruction reordering in a VLIW processor. For example, a correct execution of a load operation in the called function concurrently with a memory write operation in the calling function cannot be guaranteed (especially if the called function has been defined dynamically). In addition, VLIW implementation requires a large-size name memory, multi-input register files, and a large number of cross-links. An execution can also be halted if a situation arises that is different from the state when the execution scheduling was generated (a cache miss during execution, for example).

Superscalar microprocessors and VLIW processors are products of modern electronics, and their efficiency is constantly rising. However, using these processors entails a thorough examination of the architectural methods used for obtaining high performance. At the same time, these methods should be checked for being adequate for the problem area whose tasks the computer system being built is to solve.

Currently, further increase of microprocessor performance is linked to the static and dynamic code analyses directed at locating program segment-level parallelisms. These analyses are carried out using information made available by high-level language compilers. Research in this area has led to the development of multithread processor architecture, which is a next stage of the superscalar architecture development.

At the present time, the work in this direction is at the theoretical research and simulation stage. However, Intel has already built the first multithread microprocessor. IBM and Sun are also working on developing processors that use all the benefits offered by the multithread architecture. Therefore, the main aspects of this architecture will be considered in sufficient detail in Section 1.4.

Another potential approach is a move to single-chip multiprocessor computational systems. In this case, multisequencing high-level language compilers are involved.

1.3.3 Hierarchic Memory Organization

MEMORY COHERENCE PROTOCOLS

One of the uses that microprocessors are put to is their employment as computational system elements. The hierarchic memory-coherence protocol employed in a particular processor depends on the architecture of the parallel system architecture the processor is used to build. Two such protocols will be considered in the following sections. One protocol is for SMP architecture systems, which use a system bus to link processor to memory. The other protocol is employed in the distributed memory processor systems.

MESI COHERENCE PROTOCOL

Microprocessors used in SMP architecture systems employ *Modified, Exclusive, Shared, Invalid* (MESI) write-back cache organization protocol. The write-back feature of this protocol cuts down on the number of data transfers between the cache and the main memory: if the cache data has not changed, there is no need to update it in the main memory. Moreover, other improvements in reducing the number of through memory writes can also be taken advantage of [19]. Intel 80860XP processor makes use of these improvements.

Certain initial conditions must be set and definitions made to describe the MESI cache-coherence maintenance algorithm. Thus, each microprocessor has its own cache. All microprocessors share one main memory, to which they are connected by the system bus. Peripheral devices are also connected to the bus.

It is important to understand that all microprocessors can snoop on the bus transaction operations with copies of memory blocks in all caches, as well as in the main memory, that are carried out by one of the microprocessors or peripheral devices. This is caused by the fact that at any given moment only one bus client can transmit data, whereas all microprocessors connected to the bus clients can perceive it. Therefore, if a microprocessor connection other than bus is used, for the MESI protocol to remain operable, the previous principle of executing transactions must be preserved for all microprocessors to be able to snoop on them.

Each line of a microprocessor cache can be in one of the following states:

M—*Modified.* The line is available for `read` or `write` operations only to this microprocessor because it was modified relative to the main memory by a `write` operation.

E—*Exclusive.* The line resides exclusively in this cache. Its contents are the same as its corresponding block in the main memory.

S—*Shared.* The line resides in this cache. Contents are the same as the main memory. It can be shared (that is, available for reads or writes either from this cache, from the main memory, or from other caches where its copy is maintained) by microprocessors other than the owner microprocessors.

I—*Invalid.* The line cannot be used (either for reads or writes) because it contains no valid memory block copy.

Line states are used by the microprocessor to determine whether data can be accessed locally in the cache line without going to the bus to obtain it. They are also used to control the coherency maintenance mechanism.

The coherency maintenance mechanism is controlled by bit WT/WB. When this bit is set to 1, the write-through mode is enabled; when it is set to 0, the write-back cache mode is enabled.

The state of a cache line being accessed for a `read` or `write` operation is determined by Table 1.3.

TABLE 1.3 MESI Cache Line States Transitions for Memory Read/Write Operations

Initial Line State	Post-Read State	Post-Write State
I	If WT = 1 then E, else S. Refreshing cache line by reading it from the main memory	Write-through to the main memory; I
S	S	Write-through to the main memory; if WT = 1 then E, else S
E	E	M
M	M	M

A `read` cache miss causes the line to be brought from the main memory and either an E or S state assigned to it. Only `read` cache misses cause cache updates. When a `write` cache miss is made, the `write` transaction is placed into the buffer and is sent to the main memory when the bus becomes available.

When a line is invalid, as indicated by the I state, a `read` instruction from the line causes a line to be read from the main memory, placed into the cache, and the line state modified to either E or S. The E state is set for the write-through cache mode; for the write-back mode, S state is set. Until this line has been accessed by other microprocessors or peripheral devices, no data will be written back to the line, and the microprocessor will not go to the bus to get data.

When the line state is I, a `write` instruction to this line only changes the contents of the main memory; contents of the cache line are not changed and the line state remains I.

When the line state is S, reading this line does not change its state. If the line is in the write-through mode, than the state changes to E after a write is made; if the line is in the write-back mode, a write-through is made, but the line status remains S.

When the line state is E, it is preserved after a `read` operation; a `write` operation puts the line into the M state.

Finally, when the line state is M, neither `read` nor `write` operations change it.

To maintain coherence of the cache line during I/O operations and access to main memory by other processors, special cache-type-status polling cycles are generated. These cycles poll the caches' status to determine whether the line specified by the address used in the operation that had initiated the polling cycles is stored in them. A line can also be forced into the I state; this is done by the INV signal. The line states in this case are determined by Table 1.4.

TABLE 1.4 MESI Cache Line States Transitions for I/O Read/Write Operations

Initial State	INV = 0	INV = 1
I	I	I
S	S	I
E	S	I
M	S; line write-back	I; line write-back

DASH PROTOCOL

There are multiprocessor systems in which memory is physically distributed between processor nodes and memory transactions of one processor cannot be snooped on directly by other microprocessors. In this case, cache coherency of different processor nodes is maintained by internode messages. Several principal methods of doing this exist.

A straightforward approach to maintaining cache coherence in a distributed shared-memory multiprocessor system follows: every time there is a cache miss by any processor, a request to bring the needed line from the memory is initiated. The memory node from which this line is taken is called the *home node* to the line. Through the switch, the request is sent to the home node, and the needed line, again through the switch, is transferred to the module where the cache miss occurred. In this way, the cache is filled for the first time. For each block in the memory node a directory of computation modules in which it is cached is maintained. A distributed-by-computation-modules directory of the home blocks can be also organized. A block cached in more than one module is called *shared.*

The actual cache coherence-preservation process follows: when a cache is accessed during a `write` operation, after the `write` operation completes, the processor halts until the following sequence of actions has been performed. First, the changed cache line is sent to the home memory node. If the line is shared, from the resident node it is sent to all nodes listed as sharing this line. After receiving acknowledgements that all copies have been changed, the home memory node sends to the halted processor a permission to resume computations.

Although the described coherence-maintenance algorithm is logically operable, it is seldom used in practice due to long processor idle time during `write` operations. In practice more sophisticated algorithms are employed that reduce processor idle time. One such protocol is directory architecture for shared memory (DASH) protocol. It is implemented as follows.

For each of its memory blocks, each memory module maintains a list of nodes in which copies of this block are also cached. Each memory block has three global states associated with it in its home memory module:

Uncached—The block is not cached by any other node than home.

Shared Remote—The valid copies are cached in other nodes.

Dirty Remote—The block has been modified in some other node's cache.

Additionally, each cache line may be in one of three local states:

Invalid—Not cached in this node.

Shared—Valid copy of the block in this cache and perhaps in other nodes' caches.

Dirty—The line has been modified by a `write` operation.

Each processor can read its own cache if the local cache copy of the memory block it attempts to read is in the *shared* or *dirty* state. If the referenced memory location is not cached or if its cache line is in the *invalid* state, a read miss request is generated and sent to the home module.

If the global state of the requested memory block in its home module is *uncached* or *shared-remote*, the block is cached in the requesting local node, and the home module updates the list of the nodes in which this memory block is cached.

If the requested memory block is in the *dirty-remote* state, a read miss request is redirected to the cache of the node containing the dirty cache line. This node sends the requested cache line to the requesting local node and to the line's home module; the remote node then updates the state of its cache copy to *shared*. The final global state of the memory block is *shared-remote*.

For write operations, if the local state is *dirty*, writes can be carried out locally and computation continues. If the local state is *invalid* or *shared*, the local node sends a read request to the home module for exclusive use of the block and suspends the write until it receives acknowledgments from the nodes sharing the requested line that they have changed local state of their copies of the block to *invalid*.

If the global state of the block is *uncached*, the block is cached exclusively in the requesting node, which resumes the suspended computations.

If the global state of the block is *shared-remote*, the home module sends a request for placing local copies into *invalid* state to all nodes caching this block. Upon receiving this request, all remote nodes change the local state of their copies of the block to *invalid* and send acknowledgments to the node that had initiated the write operation. After completion of the read operation, the global state of the block is updated to *dirty-remote*.

Efforts to increase the coherence algorithm effectiveness are made, in particular, by taking into consideration the specific character of parallel programs. In these, the same data is used asynchronously exclusively by one processor in each time period, the processing subsequently being handed over to another processor. In such a case, a cache line can be sent from one processor's cache to the other's cache more effectively.

APPROACHES TO SELECTING COHERENCE PROTOCOLS

Discussions about what cache organization methods are preferable must take into account the specific ways compilers generate object code. Moreover, in writing their programs, programmers must make use of the specific compiler and cache controller characteristics. In other words, better results can be obtained from a simpler cache organization supported by the compiler and executing programs written following certain rules arising from the compiler and cache organizations than from a complex cache.

MICROPROCESSOR CACHE ORGANIZATION

Although instructions and the data they operate on are accessed simultaneously, they are different types of data. To increase the degree of parallelism in working with memory, instructions and data are cached separately.

1.3.4 Raising Processor Context Switching Speed

APPROACHES TO REDUCING PROCESSOR CONTEXT SWITCHING TIME

Modern operating and programming systems extensively use processor context (contents of its registers and individual control triggers) switching when entering and exiting interrupt handlers or subroutines and also when multitasking. The time used to switch the contexts must be minimal, because the switching expenditures are the price paid for concurrent execution of a collection of interacting computation processes.

Reducing processor context switching time can be achieved in several ways. One is reducing the number of registers whose contents are stored in the memory. The other way is using hardware to perform the register content-saving operation. Finally, special conventions can be introduced that regulate register use by programs; this allows the processor context to be saved only partially, producing time savings.

REDUCING THE NUMBER OF SAVED REGISTERS

Reducing the number of saved registers causes efficiency to drop. Generally speaking, this method is at odds with the drive to raise the efficiency by employing fast register memory and concurrent functioning of processor devices, each of which has its own registers. However, some architectures do use this method, in particular INMOS Ltd transputers [21] and Java processors [22]. These architectures are based on stack operations. The registers saved are the pointers to the current stack position, to the memory area used, and so on. The number of these pointers is limited; there are only eight of them in the transputers, which allows the processor environment to be switched in 2 mcs at the clock frequency of 10 MHz.

HARDWARE REGISTER-SAVING SUPPORT

Hardware support of register saving can be implemented in various ways. At one extreme, this may be an accelerated transfer of the register contents to the memory. At the other extreme, each newly activated program may be allocated its own register set.

For example, Sun's SPARC architecture microprocessors use eight groups (windows) of 32 registers with adjacent windows sharing eight of these registers. Register windows overlap in such a way, that registers 24–31 of the previous window are at the same time registers 0–7 of the next window. In the developers' opinion, this arrangement makes parameter passing by subprograms more efficient.

USING FAST INTERRUPT HANDLER REGISTERS

Some interrupt handling functions are executed by specialized programs that can make do with a limited number of registers. Therefore, some processors are equipped with special registers employed for processing fast interrupts, such as an interrupt triggered by receiving the next message character that transports the character to the memory. Operations to start receiving the entire message and to terminate its reception require the memory allocation mechanism to be launched, which usually requires a regular interrupt that saves all processor's registers. However, because more often only a next signal is received, using fast interrupts substantially increases the efficiency.

1.3.5 Expanding Microprocessor Functional Capabilities

DIRECTIONS OF FUNCTIONAL CAPABILITIES EXPANDING

Along with increasing the number of ALUs, microprocessors are equipped with additional functional capabilities. Some of these follow:

- Integrated memory and peripheral device control functions that in traditional microprocessors are implemented in chipsets
- Multimedia data-processing blocks, such as those used earlier in digital signal microprocessors
- Integrated network and telecommunications system interfaces, such as Ethernet, ATM, FDDI; this makes it possible to interconnect microprocessors among themselves and with telecommunications and computational networks without using additional adapters

It must be noted that the idea of building a single-chip computer has always been popular. Currently, the task of placing a sufficiently large capacity embedded DRAM (EDRAM) and a microprocessor kernel on one chip is about to be solved. IBM produces a compact dynamic memory cell whose size (0.62 micron square) is only 1.5 times larger than a memory cell in a 64-Mb DRAM microchip [25]. A 16-Mb EDRAM block occupies only 20.8 mm^2, and its bandwidth reaches up to 50 GBps.

The era of system on chips (SOC), in which the microprocessor is only one of the components (kernels), has practically dawned. Motorola, for example, offers two families of chips in which PowerPC 603e is used as the kernel. These chip families are based on the AltiVec and PowerQUICC technologies.

INTEGRATING MEMORY AND PERIPHERY CONTROLS

The importance of integrated memory and periphery controls is obvious. For example, to build a computer using Pentium microprocessor, Intel 440XX chipset is needed; to build Pentium III, Intel 440BX AGPset is needed.

Chipsets Cobalt and Lithium were developed by SGI for Pentium II based SG 320 and SG 540 computers. The former chipset has about 10 million transistors and functions as an integrated memory and graphics-processor controller. The Lithium chipset implements an I/O processor controlling S-video, IEEE 1394, PCI, and Ethernet interfaces.

As a rule, chipsets are developed by the same companies that develop microprocessors. Integrating chipset functions into a microprocessor makes it possible to substantially simplify motherboards. Of course, using a separate microprocessor with different chipsets allows computers of varying types and price ranges to be built. But this is relevant only when hardware resources need to be conserved. The first representative of this trend is Ultra™ SPARC IIi microprocessor.

MULTIMEDIA EXPANSIONS

Many developers expand the functional capabilities of their processors by equipping them with specialized blocks for processing multimedia applications. It must be noted that a similar graphics processor, which substantially increased performance in some applications, was already incorporated into Intel 80860 second-generation microprocessor.

Microprocessors from the following companies have multimedia processing blocks: Intel (MMX instruction set expansion and 70 new SIMD instructions for Pentium and Pentium III, respectively), AMD (3DNow!®), Sun (VIS™ SPARC), Compaq®/DEC (Alpha MVI), HP (PA RISC MAX2), SGA/MIPS® (MDMX), and Motorola (Power PC AltiVec™) [23].

Multimedia processing instructions can be built into the microprocessor instruction system in various ways. It can be done as a functional block sharing the register file with the other blocks, as it is done with MMX in Pentium. Or it can be done as a separate decoupled architecture processor with its own register file. The latter method is used in Pentium III and PowerPC AltiVec.

Multimedia processing instructions, as a rule, specify parallel processing of a few short (8-, 16-, or 32-bit) fixed-point data in the processor's SIMD mode. However, this does not address all current needs and parallel processing of four floating-point 32-bit operands in the SIMD mode was introduced in Pentium III.

COMMUNICATIONS EXPANSIONS

Communications interfaces were first integrated into microprocessor chips in transputers. Modern microprocessors are equipped with integrated communications interfaces to build parallel systems. Alpha 21364 and Power 4® are good examples of such integration. IBM, Toshiba®, and Sony® conduct joint development of the Cell microprocessor. This should be a supercomputer on a chip with communications interfaces.

There is a general transition going on from shared buses, such as PCI and VME, to point-to-point channels such as PCI Express™, Rapid IO, HyperTransport™, and InfiniBand®. This process will conclusively consolidate distributed computational system interfaces in the microprocessor architecture.

In addition to the specialized interfaces for constructing parallel systems, telecommunications interfaces are being integrated into microprocessor crystals. Motorola MPC8260 processors support a variety of telecommunications protocols, such as 10/100 Mb/s Ethernet, 155 Mb/s ATM, and 256-channel 64 Kb/s HDLC.

1.3.6 Standardization of Computer Architectures

STANDARDIZATION PROBLEMS

During the entire history of computer engineering development, attempts have been made (by software developers, first of all) to standardize computer architectures, an action that would significantly expand the application area of the produced software. Having realized the futility of achieving compatibility on the machine-instruction level, developers then tried to standardize assembler languages, high-level programming languages, and the application software-to-operating system interface languages. The incentive for these attempts has been the constantly growing complexity of the processors themselves as well as of the programming systems constructed with their use.

In addition to all other considerations, developing new complex systems requires two mandatory stages: an adequate definition of the system requirements and an exhaustive testing for compliance with these requirements. Testing must be conclusive. Without citing examples from the area of large-application system development, a few widely known mistakes in microprocessors from some well-known companies will be shown along with some undeclared capabilities of microprocessors and operating systems.

The lack of standardization does not allow development of new systems by constructing them from already existing systems that have undergone approbation in various application environments by a great number of independent users.

The open-system concept is an attempt at a comprehensive solution of the standardization problem [24]. An *open system* is a collection of interfaces, protocols, and data formats that are based on the commonly available, universally accepted standards providing software portability, systems interaction, and scalability.

Portability is a property of a source code program to be executed on different hardware platforms under different operating systems.

System interaction is the ability of systems to exchange information automatically, recognizing the format and semantics of data.

Scalability is a property of a program to be executed using different resources (memory size and number and efficiency of processors) with the efficiency rate being proportional to the resources available. It is important to understand that resources may not only increase but also decrease. For example, a program can be executed in a memory area randomly allocated to it.

Within the framework of the open-system concept, processor architecture must be describable in sufficiently simple formal terms with data types, registers, and performed transformations specified and without producing undesirable side effects.

There are at least two known attempts to implement this approach.

ARCHITECTURE-INDEPENDENT PROGRAM SPECIFICATION

Currently, preparation of a draft Architecture Neutral Distribution Format (ANDF) standard is being conducted by the Microprocessor Systems Committee of the International Standard Organization (ISO/IEC). In the opinion of developers from X/Open Company, Ltd., this specification format will solve the program portability problem. Source code is supposed to be compiled in two stages. During the first stage, the source code is translated into general declarations of the application program interface (API), along with a general data type description made. The actual translated program obtained is an expression of the abstract algebra as defined by the ANDF. As a result, the program text can be subjected to formal verification and modifications. In the second stage, the object code for a specific architecture is generated.

JAVA TECHNOLOGY

This technology was introduced by Sun. It is based on the concept the virtual Java™ machine whose specification includes the following data types:

`byte`

`short`—A two-byte whole number

`integer`—A four-byte whole number

`long`—A two-byte real number

`float`—A four-byte real number

`double`—An eight-byte real number

`char`—A two-byte character

`object`—A four-byte pointer to an object

`ReturnAddress`—A four-byte return address

Java machines have the following registers:

`PC`—Program counter

`Vars`—Register for accessing local variables

`Optop`—Pointer to the operand stack

`Frame`—Pointer to the run-time environment

Most of a Java machine's instructions are one-byte long. This fits the processor stack architecture, which uses a small number of registers and data pointers.

Using one-byte-long instructions in Java machines makes it possible to reduce program size. The average instruction length is 1.8 bytes. Lately, to all existing arguments in favor of standardizing architectures, the practical needs of Internet work, which demands short programs, have been added. Open Internet systems allow accumulation of software products and construction of new systems from those already existing.

Note that the architecture of a virtual Java machine is quite similar to the architecture of the INMOS transputers architecture. The practical difference is the addition of the object-oriented technology elements. One of the obstacles in the way of the development of Java technology is the low efficiency of Java code execution. However, all the needed prerequisites for surmounting this shortcoming are available. For example, modern Intel x86 processors have a special block that translates complex instructions into a collection of simple instructions for a RISC processor. Afterward, the RISC processor executes these instructions using all the advantages of the RISC approach to achieving high efficiency. It is quite conceivable to develop a similar translator for Java code that would translate byte-code into instructions for the real processor.

Another approach to increasing efficiency was used in the T-9000 transputer [25]. It attempts to raise the execution speed by concurrent processing of a great number of instructions on parallel processing devices while preserving the byte instruction system of the T-2xx, T-4xx, and T-8xx families' transputers.

1.4 MULTITHREAD MICROPROCESSORS

1.4.1 Multithread Architecture Fundamentals

Superscalar and VLIW microprocessors have only one program counter and, therefore, are single-thread processors. In these microprocessors, instructions examined for a possibility of their concurrent execution are tied to the processor's program counter either by the execution window, as in superscalar processors, or by a long instruction, as in LIW processors. To more aggressively select instructions from one or more programs for parallel execution, processors are equipped with several program counters and, possibly, some other hardware. Microprocessors with more than one program counter are called *multithread* processors [26–28].

Multithread architecture solves the problem of combating downtime of the processor functional units caused by the processor being unable to execute the next instruction by switching to another register file. This provides the processor with another environment and allows it to continue calculations by going over to the execution of another thread (process).

Processor switches to another register file either upon the occurrence of some event causing the processor to idle (a cache miss, main memory access, or an interrupt) or as a compulsory action, for example, for every cycle, as in Tera MTA [27].

In Tera MTA, the problem of the gap between the internal processor processing speed and the main memory access time is solved by compulsory switching to work with the next register set in every cycle. The structure of a multithread processor is shown in Figure 1.21. Each register set services one computation process, called a *thread*.

Altogether, the processor has n sets of registers. Therefore, a main memory request by a process can be serviced during $n-1$ cycles, up to the moment when the processor again switches to the same register set. Therefore, instruction execution of one thread is slowed down by n times. The value of n is selected in such a way as to provide memory access time shorter than the length of $n-1$ processor cycles.

With all the differences in approaches to constructing multithread microprocessors, their common trait is the introduction of a collection of processor elements containing an instruction-fetching unit that organizes the execution window for one thread. Within the boundaries of one thread, branches can be predicted, registers renamed, and instructions dynamically prepared for execution. Therefore, the overall number of instructions being processed is significantly larger than the execution window of a superscalar processor on one hand, and, on the other hand, the clock rate is not limited by the size of the execution window.

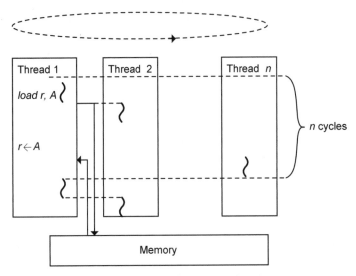

FIGURE 1.21 Multithread architecture.

Threads can be identified by the compiler during the high-level source code or object-code analysis. However, compilers cannot always solve the dependence problems arising from sharing registers and memory cells by threads. These dependencies must be solved in the progress of thread executions.

For this purpose, microprocessors are equipped with special conditional thread-execution hardware, which allows return of execution control and discards the results when a violation of thread dependencies is detected. An example of a dependence violation is a `write` operation attempt by one thread to a cell from which another thread is performing a `read` operation that it should do in the next cycle. In this event, a violation is recorded, and the execution of the second thread is returned to the point where the correct value can be read.

The interface between the part of the multithread processor hardware that supports processing of each individual thread and the part common to executing all threads can be put in at different levels. It can be added after the thread instruction-fetching units and the thread's functional units, as well as on the shared memory-access level. In the former case, each thread uses both its own functional units, such as integer ALU, and some functional units shared by all threads, such as floating-point operation units. Close connection in terms of resources allows sequential programs with strong dependencies between threads to be executed efficiently. In this case, we have an instance of a simultaneous multithread (SMT) processor implementation.

In the latter case, a functionally complete processor is actually allocated to execute each thread. Overall, this structure is oriented at the execution of independent or weakly connected threads created either by one program or a collection of programs. In this case we have an instance not of a processor but of a single-chip system (CMP, or chip multiprocessing).

A multithread processor can execute threads from one or several programs. If the processor executes only one program, it is evaluated in terms of throughput; if it executes more than one program, it is evaluated in terms of its bandwidth.

It is estimated that the multithread processor Alpha 21464 will be 14 times as efficient in processing transactions as the contemporary Alpha 21264 [29]. Here, a two-fold increase is expected exactly from the use of the multithread architecture. Another two-fold increase is expected to be delivered by raising the use of the instruction-level parallelism. The remaining 2.5-fold increase is supposed to be achieved by the modernized technology and increased clock rate.

Multithread microprocessors and single-chip systems absorb methods of raising microprocessor performance accumulated during the evolutionary development; they use compiler-hardware symbiosis for respective static and dynamic detection of parallelism in the sequential source programs. The orientation of the execution of threads having a certain degree of mutual dependencies stipulates that specific solutions concern shared use of functional devices by the threads. Using a collection of register files in a processor is the essence of the multithread architecture. That is, transistor resources are spent not to build cache but to construct a collection of register files, including program counters, status registers, and work registers.

Although Intel has already introduced two-thread architecture in its Xeon® and Pentium 4 processors, there is still significant research into multithread architectures to be done. Tera [27] announced a project development of the multithread microprocessor Torrent. Level One, an Intel daughter company, produces a multithread network microprocessor IXP1200 that contains six four-thread processors [28]. IBM announced a project development of Blue Gene® computer [26] with the throughput of 10^{15} floating-point operations per second (FLOPS). The Blue Gene crystal has 32 eight-thread processors and is accessed by a 256-bit bus. Because EDRAM has high bandwidth and low latency, the eight-thread processor structure makes it possible to forgo cache, instead of which a small buffer memory is used between the processor and the main memory. IBM, Sony, and Toshiba are conducting a joint development project of the multithread processor Cell; its very name quite expressively tells its purpose. (Could this be a second try to introduce a name for the new class of microprocessors after the flop with the term transputer?) It is claimed in [54] that this processor will be produced at the end of 2004. Its throughput is supposed to be more than 100 times that of 2.5 GHz Pentium 4.

1.4.2 Thread Detection

Thread detection can be based both on the control streams (control driven) and on the data streams (data driven) of programs. Threads can be conditionally or unconditionally executable.

Access to the shared register and memory variables must be synchronous in unconditionally executed threads. Threads should be generated by a *fork* operator only when they are truly necessary, because the processor hardware does not watch access order violations. Shared variables must have the same state, as with sequential program execution. Unconditionally executed threads have explicitly defined synchronization and interaction operators. The difference in executing unconditional threads on a multithread processor from executing them on a parallel system processor is that in the latter case, processes interact by exchanging data and synchronization messages without any restrictions, as they are one parallel program. In the former case, however, data can be transferred only from the earlier generated threads to the later generated threads, because in this case, a sequential program is executed.

The access order to the shared variables for conditionally executed threads is maintained by hardware, which detects access-order violations. Access to the variable is carried out under the assumption that it is allowed; when it turns out that there was an access-order violation, the erroneous data is discarded. Buffers can be used for this purpose; the results are stored in them without changing the thread's architectural state until the correct access order is confirmed.

When employing control-driven thread detection, such control structures of a sequential program as cycles, functions, and subprograms access iterations are analyzed. Attempts are made to detect code segments containing minimal interdependencies.

On the conceptual level, threads detected by data-driven methods consist of one instruction. A thread is generated when an instruction receives one operand. A possible expansion of this concept is the execution not of one instruction upon all ready operands but of a sequence of instructions headed by this instruction. Data transfers among threads are defined explicitly.

Automatic conversion of programs written in imperative languages into data-stream programs is in the research stage. Data-stream programs are written using functional languages or special data-stream languages.

1.4.3 Multithread Processors with Threads Detected by Analysis of Program Control Flows

MULTITHREAD PROGRAM EXECUTION MODEL

Conditional thread-execution multithread processors [31] employ an aggressive code-execution model to extract speculative execution threads from a sequential program. According to this model, a program is broken into a collection of processing units by software and hardware means; these units are thereafter called segments. A *segment* is a part of a program that is executed by a thread. A segment is a contiguous instruction sequence area (for example, a part of a base block, a base block, a set of base blocks, a single iteration of a cycle, a cycle, a function call, and so on). Various methods exist to form segments and to organize their execution. The following is one of the possible methods [31], giving an object illustration of the essence of a conditional thread-execution multithread microprocessor.

For execution on a multithread processor, program segments are statically separated by annotations. Control dependencies among program's operators are represented as a control-dependence flowchart (CDF); segments are represented as nodes; and the order of their execution is shown by the curved lines that connect them. The dynamics of program execution can be viewed as a traversal of the CDF program. At each step of the traversal, the processor assigns one segment to one of the processor units designated for thread execution and continues the traversal. The assignments are made without considering the actual content of the segments.

A segment is assigned for execution to a processor unit by handing the initial value of the program counter over to it. Many segments initialized in this way are executed in parallel by processor units, resulting in execution of a set of instructions in one processor cycle.

Each of the processor units fetches and executes instructions pertaining to the segment assigned to it. The values of the shared registers are copied into each processor unit. The modified register contents are dynamically sent to the collection of the parallel processor units according to the masks generated by the compiler.

Memory is accessed speculatively, the sequence of the preceding load or save instructions not being known. Data is accessed in parallel by many processor units; the processing is suspended only in case of real data dependence.

Figure 1.22 shows an example of multithread processor architecture.

A multithread processor can be viewed as a parallel computational system consisting of a collection of processor units and a scheduling program that assigns segments to the processor units. As soon as a segment is assigned to a processor unit, this unit starts fetching and executing a segment's instructions until it reaches the end of the segment. The collection of the processor units, each of which has its own internal mechanism for sequential instruction execution, makes it possible to execute a collection of segments. Instructions inside the dynamic execution window of a multithread processor are delimited by the first instruction in the first executed segment and the last instruction in the last executed segment. Provided that each segment can have loops and function calls, the effective size of the window can be extremely large. The important thing is that not all instructions inside this broad range are considered for concurrent execution—only a limited set of instructions within each processor unit.

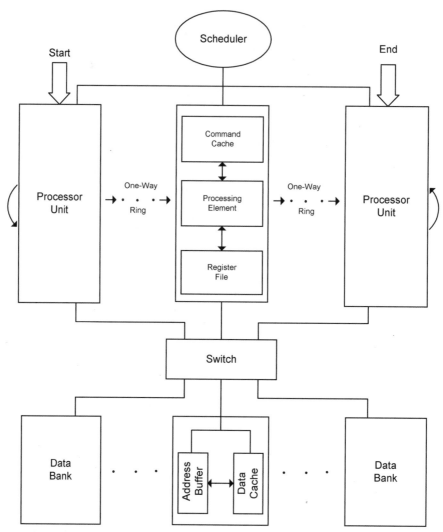

FIGURE 1.22 Multithread processor architecture.

Consider a CDF for a fragment of program with five basis blocks—A, B, C, D, and E—shown in Figure 1.23. One of the possible dynamic basic-block execution sequences looks like this:

$$A_1^1, B_1^1, C_1^1, B_2^1, B_3^1, C_2^1, D_1^1, A_1^2, B_1^2, B_2^2, C_1^2, D_1^2, A_1^3, B_1^3, C_1^3, B_2^3, C_2^3, D_1^3, E$$

In this sequence, the upper and lower indices correspond to the outer and inner cycle iterations, respectively.

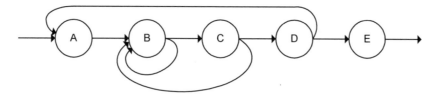

FIGURE 1.23 A flowchart of control dependencies of a program fragment.

An iteration of the CDF outer cycle that is shown in Figure 1.23 as a segment will be considered. In other words, the basic static blocks A, B, C, and D (as well as the control flow going through them) are part of the segment. The segment corresponding to the first iteration of the outer cycle can be assigned to the first processor unit. The segment corresponding to the second iteration of the outer cycle will be assigned to the next processor, and so on. The processor that is assigned the first iteration sequence as a segment executes dynamic instructions of the basic blocks $A_1^1, B_1^1, C_1^1, B_2^1, B_3^1, C_2^1, D_1^1$.

Likewise, the next-in-line processors execute dynamic instructions of the basic blocks $A_1^2, B_1^2, B_2^2, C_1^2, D_1^2$ and $A_1^3, B_1^3, C_1^3, B_2^3, C_2^3, D_1^3$ according to the second and third iteration, respectively. In this example, executing three productive instructions within the cycle is a potential result of the multiscalar approach. For example, in the given cycle, the processors could execute instructions from the dynamic basic blocks B_2^1, C_1^2, and B_2^3 concurrently.

It is important to note that even though segments are divided into instruction groups, they are not independent. Because segments are parts of the sequential instruction flow, the data and instruction relationships between individual instructions must be maintained during the execution. The key issue in the multithread implementation is providing data and control links among the parallel processors, that is, how can a CDF be traversed sequentially if, in fact, a nonsequential traversal is performed?

Sequential traversal is done in the following way: first, each processor is provided with a sequential execution model of the segment assigned to it. Second, for all processors, a sequential execution order, which is supported by a cyclic processor unit queue, is specified. The start and end of queue pointers identify the processor units that execute the very first and the very last segments, respectively. For example, according to the considered CDF (Figure 1.23), the processor unit at the beginning of the list that executes the first iteration precedes the processor unit that executes the second iteration, which in turn precedes the processor unit at the end of the list that executes the third iteration.

As a segment's instructions are being executed, the values of the program variables are generated and utilized. These values are associated with the storage area, namely with the registers and the memory. Because for sequential program executions, variable storage area is considered as a single register and memory collection, multithread execution must support the same model. Moreover, it must be ensured that variable values are produced and used in the same way as they are in a sequential execution. In the considered example, the values used by the instructions in the dynamic basic block B_2^2 must be produced by execution of the block sequence $A_1^1, B_1^1, C_1^1, B_2^1, B_3^1, C_2^1, D_1^1, A_1^2, B_1^2$ and the preceding instructions in B_2^2. To ensure this, exchange among the segments must be synchronized.

When register memory is used, the control logic synchronizes generation of register values in preceding segments with utilization of these values by the successive segments. Register values produced by a segment can be defined statically and flagged in the segment-generation mask. At the moment a corresponding register value is generated, if there is a flag in the generation mask, this value is sent to the successive segments through the one-way ring channel (see Figure 1.22), that is, to the processor units that are logical successors to the processor unit that has generated the value. The values loaded into registers from the ring channel and intended for the successor segments are defined in the accumulator mask, which is a union of the generation masks of the active predecessor segments. As soon as the values are received from the predecessor processor units, the storage flags in the successor processor units are cleared. If a segment has to use one of these values, the using instruction can be executed only if it receives the value; otherwise, it must wait for the value it requires.

Unlike register values, determining exactly which values are generated and which are used by a segment cannot be done in advance for the values stored in memory due the addresses being generated dynamically. If it is known that a segment uses a value (bringing it in by a load instruction) that was generated (by a store instruction) in an earlier segment, it is possible to synchronize the generation and utilization of this value. In other words, the load operation in the successor segment can be postponed until the store instruction has been executed in the predecessor segment. (This is similar to the situation with the registers; nevertheless, the synchronization mechanism is different due to the name range sizes being incommensurable.)

In the general case, when this knowledge is not available, a conservative or aggressive approach can be undertaken. The conservative approach implies waiting until such time when it becomes certain that the load instruction will read the correct value. This approach usually involves delaying execution of load instructions inside the segment until executions of memory-store instructions, whose results can be used by the next instruction, have been completed by all predecessor segments. With the aggressive approach, loads from memory into register of processor units can be performed speculatively, with the assumption that no predecessor segments store values into the same memory cell. To guarantee that no predecessor segment writes to a memory cell that has been read from by a successor segment, a check must be performed during the computation process. If this check detects a contradicting load and store (not done in the proper sequence), the later processing unit must be interrupted and a corresponding restoring procedure must be initiated. Multithread processors utilize the aggressive approach.

Because of its speculative nature, multithread execution must have both the correct execution confirmation means and the erroneous execution correction means. Instruction execution inside a segment can be considered speculative from two aspects:

■ Control speculative
■ Data speculative

If a speculative-control operation resulted in prediction of a wrong next segment, the next segment or segments must be cancelled and the correct segment sequence restored. Similarly, a segment using incorrect data must be cancelled and the correct data values restored. In any case, canceling one segment causes the cancellation of all segments executed after the cancelled one. (It would be difficult to preserve sequential semantics otherwise.)

To make preserving sequential semantics of a program execution simpler, a multithread processor removes segments from the cyclic queue in the same order in which it had placed them in there. In the process of speculative execution, a segment produces both correct and incorrect values. Only correct results produced by a segment can be safely used by the other segments. Nevertheless, in a multithread processor, values are optimistically sent for speculative execution for processing by other segments. Because a segment sends values to other segments in advance as soon as it generates them, most, if not all, of the values will be sent by the time a segment gets to the head of the queue. Consequently, a segment can be cancelled and a new segment launched simply by modifying the queue-head pointer.

To illustrate the power of the multithread model, consider the example shown in Listing 1.6. In the process of executing a given program, a character is repeatedly fetched from the buffer and checked for matching elements of the linked list. If there is a match, the function that handles character processing is called. If there is no match, the character is added to the list. As the program is executed, adding new items to the list will become infrequent, because most of the fetched characters will match those already in the list.

When the program is executed on a multithread processor, the segment assigned to a processor unit performs one complete list search for a matching character. Processor units perform linked list searches in parallel, each for its own character; as the result of this, many instructions are executed in one cycle.

LISTING 1.6

```
for (indx = 0; indx ← BUFSIZE; indx ++) {

/* Get a character to search for*/

symbol = SYMVAL (buffer [indx]);

/* Sequential list search for the character */

for (list = listhd; list; list = LNEXT (list)) {

/* if the character is on the list, process it */

if (symbol == LELE (list)) {

process (list);

break;

}

}

/*if the character is not on the list, add it to the end of
the list */

if (! list) {

Addlist (symbol);

}

}
```

MULTITHREAD PROGRAMS

A program intended for execution on a multithread processor must have a CDF that can be traversed rapidly, distributing the segment collection among the processor unit collection.

Code specifications for all segments are the same. A segment is defined as a part of a sequential program. Even though the instruction system in which the code is presented exerts influence on the construction of each processor unit, this has no effect on the construction of the rest of a multithread processor.

To speed up traversal of the CDF, the scheduler needs information about the structure of the program's control flow. In particular, it needs to know which segments are potential successors to any segment in the CDF. The scheduler uses this information to predict one of the possible successor segments and to continue traversal of the CDF, starting from the current mark. This information can be defined statically and placed into the segment's descriptor. Segment descriptors can be interspersed within the program text (before each code segment, for example) or be grouped together and placed separately (that is, at the end of the program).

To coordinate the execution of different segments, each segment must be described according to the set of the values used and produced by the segment.

The procedure to process register values is simple. Static analysis of the CDF by the compiler produces a creation mask. Using values, the segment awaits them only if they have not yet been produced by the predecessor segment. Otherwise, it finds the values it needs within the local memory, transmitted there by the predecessor segment over the ring channel.

It is only natural to expect the creation mask to be located in the segment descriptor. Because a segment can contain multiple basic blocks whose executions depend on the data being processed, it is impossible to make a static determination of what register values will be created during the computational dynamics because of potential conditional branches. A creation mask must be conservative and, consequently, include all possible register values that can be produced.

As segment's instructions are executed by the processor unit, the produced register values are sent to the successor segments. Because a processor unit cannot determine beforehand the type of instructions contained in the register assigned to it, it does not know which instructions modify the registers whose values must be passed on to other segments. In conformance to the sequential semantics, other segments can be sent only the result of the last modification of a register by the segment. The strategy of awaiting execution completion of all segment's instructions (when no further register modifications are possible) is inexpedient, because it often leads to other segments awaiting a value that is already available.

Not all values produced by a segment must be passed on to successor segments. It will be sufficient to send only those values that will be used outside of the segment that produced them.

A compiler can determine the last instruction in a segment that modifies the corresponding register. It can mark this instruction as special (execute and pass on), which, in addition to performing a certain operation, passes on the results to the successor processor units. Moreover, because the processor unit executes the segment's instructions, it can identify those registers for which no values will be produced.

For the same reasons a processor unit cannot determine which instructions the segment assigned to it actually performs, the same as it cannot determine in advance the segment's terminating instruction, that is, at what point the segment hands off the control. When the compiler is breaking the CDF into segments, segment boundaries and control-transfer nodes are defined. In a control-transfer node, an instruction can be marked with special halt conditions in such a way that these conditions could be evaluated by the time this instruction is fetched by the processor. If the halt conditions associated with the instruction are fulfilled, the segment is completed.

Transfers and halts can be specified by marking each instruction in a segment with a bit tag (transfer bits and stop bits). Marking by bits can also be done in a different way. For example, a tag bit table can be associated with each static instruction. Instructions and the corresponding bit tags from the table are selected from the program text by hardware; this pair is then combined into a new instruction and placed into the instruction cache. The register can be released by adding a special release instruction to the basic instruction system.

A multithread program can be generated from an existing binary program by adding descriptor segments and bit marks. This information can be interspaced in the program code or precede or follow it.

Separating the executable code from the descriptors makes it possible to simplify porting programs to other hardware platforms.

ESSENTIAL HARDWARE FOR MULTITHREAD ARCHITECTURE

The tasks of CDF traversal, assigning segments to processor units, and executing these segments while preserving sequential semantic of the program is placed on the hardware. The software scheduler is entrusted with defining the segment execution order. The scheduler selects a segment descriptor by its address and assigns the segments to a processor unit, issues the address of the first instruction, and sets the creation and accumulation masks for the segment. By using a static or dynamic prediction method, the scheduler also predicts the successor segment based on the information from the segment's descriptor. Each processor unit fetches and executes a segment's instructions until it encounters a halt instruction, which identifies the segment's completion.

The main purpose of distributing segments among the processor units of a multithread processor is to create a possibility of executing several instructions in one cycle. In multithread processors, efficiency loss can be caused by unsuccessful computation, wait, and inactive cycles.

Unsuccessful computation cycles are the work that will be discarded in the future because of incorrect data or predictions used in them. Wait cycles occur when waiting to receive a value produced by an instruction in the predecessor segment or by an instruction in the same segment (for example, when executing a long-execution-time operation or because of a cache miss), as well as when waiting for the segment rescheduling to be completed. Inactive cycles are the time during which the processor has no assigned segments. Their main cause is an imbalanced processor load.

To reduce losses caused by inactive cycles, the possibility of discarding results of previous operations should be reduced by synchronizing processor units in terms of data, as well as establishing the fact of a discarding having to be done (if such exists) as soon as possible.

ADVANTAGES OF MULTITHREAD ARCHITECTURE

Certain features favorably set multithread processors apart from traditional superscalar microprocessors.

With the superscalar approach, the accuracy of branch predictions limits the degree of parallelism. If the average probability of a correct branch prediction is 0.9, the probability of a correct prediction five branches ahead is only 0.6.

Multithread processors have a greater prediction depth and provide a greater probability of selecting the correct direction of computations. This property stems from the branch-selection discrimination. Multithread processors break a sequential instruction stream into segments. Even though segments can contain internal branches, the scheduler must predict only those branches that separate the segments. The branches within segments are not predicted (if they are not predicted separately internally to the processor).

A wide execution window in superscalar processors increases the number of postponed instructions and makes it more complicated to keep track of the execution results of all instructions in it.

With multithread implementation, the execution window can be very wide; however, at any given moment only a few instructions must be examined for having produced results (only one instruction for each processor unit).

The boundaries of the postponed instructions window can be identified by the first and last instructions in the execution queue.

To produce n results simultaneously, a processor must use n^2-degree complexity logic so that the crosscheck of instruction interdependencies could be performed. This limits the bandwidth of the superscalar processor output logic. In multithread processors, each processor unit issues instructions independently, that is, its logic is of n-degree complexity.

Before memory accesses can be reordered, all addresses of the value loads and stores must be identified and calculated.

In the superscalar implementation, load and store instructions are ordered (or stored in the initial sequence) and are placed into the buffer, along with the memory-access address. When an instruction loading a value from or storing it to the memory is executed, the buffer is checked to ensure that there is no earlier store or load instruction from the same or not-yet-determined address. In the multithread implementation, load and store instructions can be executed independently, without knowing in what sequence load and store instructions were executed in the successor or predecessor segments.

A quite wide execution window with a large branch-prediction depth can be, in principle, generated in superscalar processors. A flexible instruction-execution schedule can also be generated. For example, a load instruction in a called function can be executed concurrently with a store instruction in the calling function. However, superscalar processors have no concept of program CDF. That is why it is necessary to predict each branch that, in the long run, causes prediction accuracy and computation efficiency to drop.

In many aspects, multithread processors are like multiprocessor systems with shared memory and very low-level scheduling overhead [31–33]. The main difference between them lies in that a multiprocessor system requires that the compiler divide programs into segments in which all intersegment dependence relationships are known (taken care of by the programmer by using synchronization operators and interprocessor communications); a multithread processor, on the other hand, does not require any prior knowledge concerning control and data instruction interdependencies.

Multithread architecture combines the principles of low- and high-level paralleling and methods of analyzing static and dynamic program structure. This makes it possible to achieve the higher efficiency of using processor computational resources. Actually, multithread processors implement a symbiosis of an automatic paralleling compiler, which gives instructions to the processor hardware using instruction marks and special instructions, and the processor hardware, which receives these instructions.

Of course, the presented approach is not the only one possible that can be used to implement this productive idea: to recruit the compiler to schedule tasks and balance the load among the processors in multiprocessor systems.

1.4.4 **Multithread Processors with Threads Detected by Analyzing Program Data Flows**

An example of a multithread processor with treads detected by analyzing program data flows is the Kin microprocessor [34]. This microprocessor is built as a collection of functional blocks that intercommunicate using FIFO queues. Internally, each functional block can be both synchronous (which is quite possible on account of the limited size block) and asynchronous. The latter circumstance is quite important, because there are claims cited in [35] that crystal microprocessors with a billion transistors and frequencies in the gigahertz range cannot be synchronous. In large crystals, the propagation time of any type of signal, including clock, from one side of the crystal to the other is about 3 ns, or 5–7 clocks at 2 GHz. The traditional solution based on using a clock-signal distribution network requires a great amount of resources and energy. For example, the Alpha microprocessor uses 40 percent of the power to eliminate skew and jitter from the clock signal.

To understand how the Kin microprocessor functions, it is important to know that a block fetches instruction items from the input queue as they arrive, performs the specified actions, and places the results into the output queue. With this type of organization, the instruction flow is distributed among the blocks as a stream of packets that contains tags and other information necessary for controlling functional blocks. The ILP computation organization model used in Kin is a modern development of the data-flow computations conception.

The main architectural concept of the Kin microprocessor is a basic block: an instruction sequence transforming data in registers and memory and completed by a control-transfer instruction.

The instruction-fetching unit fetches and decodes instructions of the next basic block and places the results into a code cache line. Each instruction is given its own dynamic tag. Then, register-renaming blocks remove all extraneous WAR and WAW instruction interdependencies, and the instructions are placed into the out-of-order execution-organization block. From the out-of-order execution unit, instructions are moved to the reservation stations, where they wait until operands become available to start executing.

The reservation stations send the instructions, which have their operands ready, to functional units to be executed: control-transfer and arithmetic, logic, and shift instructions interpreted, memory-access instructions executed. Functional blocks place the results of their work into reservation stations and the out-of-order execution block; in case of control-transfer instructions, the results are placed into the instruction-prefetch block.

The described functional organization of the Kin microprocessor is typical for out-of-order instruction-execution processors. The Kin's architectural special feature is its use of avid mechanisms for basic block instruction fetching and for removing instructions that were fetched as a result of a wrong computation direction prediction.

The instruction-prefetch block that predicts control transfers and performs avid instruction prefetch fetches instructions from the decoded instruction cache and adds a dynamic tag to them; one of the items included in the tag is the trace to which the instruction belongs. The same instructions can be fetched more than once. For example, in the case of a conditional branch-instruction cycle, instructions of the cycle's body are fetched repeatedly. Nevertheless, they are considered to be self-contained. Instructions from different decoded instruction cache lines can be fetched simultaneously. This is exactly the gist of avid instruction prefetching.

With avid instruction prefetching, instructions are fetched based on both the direction of the branch and the direction where there is no branch (that is, the control flow continues through). Therefore, when determining the value of a branch instruction's predicate, the results of the incorrectly prefetched instructions need to be discarded as well as the instructions themselves. For this purpose, a dynamic tag that is generated when branch instructions are executed is used. If there is an m trace in a branch instruction's tag, then the prefetched instructions of one direction are given trace $m0$, and those of the other direction are assigned trace $m1$. Consequently, it becomes possible to detect all fetched wrong instructions every time when the branch direction is being determined. The fetched wrong instruction-discarding block sends messages about traces of the instructions that are to be discarded to the other blocks.

1.4.5 Specifics of Multithread Paralleling Models

When implementing multithread architectures, it is important to delineate the tasks performed by the programmer, compiler, operating system, and processor. Threads must be executed by the processor; but there also is thread control: thread selection and generation, establishment of the execution order, resource allocation, and interthread communication

The most difficult task of thread selection must be done by the compiler or programmer. Here, auxiliary instructions for the hardware must be made that will make the processor's work easier further ahead.

A program intended for execution on a multithread processor must consist of a collection of threads, each of which is one data-processing unit. An important feature of multithread processors is the ability to effectively implement not only memory access but also access to I/O devices, as well as to the controllers of the communication environment network adapters. A multithread program makes it possible to achieve the maximum combination.

Combining computations in multithread processors with processor interexchange can be achieved to an extreme degree in multithread programs. A thread referencing a network adapter will wait for the completion of the exchange, while the other threads continue execution. This property of multithread processors makes them oriented toward the task of constructing massively parallel systems.

It should be noted that transputers can also be included into the multithread microprocessor category. Operating at appropriate clock frequency, their processor context-switching time is supported by hardware and is extremely low; they also have virtually implemented channels (T-9000) [21, 25].

Therefore, after superscalar architecture, multithread architecture is the next step in the direction of developing methods of dynamic program paralleling while fully retaining the results of static paralleling. This architecture does not make any demands on the localization of processed data and may, in fact, eliminate these demands altogether.

Multithread architecture offers solutions to the following problems, which are in the way of the superscalar processor development:

- Efficient loading of a greater number of functional units
- Bridging the gap between the execution speeds of register instructions and memory-reference instructions
- Reducing the interrupt servicing time as compared to superscalar processors by singling out a separate thread that executes interrupt handlers, which makes it possible to work in a rigid real-time mode
- Efficient combining of computations in microprocessors with interprocessor exchanges

1.5 MICROPROCESSOR ARCHITECTURE DEVELOPMENT

1.5.1 Influence of the Elementary Base on Architecture Realizability

Using discrete elementary electronic components together with small- and large-scale integration circuits, it was possible to design and build a processor of any architecture, the only limit being the overall amount of the equipment used. Processors were built from cards onto which electronic components were soldered. Using connectors, cards in turn were assembled into blocks; blocks themselves were joined by cables with interblock connectors and mounted into racks and other higher-level constructs. Therefore, after a processor was designed, a large number of assembly workers and setup technicians were needed to build and adjust cards, blocks, racks, and the connections between them.

To obtain high performance, it was necessary to have the appropriate number of registers and functional devices; however, the extent of electronic components and soldered and detachable connections necessary to build them was circumscribed by the reliability, cost, and power consumption of the processor under construction. In these circumstances, processor designers came up with a great variety of processor architectures, each of which reached the highest efficiency/cost ratio benchmark when executing programs oriented toward its task area. To these architectures belong vector-pipelined processors and SIMD architecture associative processors, all of whose processor units execute one stream of instructions produced by one common control unit. During the era of SSI and MSI circuits, several processors of this type were built; for example STARAN [36], DAP [37], and Connection Machine (CM-1, CM-2) [38]. These processors consisted of a large number of one-bit processor units, each with built-in local memory. Using a communications network between the processor units, these systems were employed for solving various types of problems, including equation-set solving, signal and image processing, and associative data processing.

Increasing the integration level of VLSI circuits accompanied by little change to the output pin numbers led to the situation where, on one hand, quite a few processor units could be built on one VLSI circuit, whereas, on the other hand, this number was markedly not enough to build a full-fledged SIMD processor. Several VLSI multiprocessor chips could not be connected into one system because of the small number of output pins on them. Because of this, memory could not be connected to processor units nor effective communication network between processor units created.

The same problem of a shortage of pins stood in the way of constructing vector-pipelined processors: their blocks that could be implemented in LSI chips would have a number of connections with other LSI chips greatly exceeding the number of chip output pins.

Therefore, as was noted earlier, vector-pipelined and SIMD architecture processors yielded the arena to the microprocessors, which at one time were even called killer microprocessors [7].

However, when the integration level reached 10^8 to 10^9 transistors on one chip, it became possible to build full-fledged single-chip vector-pipelined and SIMD processors.

Pentium 4 and Xeon microprocessors with MMX, SSE, and SSE2 instruction system extensions are actually SIMD processors. In addition to being able to execute up to eight short-operand operations in one cycle, they can also concurrently process several double-precision floating-point numbers. Thus, the capability to execute two 64-bit-long operand operations in one cycle has significantly raised the performance of these microprocessors in such benchmark tests as Linpack.

1.5.2 Single-Crystal Vector-Pipelined Processors

When the number of transistors on a chip reached the level sufficient to implement a full-fledged vector-pipelined processor, a striking representative of this class appeared: SX-6 microprocessor [39] built by Japanese NEC®. The microprocessor uses 0.15-micron CMOS technology with copper conductors and has approximately 57 million transistors. The microprocessor's structure is depicted in Figure 1.24.

The microprocessor's main components are a scalar processor and eight identical vector units. The scalar processor has superscalar architecture and produces four results per cycle using 128 64-bit registers. At 500 MHz, the peak performance of the scalar processor is 10^9 FLOPS (1 GFLOPS).

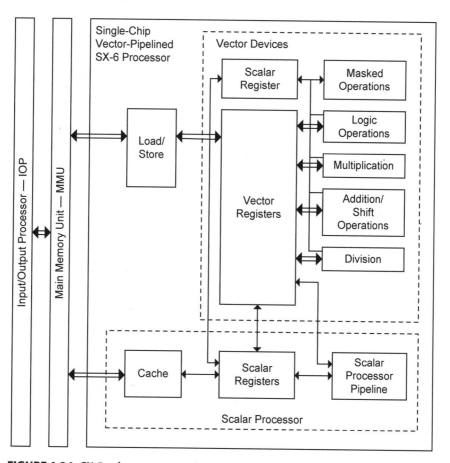

FIGURE 1.24 SX-5 microprocessor structure.

Each of the eight identical vector units has five data-processing pipelines that perform logic operations, masking, additions and shifts, and multiply and divide operands located in the vector registers. They also have one pipeline for exchanging data between the registers and the main memory. The total register capacity of the eight-vector units is 18 KB (an equivalent of 288 64-bit registers).

Multiplication and division operations can be concatenated in the vector unit; this raises peak performance of one device up to 1 GFLOPS at 500 MHz.

The memory interface bandwidth is 32 GBps, which allows each of the eight vector units to load from or store to the memory one operand per clock.

The performance of the SX-6 chip is 8 GFLOPS.

Chips containing several vector-pipelined processors forming a single-chip system with shared memory can be expected in the future.

1.5.3 Reasons for Moving to Building Single-Chip Multiprocessor Systems

The most noted tendencies causing emergence of single-chip multiprocessor systems follow:

- Transferring to the compilation stage the task of extracting from sequential programs instructions capable of parallel execution in particular and branches of parallel programs in general. If superscalar processors extract parallel execution instructions by themselves, additional functions of extracting parallel branches, already in multithread microprocessors, are placed on the compiler, and the LIW microprocessors place all tasks of loading parallel functional units on the compiler. Under these circumstances, the task of building a paralleling compiler for multiprocessor systems does not appear unsolvable.
- The extent of equipment for loading functional units is quite voluminous in superscalar architecture processors; it grows quadratically, depending on the number of instructions being processed. An increasing number of functional units leads to increasing the number of instructions fetched for an execution, which leads to an increase in equipment that does not directly process data. The total amount of control circuits in a multiprocessor system built from discrete regular processors can be significantly lower than in a superscalar architecture processor, with their total number of functional units being equal, that is, their efficiency with fully loaded functional units being the same. It must also be noted that regular discrete processors in a multiprocessor system can operate at higher clock frequencies.
- Owing to its inherent redundancy, a multiprocessor can continue functioning when part of its equipment fails. Failures can be both internal, caused by the silicon wafer or a manufacturing technological process defect, or arise during

the course of system use. A multiprocessor system can be built as a single-chip or a multi-chip assembly. The reality is that the efficiency and cost ratio of a multiprocessor system microassembled using regular processors in one casing can be substantially higher than that of a single-chip system, with their total chip areas being the same [40, 41]. Microassemblies do not significantly differ from VLSI circuits. The choice in favor of either single-chip or microassembly implementation is made on the basis of the technical and economic characteristics of each. Such examples as memory microassemblies and the Pentium Pro microprocessor demonstrate the capability of the microassembly technology. A most interesting engineering project, in which architectural and technological advances are concentrated, including single-chip systems and microassemblies, is Power 4 microprocessor from IBM [42].

■ By various estimates, traditional computers, consisting of a 108 transistor microprocessor and memory chips with total of 109 transistors, use only 104 to 105 transistors in each clock. In other words, a substantial part of equipment that could potentially process data idles. Of course, idling is of some benefit when using CMOS technology: equipment consumes little power and produces little heat. With the current constructions of chip casings, the issue of heat removal can become the decisive factor in the choice of chip's architecture. However, at present it appears that a multiprocessor system containing a large number of processors, each with its own small built-in memory, can be implemented on one chip quite effectively. Computational structures of this type are called processing memory, multifunction memory, smart memory, and several other terms.

1.5.4 Approaches to Building Single-Chip Multiprocessor Systems

Multiple instructions-multiple data (MIMD) architecture multiprocessor systems can be built on a single chip. These systems process many data streams by many instruction streams. They can be implemented as both symmetrical multiprocessor (SMP) and massively parallel processor (MPP) systems [43].

The closest to the multithread architecture microprocessor will be the SMP implementation constructed from 8 to 32 regular microprocessors that are architecturally close to first- and second-generation RISC processors, with a simple instruction-fetching control unit but with a built-in L1 cache. Moreover, all chips' processors have a shared on-chip L2 cache. In essence, this multiprocessor differs from a multithread architecture processor by the use of a collection of simple, independent control units for each processor instead of one complex control unit.

The MPP constructed from nodes (similar to the described regular processors) interconnected by point-to-point channels also is a possible version of single-chip multiprocessors. These multiprocessors use up to 30 percent of their hardware to create channels for exchanges with the other nodes and peripheral devices.

An important feature of the considered MPP is the complexity of a processor node and the number of nodes on the chip. Thus, decreasing the node complexity to one-bit ALU and one-bit register (one trigger) produces reconfigurable microprocessors; depending on the implementation they can be considered either homogeneous computational systems [44] or PLIC.

A unique capability of reconfigurable systems is their ability to be configured by software to implement a given algorithm by hardware; in logic-symbolic processing tasks, this feature makes it possible to achieve substantial processing acceleration compared to the software implementation of the same algorithm.

In reality, the organization of SIMD architecture single-chip systems built from one-bit processor unit is similar to computation environments and programmable logic integrated circuits (PLIC) and [44, 45].

The main trend in the development of VLSI circuits is building single-chip systems from a large number of simple processors. These systems can have multithread or vector-parallel architecture; they can also be MIMD or SIMD systems built from simple processors or reconfigurable arrays of simple (one-bit) processor units. Along with processor logic circuitry, these chips must have built-in memory.

Single-chip systems, in turn, are combined into a single system with MIMD or SIMD architecture or with some other architecture that is a combination of the previously mentioned.

1.6 METHODS TO MEASURE PROCESSOR PERFORMANCE

1.6.1 Computer Peak Performance

Historically, the main method to measure performance has been determining the peak or technical performance, which is the theoretical maximum operating speed under ideal conditions. This maximum is defined as the number of operations executed within a unit of time by all ALUs of a computer. Peak performance is measured in instructions per second (IPS); for example, in modern computers peak performance is measured in millions of instructions per second (MIPS) or millions of floating-point operations per second (MFLOPS).

Peak operating speed is achieved by executing an infinite sequence of instructions that are not interlinked and have no memory-access conflicts (that is, when the result of any operation does not depend on the operations performed by the other instructions). In this connection, it is assumed that the operands are fetched from the on-chip data cache and the instructions are fetched from the instruction cache. Of course, this type of situation is purely hypothetical and in reality, no computer can work for any significant length of time at peak performance, although it can get close to it.

Peak performance is the only truly objective evaluation (only a few computer characteristics need to be known to determine it) and does not depend on the type of programs executed. The first characteristic in question is the processor clock rate, which for the majority of modern processors defines the rate with which results are generated at the output of the arithmetic pipeline. The other characteristic is the number of arithmetic pipelines the processor has. Peak performance is determined by multiplying the clock rate by the number of concurrently executed operation. Here, the architectural capabilities to combine operations executed in one cycle must be taken into account.

For example, an arithmetic unit of Pentium can generate one 64-bit floating-point operation result or two 32-bit integer operation results every clock cycle. Subsequently, Pentium 90 (operating at 90 MHz clock frequency) has a peak performance of 90 MFLOPS (1 MFLOPS = 10^6 FLOPS) when doing floating-point calculations and 180 MIPS when processing 32-bit integers.

Determining peak performance of Intel Itanium® and IBM Power 4 processor is another example. These processors can simultaneously execute an accumulation instruction $a + b \times c$ in two floating-point processing units in one cycle. Therefore, their floating-point operation peak performance is four times the clock frequency. However, this is true only in the case when the numbers of additions and multiplications that need to be carried out are the same; this situation is typical for many scientific and technical algorithms, for example when numerically solving differential equations. In situations where only additions or multiplications need to be carried out, the peak performance of the processors is considered to be a double of the clock frequency.

Notwithstanding the simplicity in obtaining them, performance figures can be exceedingly useful when discussing computers' merits. The peak performance figures are especially convenient when making rough comparisons of processors' capabilities. American arms-export-control specialists use the composite theoretical performance (CTP) index, measured in millions of theoretical operations per second (MTIPS), to assess computer efficiency level [46]. CTP is determined only by computer hardware characteristics (clock rate, set of functional units, internal bus collection and their bandwidths, bit array width, and so on).

However, using peak performance figures to compare computer efficiencies does not always produce adequate results. Peak performance expressed in the number of instructions executed within a unit of time does not take into account the functional content of instructions. The same result, such as adding two memory arrays, for example, can be obtained using different numbers of instructions in RISC and CISC processors. In RISC processors, for example, addresses of the array elements are generated by separate instructions, whereas in CISC processors this is done by a part of a complex instruction. Different instruction formats take different amounts of time to execute, and the execution period for a sequence of instructions depends on where in the sequence each instruction is specifically located. The effect of the width of the processed number must not be forgotten either, so as not to obtain incorrect results by comparing performances of processing 32- and 64-bit data.

With the appearance of CRAY-1–type supercomputers, a new unit for measuring performance was introduced; it is expressed by the number of floating-point results per unit of time. The reasons for introducing this index were based on the following. There are main necessary floating-point operations that produce the results, and there are auxiliary operations for setting up the computations that must not be considered when evaluating useful performance. For example, when multiplying matrices, only floating-point operations to obtain the element values of the resulting matrix are of importance for reflecting its computational efficiency; fixed-point operations to generate addresses of the matrix elements, update cycle counters, and so on depend on the computer's internal organization. However, the obvious orientation nature of this performance index toward the scientific and technical calculations tasks has not made it widely used to any extent.

Modern processors, such as Intel Itanium and IBM Power 4, combine executions of fixed-point and floating-point operations; therefore, their floating-point operation performance more often than not corresponds to the performance expressed in the results.

1.6.2 Real Performance

As a rule, when assessing microprocessor efficiency, the microprocessors themselves are not evaluated; the computers built on their base are. This is due, above all, to the user's approach to the assessment; he is not interested in the processor's characteristics per se but in the overall task-solving efficiency of the computer system built using this processor.

When executing real application programs, the effective (real) computer efficiency can be substantially lower (up to several times) than the peak. This is explained by that fact that modern high-performance microprocessors have complex architecture (pipelined and superscalar processing, hierarchic memory, and so forth). Their functional characteristics at the internal unit level substantially depend on the programs executed and data processed. Therefore, it is impossible to assess the efficiency to the necessary degree of precision based only on the operating clock rate, number of clock cycles it takes to process one instruction, and number of processing units. Moreover, efficiency depends on the operating system used, compilers, software component libraries, and other software. For example, execution efficiency of object codes of the same program generated by different compilers can differ substantially.

Historically, the first attempt to measure real efficiency was the creation of the synthetic benchmarks, consisting of a mixture of instructions with a given percentage for each instruction type. However, it soon became clear that the results received using synthetic tests are difficult to apply for assessing the real computer efficiency in a specific task.

Therefore, using fragments of real tasks that are characteristic for one or another application area of computational equipment has received wide acceptance in the world practice for assessing efficiency of various computational means. The time taken to execute each of the collection's tasks is used as the base for calculating the efficiency index of the computing machine being evaluated. The efficiency index is a relative estimate providing information about how many times faster or slower the machine being evaluated performs this type of task as compared to some common computer (often called the base or standard computer).

If the efficiency of the standard computer expressed by the number of computation operations per second can be established in some way, then it will not be difficult to move from using the efficiency indices to absolute values to express the operating efficiency of computing machines.

1.6.3 Methods for Measuring Real Performance

The following three problems associated with analyzing the results of the benchmark performance testing must be solved when evaluating performance [47]:

- Separating unconditionally trusted figures from those that must be considered with a certain degree of circumspection (the reliability of the estimate problem)
- Selecting benchmark tests that most closely characterize the efficiency of processing user's typical tasks (the adequacy of the estimate problem)
- Correct interpretation of the results of the performance testing, especially if they are expressed in rather exotic units of the type, such as Mega Whetstone Instructions Per Second (MWIPS) [48], Dhrystone/s, and so on (the interpretation problem)

The existing benchmarks can be divided into three groups. To the first benchmark group belong manufacturer tests developed by the computer manufacturers for in-house use: evaluating the quality of their products. The main feature of these tests is that they are oriented toward comparing a limited number of one-type computers, often belonging to one line. These tests let the computer developers optimize the structural and technical solutions of their products. For example, in 1992, Intel introduced Intel Comparative Microprocessor Performance (iCOMP®) index for measuring the performance of processors with the x86 architecture. The 486 SX-25 microprocessor was accepted as the standard processor, with the performance index of 100. The iCOMP index is determined by running a synthetic test that generates a combination of operations. Of them, 67 percent are 16-bit integer operations, 3 percent are 16-bit floating-point operations, 25 percent are 32-bit integer operations, and 5 percent are 32-bit floating-point operations. For example, iCOMP indices for 486 SX2-50, Pentium-100, and Pentium-166 microprocessors are 180, 815, and 1,208, respectively. It must be noted that the iCOMP index evaluates performance of the microprocessor itself and not of a computer system built using it with the system memory and peripheral devices added.

IBM has developed specialized benchmark packages to measure computer performance of the System/370 and System/390 architecture mainframes, as well as tests for the AS/400 architecture computers.

Manufacturer benchmarks are an almost ideal means for evaluating operating speed and technical and economical characteristics of processors and computing systems having the same architecture but implemented in different ways. However, in their original form they cannot be used for testing other computers because of the too-obvious orientation toward the specific company architecture. These benchmarks are used by the system designers for evaluating different implementation versions.

To the second group belong standard benchmarks. Standard benchmarks, which are developed for the purposes of comparing a wide range of computers, lay claim to being totally universal performance-assessment means. These ambitions are based on developers of this category of tests being either independent analysts (such as, for example, Jack Dongarra, who jointly with a group of other researches introduced the Linpack benchmark suite [49]) or groups of the largest computer manufacturers (such as Standard Performance Evaluation Corporation [50]). The collective nature of the development process practically eliminates the possibility of a standard benchmark being oriented toward a specific computer manufacturer.

The third group of benchmarks consists of user benchmarks that take into account the specifics of the particular application of the computer or computing system. User benchmarks are developed by large companies that specialize in introducing computer technologies or by joint efforts of group of users united by the similarity of the tasks they work on.

User benchmarks are intended specifically for choosing computers and software most suitable for concrete application problems. In principle, this approach makes it possible to obtain the most precise performance assessment for a specific application class, although it requires significant efforts on the user's part to develop benchmark programs and to conduct computer testing.

1.6.4 Benchmarks for Evaluating Memory Bandwidth and Processor Performance

There are several rather simple, small-size benchmarks that can be used to evaluate processor performance and its memory operating characteristics for floating-point arithmetic operations. Some of them follow:

Stream [51]—A synthetic benchmark that measures the bandwidth of system memory in MBps by repeatedly cycling through four floating-point arithmetic operations. *Stream* can be used to evaluate memory bandwidth as well as maximum, minimum, and average memory exchange time for each of these four operations.

Cpu-rate benchmark—A test for evaluating performance of processors and system memory for floating-point tasks. Unlike the *Stream* benchmark, *Cpu-rate* distributes the processed vectors dynamically to receive information concerning the effect of the memory speed when repeatedly cycling through arithmetic operations. Measurements can be conducted both for statically and dynamically allocated memory; this provides more objective evaluation of the processor operating speed than that of *Stream* when processing various format floating-point numbers. Results of the *Cpu-rate* benchmark are presented as the processor/memory operating speed measured in FLOPS.

1.6.5 Linpack Benchmark

Linpack benchmark is widely used to measure performance of computer systems of various architectures. Performance evaluation results are published periodically in the report, "Performance of Various Computers Using Standard Linear Equations Software," prepared by Jack Dongarra from the University of Tennessee.

Linpack benchmark is a dense matrix of linear equations with n unknowns over the field of real numbers. For each n is known how many floating-point operations must be performed to solve the system of linear equations using the Gaussian exclusion method. The number of these floating-point operations is computed as $2n^3/3 + 2n^2$. No matter what method and what program is used to solve the system of linear equations with n unknowns, the number of operations is assumed to equal the number of results and is expressed by the previous formula. It must be noted that the benchmark presupposes equal duration of the addition and multiplication floating-point operations, which is an uncertain assumption because multiplication can take more cycles to carry out. Moreover, when the difference in the execution times of addition and multiplication operations was significant, Linpack barred using the Schrassen algorithm, which speeds up calculations by substituting some multiplication operations with addition operations. The performance assessment in FLOPS is obtained by dividing the number of operations $2n^3/3 + 2n^2$ by the benchmark execution time in seconds.

Benchmarking is performed using three methods that determine solution time for different values of n and the given restrictions. They follow:

$n = 100$; 100×100 (**LINPACK Benchmark**)—The original benchmark written in FORTRAN using Level 1 BLAS ($y = y + a \times x$) component libraries, where y, a, and x are scalars, for solving linear algebra problems for dense matrices. No modifications to the benchmark can be made in FORTRAN; however, there are C versions of the test available. The benchmark performs two operations for each three memory references; a DAXPY subprogram from Level 1 BLAS library is used for main computations.

$n = 1000$; $1,000 \times 1,000$ (**Towards Peak Performing [TPP]**)—Has no restrictions on program modifications and using an assembler to achieve the highest level of performance. Block processing can be used; for example, using DGEMM subprogram from the Level 3 BLAS ($C = C + A \times B$) library, where C, A, and B are matrices. This increases the locality of processed data (number of computation operations per one memory reference).

$n = N_{max}$ (**Highly Parallel Computing**)—Has no restrictions on program modification and using an assembler to achieve the highest level of performance by choosing such matrix size N_{max} at which the highest level of performance R_{max} is achieved.

When running the Highly Parallel benchmark, such $n = N_{max}$ is chosen at which the highest performance level R_{max} is attained. (The value of n is chosen by trial method to the point at which the matrix can be located in the system memory, because disk accesses substantially lowers the efficiency.) Afterward, when $N_{1/2}$ is determined, the performance level is half the R_{max}. These are the values that are published for the 500 fastest computer systems in the Top 500 report, along with the theoretical peak performance R_{peak}, number of processors, and their clock rates. Computer systems are listed in the Top 500 report in decreasing order of their R_{max}.

Consequently, the LINPACK benchmark suite is a collection of programs for solving linear algebra problems. The following are used as parameters: the degree of matrix (for example, 100×100, 1000×1000), the format of the values of matrix elements (single- or double-precision), compilation method (optimized or not), and the possibility of using an optimized library of standard functions.

When measuring the performance level of modern microprocessors, program and data of Linpack 100 × 100 test (LINPACK benchmark) can be located in cache, which makes it possible to practically evaluate peak performance in results per second.

1.6.6 SPEC CPU Benchmark Suites

REVIEW OF BENCHMARKS

Standard Performance Evaluation Corporation (SPEC®) develops benchmark suites for measuring computer performance in various modes of operation. The corporation's goal is to correlate parameters objectively, in such a way so that neither the benchmarks themselves nor the methods used to compile them can be oriented toward a particular manufacturer. Over the years, SPEC has produced the following benchmark suites for measuring processor performance: SPEC CPU 89, SPEC CPU 92, SPEC CPU 95, SPEC CPU 2000. At the present time, SPEC CPU 95 and SPEC CPU 2000 test results for computers from different manufacturers are available at the site *www.specbench.org/osg/cpu2000/results.* Benchmarks have been replaced mainly due to the microprocessor development and the progress in understanding benchmarking technologies. Taking a look at the development history of SPEC CPU benchmark suites will be conducive to understanding benchmarking methods.

SPEC CPU 89 SUITE

The SPEC CPU 89 suite comprises two benchmarks. These are Cint89, consisting of four integer-processing programs, and Cfp89, combining six programs that have a significant number of double-precision floating-point operations. All ten programs are quite complex, written in C and FORTRAN, and can be applied to solve a wide range of problems, from optimization of Boolean logic function representations in programmable logic circuits to atom replacement modeling in quantum chemistry.

The SPEC 89 benchmark method presupposes generating ten $SPECratio_i$ differential assessments. Each of these is defined as the ratio of the execution time of #i program from Cint89 or Cfp89 suite on the benchmarked computer to the execution time of the same program on DEC VAX 11/780 computer.

An integral characteristic of computer performance is the SPECmark® index, which is a geometrical average of the ten individual SPECratio estimates. Two more estimates are added to the SPECmark parameter—SPECint89 and SPECfp89—which provides separate estimates of computer performance for integer and real-number processing. These indices are calculated is the same way as the SPECmark: SPECint89 is a geometric average of the individual SPECratio estimates for four programs from the Cint89 suite; the SPECfp89 is an analogous value for six programs for the Cfp89 suite.

SPEC CPU 92 BENCHMARK SUITE

This suite expands the collection of the problem classes represented in the benchmark as compared to SPEC 89. The calculation method of the main performance characteristics has not undergone any principal changes in SPEC 92. As before, the SPECint92 and SPECfp92 indices are defined as geometric averages of the individual SPECratio estimates obtained by running benchmark programs from Cint92 and Cfp92 suites (taking into account the increased number of the benchmark programs, of course).

The Cint92 benchmark suite is intended for measuring performance of computer systems when performing integer operations, prevalently in the commercial application area. It comprises six standard benchmarks in C language, representing a network theory problem, a LISP interpreter, a logic design problem, a Unix utility to pack a 1 MB test file 20 times, electronic spreadsheet row and column operations, and a C compiler.

The Cfp92 benchmark suite is intended for measuring performance of computer systems when performing floating-point operations (mostly in technical and scientific application areas).

It consists of 14 real application programs, two of which are in C and 12 in FORTRAN. Among the package programs are those for circuit simulation, nuclear reactor thermodynamic simulation using the Monte Carlo method, quantum chemistry and physics problems, solving Maxwell equations, coordinate conversion, optic-ray tracing, robotic technology and neural networks problems, human-ear modeling, solving Navier-Stokes equations for determining characteristics of intergalactic gases, seven library functions for matrix processing (multiplication, transformation, and so on), and several others.

One qualitative innovation was introduced in SPEC 92. It consisted in the characteristics of multiprogram processing of SPECrate that are generated under the uniform load method. The essence of the latter consists of the following: a computer system performs a task consisting of multiple copies of the same program. The number of program copies executed within a certain time period is the index of multiprocessor processing performance. The same programs as for calculating SPECint92 and SPECfp92 indices are used to obtain the SPECrate estimate. The only difference is that the benchmark module is implemented as several copies forming one task, and the benchmark result is the normalized total time taken to execute all copies of the task. Each of the benchmark's 20 programs are subjected to this procedure, which allows six individual SPECratio estimates to be obtained for integer-processing programs and 14 for real-number processing programs.

Consequently, SPECrateint92 and SPECratefp92 measure the average task execution speed in the multiprocessor operating mode of a system. Moreover, these indices allow an idea about the compiler capabilities concerning producing parallel multitask code as well as about the operating system capabilities concerning its dynamic system-resource allocation efficiency (processors in particular) among the parallel programs. This makes the SPECrateint92 and SPECratefp92 estimates especially representative for collective-use SMP systems operating in batch mode.

Table 1.5 gives some brief information about programs comprising SPEC89 and SPEC92.

TABLE 1.5 List of SPEC 89 and SPEC 92 Programs

Program	*Benchmark, Language, Data Type*
PLA Sumulation	89 & 92, C, FX*
Lisp Interpreter	89 & 92, C, FX
Generating logic truth tables	89 & 92, C, FX
UNIX utility to pack a 1 MB test file 20 times	92, C, FX
Operations with spreadsheet rows and columns	92, C, FX
GNU compiler, translating 19 C programs into optimized assembler code	89 & 92, C, FX
Simulation of analog circuits with high intensity memory exchanges	89 & 92, FORTRAN, FL**, DP***
Simulation of a nuclear reactor thermodynamics by Monte Carlo method; contains many branches and short cycles	89 & 92, FORTRAN, FL, DP
Quantum chemistry problem for a 500-atom system	92, FORTRAN, FL, DP
A single-precision Mdijdp2 benchmark version	92, FORTRAN, FL, SP****
Maxwell equation's solution	92, FORTRAN, FL, DP

TABLE 1.5 List of SPEC 89 and SPEC 92 Programs *(Continued)*

Program	Benchmark, Language, Data Type
Mesh generation in modeling flow processes; the program is oriented at testing parallel systems	89 & 92, FORTRAN, FL, DP
Simulation of robot movement control using video system	92, C, FL
Shallow-water 256 × 256 matrix problem solving	92, FORTRAN, FL
Quantum physics problem of calculating mass of elementary particles using Monte Carlo method; easily vectorized and oriented toward parallel system testing	92, FORTRAN, FL, DP
Solving Navier-Stokes equations for determining parameters of intergalactic gases; easily vectorized; oriented toward parallel system testing	92, FORTRAN, FL, DP
Seven library functions for matrix processing (multiplying, transformation, etc.)	89 & 92, FORTRAN, FL, DP
Simulation of atom replacement process in Gaussian series; cannot be parallelized; contains a large volume of I/O	89 & 92, FORTRAN, FL, DP
A synthetic test simulating various matrix multiplication algorithms	89, FORTRAN, FL, DP
Human-ear simulation	92, C, FL, DP

  * FX – fixed point

 ** FL – floating point

 *** DP – double precision floating point

**** SP – single precision floating point

SPEC CPU#95 BENCHMARK SUITE

Introduction of the new benchmark suite was occasioned by advancements in microprocessors (increased performance and on-die memory), improvements in compilers, desire to take into account open-system standards, and corrected understanding of the importance of different application areas gained after SPEC 92 was introduced.

Modern microprocessors run SPEC CPU 92 benchmarks in from fractions of a second to a few seconds, which introduces quite a large inaccuracy into the estimates. The size of the SPEC 92 program code and data is such that the latter can be placed in the processor cache. This makes it altogether impossible to obtain performance estimates that are to any extent reliable.

The performance indices in SPEC 95 are given with respect to the standard machine SPARC station 10/40 configured with L2 cache.

There are two collections of benchmarks, CINT 95 and CRP 95, consisting of eight and ten programs, respectively.

The benchmarks generate the following indices:

■ Performance indices SPEC int 95, SPEC fp 95 and SPEC int base 95, SPEC fp base 95 for fixed- and floating-point operations with aggressive and conservative compilation optimization, respectively.

■ Bandwidth indices SPEC int rate 95, SPEC fp rate 95, SPEC int rate base 95, and SPEC fp rate base 95 for evaluating multitask modes and SMP architectures with aggressive and conservative compilation optimization, respectively.

All integral performance indices are generated as a geometric average of the individual benchmark indices.

SPEC does not make any recommendations concerning correlating values of SPEC 92 and SPEC 95 indices. Tables 1.6 and 1.7 give some information about the programs comprising the SPEC 95 benchmark suite.

TABLE 1.6 CINT 95 Benchmark Programs

Application Area	*Task Specification*
Simulation	Simulation of Motorola 88100 chip
Compilation	Compiling a C program and optimized code for SPARC processors
Artificial Intelligence	*Go* game—program plays against itself
Compression	Compressing a 16 MB text file
Interpretation	LISP interpreter
Image Processing	Compressing graphic object (JPEG) using different parameters
String Manipulation	Shell interpreter
Data Bases	Constructing and manipulating tables

TABLE 1.7 CFP 95 Benchmark Programs

Application Area	Task Specification
Hydrodynamics, geometric aspect	Generating two-dimensional coordinate grid in any area
Weather Prediction	Simulation of water surface using finite element model (single-precision real arithmetic)
Quantum Physics	Calculating the mass of elementary particles using the Monte Carlo method
Astrophysics	Navier-Stokes intergalactic gas parameters computations
Electromagnetism	Three-dimensional potential field calculations
Hydrodynamics	Solving systems of partial derivative equations
Simulation	Simulation of turbulence in a cube
Weather Prediction	Calculating temperature, air-flow, and pollution-level statistics
Quantum Chemistry	Generating an electron stream
Electromagnetism	Solving Maxwell equations

SPEC CPU 2000 BENCHMARK SUITE

The reasons for the introduction of the SPEC CPU 2000 benchmark suite are, in general, the same as for the SPEC CPU 95: progressive increasing of the performance indices and computer memory size as well as advances in compilers and operating systems. The indices measured by the benchmarks and the methods of their calculation have remained the same as for the SPEC CPU 95. Sun Ultra5_10 workstation is used as the standard machine with the performance index of 100.

The CPU 2000 benchmark suite comprises 25 programs, 19 of which were not used in any of the previous SPEC benchmark suites. Criteria for including and not including candidates into CPU 2000 benchmarks are given in [53].

A benchmark is recommended to be included into the suite if it:

- is widely used
- uses substantial hardware resources
- solves an interesting technical problem
- produces results that are published in popular publications
- expands the range of the problems solved by the suite's benchmarks

A benchmark is not recommended to be included into the suite if it:

■ cannot be ported within reasonable time limits
■ performs too many I/O operation and, therefore, cannot be considered to be solving a computational problem
■ used to be included into the SPEC CPU before and its operating load has remained substantially unchanged
■ is more of a code fragment rather than a complete application
■ solves problems that are solved by other candidate benchmarks
■ performs different tasks on different platforms

Benchmarking results are available at *www.specbench.org/osg/cpu2000/results.* Table 1.8 lists programs included in the SPEC CPU 2000 benchmark suite.

TABLE 1.8 SPEC CPU 2000 Benchmark Program List

Benchmark	Programming Language	Description
SPECint2000		
164.gzip	C	Compression
175.vpr	C	FPGA Circuit Placement and Routing
176.gcc	C	C Compiler
181.mcf	C	Combinatorial Optimization
186.crafty	C	Game Playing: Chess
197.parser	C	Text Processing
252.eon	C++	Computer Visualization
253.perlbmk	C	PERL Interpreter
254.gap	C	Group Theory Interpreter
255.vortex	C	Object-Oriented Data Base
256.bzip2	C	Compression
300.twolf	C	Processor Element Placing and Routing

TABLE 1.8 SPEC CPU 2000 Benchmark Program List *(Continued)*

Benchmark	Programming Language	Description
	SPECfp2000	
168.wupwise	F77	Physics: Quantum Chromodynamics
171.swim	F77	Shallow-Water Simulation
172.mgrid	F77	Multigrid Methods: Three-Dimensional Potential Field
173.applu	F77	Partial Differential Equations
177.mesa	C	3D Graphics Library
178.galgel	F90	Computational Hydrodynamics
179.art	C	Image Recognition/Neural Networks
183.equake	C	Seismic-Wave Propagation Simulation
187.facerec	F90	Image Processing: Face Recognition
188.ammp	C	Computational Chemistry
189.lucas	F90	Number Theory/Prime Number Search
191.fma3d	F90	Finite-Element-Method Collision Simulation
200.sixtrack	F77	Elementary Particle Accelerator Design
301.apsi	F77	Meteorology: Pollutant Dispersion

Below are listed officially published (September 2002) SPECint200 benchmark results for some processors:

- Intel Xeon 2800 MHz: 921
- Alpha 21264C 1250 MHz: 845
- Intel McKinley 1000 MHz: 807
- IBM POWER4 1300 MHz: 804

The following are the respective SPECfp2000 estimates:

- Intel McKinley 1000 MHz: 1356
- IBM POWER4 1300 MHz: 1202
- Alpha 21264C 1250 MHz: 1016
- Intel Xeon 2800 MHz: 878

REVIEW QUESTIONS TO CHAPTER 1

1. Name the main types of VLSI circuits and state their development tendencies.
2. State the benefits and shortcomings of static and dynamic memory chips.
3. State the mechanisms for lowering memory access time and the conditions under which this can be achieved.
4. Name the characteristic memory-access time indices for different types of memory.
5. Give a definition of programmable logic integrated circuits; state their main types; describe PLIC device design technology.
6. Give the classification of microprocessor instruction formats and types. Point out the differences between vector and superscalar microprocessors.
7. Name the main features of decoupled architecture.
8. State the conditions for efficient use of cache.
9. Name the structural methods for reducing memory-access time.
10. Describe direct-mapped cache organization.
11. Give a description of associative cache and set-associative cache.
12. What are write-through and write-back caches?
13. State the cache organization specifics in multimodule, multiprocessor systems.
14. Give the definition of memory interleaving.
15. State the difference between RISC and CISC processors.
16. Why is it important to switch processor context rapidly? Name the main context-switching implementation mechanisms.
17. State the main processor architecture standardization directions.
18. What is peak performance?
19. Name the main performance benchmarks.
20. What are SPECxx benchmarks?
21. Explain the difference between multithread and LIW architectures.

22. What processors use instruction-level parallelism?

23. What instruction interdependencies are there in sequential programs? Enumerate the dependencies that can be eliminated by renaming resources.

24. Give examples of methods to reduce performance losses due to resolving instruction-control interdependencies.

25. Enumerate resource-renaming methods that you know.

26. Give examples of various implementations of branch-prediction mechanisms.

27. Name the dispatching methods used in microprocessors that you know.

28. State what obstacles there are to raising the performance of superscalar and LIW processors.

29. Enumerate the characteristic features of multithread architecture.

30. State the methods used in designing multithread programs and processors.

31. How does multithread architecture differ from multiprocessor systems?

32. Enumerate the main principles of using a transistor resource in building microprocessors to reach the integration level of a billion transistors per chip.

33. Explain the speculative instruction-execution mechanism and how it differs from conditional branch instructions.

34. What is the reason for speculative-instruction execution in VLIW processors?

35. What is `if` conversion?

36. Describe the main features of IA-64 architecture.

37. What is the specific feature of the Moscow SPARC technology center NArch architecture in calculating predicates of the conditional execution instructions?

38. List the main directions in which the specialization of microprocessor functional units is going.

39. Name the main architectures of single-crystal multiprocessor systems.

40. State the main construction principles of building reconfigurable computers.

41. Explain the main strong and weak points of single-crystal SIMD systems.

ENDNOTES

1. Kartsev, M. A. The Arithmetic of Digital Machines. Moscow. *Science*, 1979, p. 575.

2. DeHon, A. The Density Advantage of Configurable Computing. *Computer*, 2000. No. 4.

3. Neumann, John von. Theory of Self-Reproducing Automata. Trans. Moscow: *Mir*, 1971, p. 382.

4. Henessey, J., Patterson, D. *Computer Architecture: A Quantitative Approach.* Morgan Kaufman Publishers, Palo Alto, CA, 1990, p. 563.

5. Kartsev, M.A. Digital Computer Architecture. Moscow: *Science*, 1978, p. 295.

6. Cray Research, Inc., *CRAY-1 Computer System Hardware Reference Manual,* Bloomington, Minn., pub. no. 2240004, 1977.

7. Bell, G. Ultracomputers: A Teraflop Before Its Time // Communications of the ACM, 1992, Vol. 35, No. 8., p. 27–47.

8. Stroganov, A. Topology Design of Custom CMOS LSIC. *Chip News*, 2003, #2.

9. Moore, G. Nothing Is Limitless, but the Limits Can Be Moved! *Chip News*, 2003, #2.

10. *The Programmable Logic Data Book.* Xilinx, Xilinx, Inc. 1999.

11. Smith, J. Decoupled Access/Execute Computer Architectures, ACM Transactions on Computer Systems, 1984, Vol. 2(4), pp. 289–308.

12. Smith J., Sohi, G. The Microarchitecture of Superscalar Processors, Proc. of the IEEE, 1995, Vol. 83, No. 12, pp. 1609–24.

13. Pratt, Y., et al. One Billion Transistors, One Uniprocessor, One Chip. *Computer*, 1997, No. 9, pp. 51–57.

14. Hwu, W. Introduction to Predicated Execution, *Computer*, 1998. No. 1, pp. 49–50.

15. Halfbill, T. Inside IA-64. *Byte*, 1998, Vol. 23, No. 6, pp. 81–88.

16. Diefendorff, K. The Russians Are Coming. Microprocessor report, 1999, Vol. 13, No. 2.

17. Volin, V., Rudometov, V., Stoliarskiy, E. Code-Spooling Organization in VLIW Processors. *Information Technologies and Computer Systems*, 1 (1999), pp. 58–64.

18. Ostanevich, A. Experimental Research into Predicate Computation Support in Architectures with Explicitly Expressed Parallelism. *Information Technologies and Computer Systems*, 1 (1999), pp. 41–49.

19. Overview of the i860XP Supercomputing Microprocessor. Intel Corporation, 1991, p. 47.

20. Lenoski, D., Laudon, J., Gharachorloo, K., Gupta, A., Hennessy, J. The Directory-Based Cache Coherence Protocol for the DASH Multiprocessor. Proceedings of the 17th Annual International Symposium on Computer Architecture, 1990, pp. 148–159.

21. *Transputer Databook.* INMOS Ltd., 1985.

22. Korzhov, V. Silicon Java. *PC WEEK*, Rus. Ed., 20 December 1996, p. 48.

23. Diefendorff K., et al. How Multimedia Workloads Will Change Processor Design. *Computer*, 1997. No. 9, pp. 43–45.

24. Scherbo, V., Kozlov, V. *Open Systems Functional Standards Instruction Manual.* Moscow: MSTRC, 1997.

25. The T9000 Transputer. Inmos. SGS-Thomson Microelectronics Group. 1991, p. 194.

26. Clark D. Blue Gene and the Race Toward Petaflops Capacity. *IEEE Concurrency*, January–March 2000, pp. 5–9.

27. Tera Computer Company Completes Design of Breakthrough Multiprocessor Chip, *www.tera.com.*

28. Level One(TM) IXP1200 Network Processor. Advance Datasheet, Revision 278298-001 September 1999, *www.level1.com.*

29. Barroso, L., Gharachorloo, K., Novatzyk, A., Verghese, B. Impact of Chip-Level Integration on Performance of OLTP Workloads, Proc. of the Sixth International Symposium on High-Performance Computer Architecture (HPCA), January 2000.

30. Sohi, G., Roth, A. Speculative Multithreaded Processors, *Computer*, 2001, Vol. 34, No. 4.

31. Sohi, G., Breach, S., Vijaykumar, T. Multiscalar Processors, Proceedings of the 22nd Annual International Symposium on Computer Architecture, ISCA '95, June 22–24, 1995, Santa Margherita Ligure, Italy, pp. 414–25.

32. Krishnan, V., Torrellas, J. A Chip-Multiprocessor Architecture with Speculative Multithreading. *IEEE Transactions on Computers*, 1999, Vol. 48, No. 9.

33. Tsai, J., et al. The Superthreaded Processor Architecture. *IEEE Transactions on Computers*, 1999, Vol. 48, No. 9.

34. Kol, R., Ginosar, R. Kin: A High-Perfomance Asynchronous Processor Architecture, Proceedings of International Conference on Supercomputing. July 13–17, 1998, Melbourne, Australia, pp. 433–40.

35. *The National Technology Roadmap for Semiconductors*, Semiconductor Industry Association, 1997, *www.sematech.org.*

36. Batcher, K. STARAN Parallel Processor System Hardware, 1974 National Computer Conference, AFIPS Conference Proceedings, Vol. 43, pp. 405–10.

37. Reddaway, S. DAP—A Distributed Array Processor, Proceedings of 1st Annual Symposium on Computer Architecture, IEEE, 1973.

38. Hillis, W. *The Connection Machine.* The MIT Press, 1985.

39. CRAY, Inc. Announces CRAY SX-6 Series of High-Performance, High-Effeciency Supercomputers, *www.cray.com.*

40. Davidson, E. Large Chip vs. MCM for a High-Performance System. *IEEE Micro,* 1998, Vol. 18, No. 4, pp. 33–41.

41. Koyanagi, M. et al. Future System-on-Silicon LSI Chips. *IEEE Micro,* 1998, Vol. 18, No. 4, pp. 17–22.

42. Diefendorff, K. Power4 Focuses on Memory Bandwith, Microprocessor Report. 1999, Vol. 13, No. 13.

43. Korneev, V. *Parallel Computer Systems.* Moscow: Knowledge, 1999.

44. Yevreinov, E.V., Kosarev, Yu.G. High-Performance Homogeneous Universal Computer Systems. Novosibirsk: *Science,* 1966.

45. Smith, D., Hall, J., Miyake, K. *The CAM2000 Chip Architecture.* Rutgers University, 2000, *www.cs.rugers.edu/pub/technical-reports.*

46. Goodman, S., Wolcott, P., Burkhart, G. *Building on the Basics: An Examination of High-Performance Computing Export Control Policy in the 1990s.* Center for International Security and Arms Control. Stanford University, 1995, p. 78.

47. Kruchinin, S. Standard Performance Benchmarks. *Computer Week.* Moscow, 5(211), 8–14 February 1996.

48. Curnow, H., Wichmann, B. A Synthetic Benchmark, *Computer Journal,* 1976, Vol. 19, No. 1.

49. *www.netlib.org/benchmark/performance.ps.*

50. SPEC (*www.specbench.org*).

51. STREAM: Sustainable Memory Bandwidth in High Performance Computers, *www.cs.virginia.edu/stream.*

52. CPU-Rate, *www.phy.duke.edu/brahma/dual_athlon/src/cpu-rate/cpu-rate.html.*

53. Henning, G. SPEC CPU2000: Measuring CPU Performance in the New Millennium. *Open Systems,* 2000, #7–8.

2 Universal Microprocessors

2.1 STRUCTURE OF THE UNIVERSAL MICROPROCESSOR MARKET

Processors with the x86 instruction system are dominant in the universal processor market; their main manufacturers are Intel, AMD, and VIA®. Annual production growth of this type of microprocessor is 10 to 15 percent. The rest of the RISC architecture microprocessor manufacturers hold about 20 percent of the microprocessor market.

Table 2.1 shows microprocessor architectures of computer systems built and used at present.

TABLE 2.1 Most Common Microprocessor Architectures

Microprocessor Architecture	Manufacturer
x86	Intel, AMD, Cyrix®, IDT®, Transmeta®
IA-64	Intel
Power PC	Motorola, IBM, Apple
Power	IBM
PA	Hewlett-Packard
Alpha	Hewlett-Packard (DEC)
SPARC	Sun
MIPS	MIPS
MAJC	Sun

Historically, x86 architecture microprocessors have been dominant in personal computers whereas RISC processors have been used in workstations, high-performance servers, and supercomputers. At the present time, x86 processors have made some inroads into the traditional application areas of RISC processors. At the same time, some workstation manufacturers, Sun for example, are trying to get their processors into the personal-computer market.

Nowadays, mainstream microprocessor manufacturers possess approximately equal technological capabilities; therefore, in the fight for speed, the architecture factor has become of foremost importance. Microprocessor architecture has been developing in two main directions. The architectural performance-increasing methods considered earlier in the book are used to a greater or lesser extent within each direction; however, each direction has its own priorities.

The first direction has been code-named Speed Daemon. It is characterized by its drive to achieve high performance mainly by using high clock frequency while keeping the internal microprocessor structural organization simple.

Brainiac, the second direction, achieves high performance by employing sophisticated calculation-scheduling logic and advanced internal processor structure. Each direction has its proponents and opponents and, as it seems, right to exist.

Manufacturers of RISC processors have created and are actively developing their microprocessor architectures. They provide software compatibility between different generations of the same family of microprocessors while using lower figure-processing technology and increasing performance.

A common feature of most RISC microprocessors is their capability to process 64-bit fixed- and floating-point operands at high speeds. To build functional units of such microprocessors requires sophisticated schematic design approaches, such as using a great number of transistors in processor circuits and many metallization layers for connecting those transistors.

In search of ways to achieve maximum performance, RISC architecture microprocessor designers more and more often allow themselves to diverge from the design's canonical principles. At the same time, CISC architecture microprocessors, whose notable representative is the x86 family, employ the solutions worked out in developing RISC processors.

In this chapter, using examples of microprocessors from different manufacturers, the main architectural and technical solutions employed in building modern microprocessors will be considered.

2.2 X86 ARCHITECTURE PROCESSORS

2.2.1 Pentium Microprocessor (P5)

The Pentium [55, 56] family of microprocessors from Intel includes processors operating at clock frequencies from 60 MHz to 200 MHz. Pentium processors are wholly compatible with the earlier Intel 80386 and Intel 80846 processors and can run software developed for those processors.

The first representatives of the Pentium family, operating at 60 MHZ and 66 MHz clock frequencies, were manufactured on 0.8-micron BiCMOS technology, had a 5 V power supply, and contained 3.1 million transistors. Processors operating at 90 MHz and 100 MHz were manufactured using 0.6-micron BiCMOS technology, were powered by 3.3 V, and contained 3.3 million transistors. Starting with the 120 MHz Pentium, processors were manufactured using 0.35-micron CMOS technology, and the power supply voltage was lowered to 2.8 V.

Unlike their predecessors sharing the x86 instruction set, the Pentium processor family has quite a few technological innovations. Some of them follow:

- Close to superscalar architecture
- Split code and data caches
- Dynamic branch prediction
- High-performance pipelined floating-point unit
- Enhanced 64-bit data bus
- Data integrity support
- SL technology power-management features
- Multiprocessor support
- Performance monitoring
- Variable memory page support

A Pentium processor block diagram is shown in Figure 2.1.

SUPERSCALAR ARCHITECTURE

The superscalar construction of Pentium processors is a natural development of the previous generations of Intel processors: Intel 80386 and the 32-bit architecture Intel 80486. (The latter already could execute several instructions per clock.)

The processor's two instruction pipelines are capable of executing instructions simultaneously. As in the case of one pipeline in the Intel486 processor, the dual pipeline of the Pentium processor executes integer instructions in the following five stages:

- Prefetching
- Decoding 1
- Decoding 2
- Execution
- Storing results

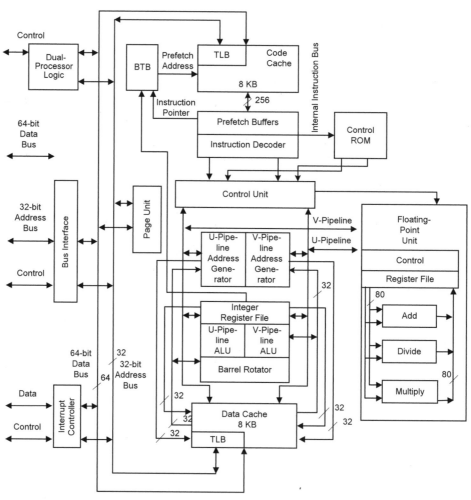

FIGURE 2.1 Pentium processor block diagram.

However, the two pipelines are not completely independent. When one pipeline stops, so does the other one, because the floating-point arithmetic unit uses the integer arithmetic unit. Consequently, these operations cannot be executed concurrently. This limits the processor's superscalar capabilities.

Execution of many of the frequently used simple instructions that used to be executed using microcode is implemented by hardware in the Pentium processors. The microcode for executing complex instructions has been improved.

SPLIT CODE AND DATA CACHES

Each of the Pentium processor caches is 8 KB. They both are of the set associative type. The necessary information is searched for in the standard 32-byte lines.

The translation look-aside buffer (TLB) converts the addresses of the main memory cells into the corresponding cache-line data-location addresses.

Data cache of the Pentium processor is of the write-back type and uses MESI cache coherence protocol. The write-back method makes it possible to modify data in cache without accessing the main memory. (Data is written to the main memory only when evicted from the cache.) Previous processor generations used write-through caches. This type of cache writes data to the main memory every time it is modified. The write-back type is more efficient and reduces the load on the memory interface bus.

Support of the MESI protocol in the Pentium processors makes it possible to ensure cache coherence: correspondence of cache data to the main memory data when working in a multiprocessor system.

BRANCH PREDICTION

The Pentium processor is the first x86-compatible processor to use branch predictions technology; until then, it was used only in mainframes and RISC processors.

The Pentium processor performs speculative branch processing using a Branch Target Buffer (BTB) and two prefetch buffers. One prefetch buffer is used to prefetch instructions with the assumption that there is no branch; the other prefetch buffer prefetches instructions using the contents of the BTB (remembered when the branch was executed for the first time).

The speculative branch-processing algorithm of the Pentium processor not only predicts simple branches but also supports more complex predictions (for example, in nested loops). It does this by storing several branch addresses in the BTB. Up to 256 executed branches are stored in the BTB, which makes for a 0.8 prediction success rate.

HIGH-PERFORMANCE FLOATING-POINT OPERATION UNIT

Intel 486DX was the first Intel microprocessor to incorporate in one crystal both the floating-point and the integer unit. In the previous generations of Intel processors, a separate floating-point processor chip was used for floating-point operations.

The floating-point unit of the Pentium processor has sophisticated eight-stage pipelines and employs internal functions. Execution of most floating-point instructions is started in one of the five integer pipelines and is then transferred to the floating-point pipelines. Moreover, simple floating-point operations (such as addition, multiplication, and division) are implemented as internal hardware functions for faster execution.

As a result of these innovations, the Pentium (815/100) processor executes floating-point operations 10 times faster than the Intel 486DX 33 MHz processor.

ENHANCED 64-BIT DATA BUS

Using its 64-bit data bus, the Pentium processor can exchange data with the memory at the rate of 528 MBps. This is more than five times the maximum transfer rate of the Intel 486DX2 66 MHz processor (105 MBps). As a result, the overall performance of the former is 2.5 times that of the latter.

In addition to expanding the data bus, bus cycle pipelining was added to increase its bandwidth. It allows the next bus cycle to start before the completion of the previous. This in turn gives memory subsystem additional time to decode addresses; consequently, slower and less-expensive memory microchips can be used.

The Pentium processor supports burst read and burst write-back cycles, which also contributes to increasing the system bandwidth. Data and address parity checking makes for enhanced reliability of the system.

The Pentium processor uses two write buffers (one for each pipeline) to raise the speed of the sequential memory write operations. This allows the processor to execute the next instructions in case the result of one of the current instructions has not been written back because the bus was busy.

DATA INTEGRITY FEATURES

With an aim toward increasing the reliability of systems built using the Pentium microprocessor, two mainframe features were implemented in the processor: internal parity checking and Functional Redundancy Checking (FRC).

Parity bits of the processor's internal buffers are used for internal parity checking. The FRC may be used for applications with enhanced requirements to the validity of the results. The FRC requires using two Pentium processors: a master and a checker. The paired-up processors execute the same instructions in lockstep. The checker processor compares its internally computed values with those output by the master processor. If these values mismatch, the checker generates an interrupt.

POWER MANAGEMENT FEATURES

Starting from Pentium 90 MHz, the Pentium processor implements new power management technology. The power-saving features are present at two levels: the microprocessor level and the system level. When executing low-intensity computation tasks (such as word processing, for example), the power-management features can lower the processor's clock frequency and reduce power consumption. The processor can even be halted altogether by being placed into the SL sleep mode. In the energy-saving mode, the microprocessor can slow down, halt, or even stop individual system components altogether, thereby minimizing power consumption.

MULTIPROCESSOR SUPPORT

The Pentium processor supports multiprocessor processing, which makes it available for use in SMP architecture systems.

As was stated earlier, the Pentium processor uses MESI protocol to assure cache coherency in multiprocessor systems.

The Pentium processors also incorporate two new multiprocessor support features: an on-chip Advanced Programmable Interrupt Controller (APIC) and a two-way L2 cache controller. The APIC supports up to 60 processors, and the L2 cache controller allows two processors to share a single L2 cache.

PERFORMANCE-MONITORING FEATURES

Pentium processor performance-monitoring features allow system designers and software developers to detect and eliminate potential bottlenecks in the program code. The developers can monitor and count cycles of internal processor events affecting the duration of `read` and `write` operations, cache hits and misses, interrupts, and so on. This makes it possible to evaluate the execution efficiency of the program code and to fine-tune the applications and systems being developed for maximum performance.

VARIABLE MEMORY PAGE SIZE SUPPORT

The Pentium processor supports both the traditional 4 KB memory page size and increased-size 4 MB pages. The memory page control mechanism is transparent to the software and was introduced to lower page switching frequency in complex graphical applications and in the operating system kernel. The larger page size allows bigger objects to be displayed, which could not be done before. Moreover, the increased page size increases the page-hit ratio, which raises the system performance.

PERFORMANCE CHARACTERISTICS

Tables 2.2 and 2.3 compare SPECint92 and SPECfp92 UNIX benchmark performance indexes of some Pentium microprocessors with those of other microprocessors.

The SPECint92 UNIX is an intensive processor-use benchmark, allowing its performance to be measured by running a representative collection of application software. The SPECfp92 UNIX benchmark measures processor's floating-point operation performance.

Table 2.2 SPECint92 UNIX Benchmark

Processor	Index	Processor	Index
Intel Pentium (815/100)	100.0	HP PA 735/99	80.6
Intel Pentium (735/90)	90.1	SuperSPARC 10/51	65.2
MIPS R4400SC-150	85.9	IBM Power PC601-66	62.6
DEC Alpha-150	84.4		

Table 2.3 SPECfp92UNIX Benchmark

Processor	Index	Processor	Index
Intel Pentium (815/100)	80.6	HP PA 735/99	149.8
Intel Pentium (735/90)	72.7	SuperSPARC 10/51	83.0
MIPS R4400SC-150	93.6	IBM Power PC601-66	76.1
DEC Alpha-150	127.7		

2.2.2 Pentium MMX Microprocessor

In keeping with the Natural Signal Processing (NSP) concept introduced by Intel in 1995, the microprocessor is entrusted with several tasks that used to be performed by separate specialized devices. This approach makes it possible to simplify computer construction and makes it less expensive to produce. Among the tasks that can be performed using a universal processor are sound and music synthesis, voice recognition, video and graphical information processing, implementing communication functions, and many others. However, the number of various applications that must be given over to the processor in accordance with the NSP principles could not be handled by the far-from-perfect x86 family architecture and required appropriate changes to be made to it. The result of embodying the NSP conception into a tangible product was the Pentium MMX processor (code-named P55C), released in January 1997. The instruction set of this processor includes 57 additional instructions that are oriented toward efficient execution of typical multimedia algorithms, which include many algorithms typical of digital-signal processing (vector operations, compression, Furrier transform, and so on). This was the first substantial change in the x86 instruction set, starting from the appearance of the Intel 80386 processor in 1985, which had 220 instructions.

The main differences between the Pentium MMX and its predecessor follow [57]:

■ Multimedia instruction set support
■ Doubled sizes of the code and data caches (16 KB each)
■ Enhanced branch-prediction logic
■ Enhanced pipelining
■ Deeper write buffering

At the same time, the Pentium MMX lost the following capabilities that Pentium has:

■ Functional redundancy checking
■ The cache controller chipset (Intel 82498/82493 and 82497/82492)
■ Split line access to the code cache

The Pentium MMX microprocessor was produced on CMOS 0.35- and 0.25-micron process technology. It has 4.5 million transistors, uses two power supply voltages (2.8 V for the core and 3.3 V for the I/O), and consumes 15.7 W.

A block diagram of the Pentium MMX microprocessor is shown in Figure 2.2.

The processor has two code pipelines (U pipeline and V pipeline). The U pipeline can execute all integer and floating-point instructions. The V pipeline can execute simple integer instructions and the FXCH floating-point instruction.

Figure 2.2 shows the processor's separate code and data caches. Each cache is 16 KB and has two ports, one for each execution pipeline. The data cache is equipped with a TLB. Caching of individual memory pages can be enabled or disabled by software or hardware.

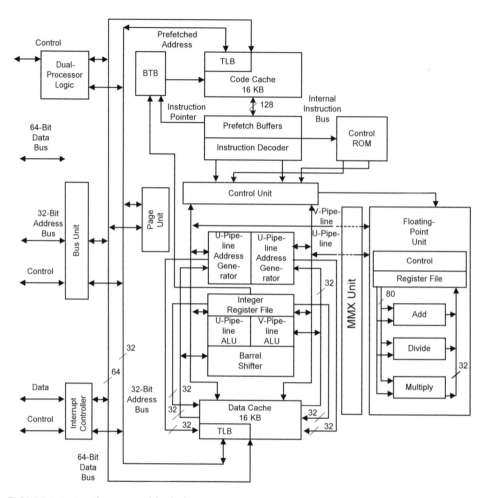

FIGURE 2.2 Pentium MMX block diagram.

The code cache, branch-address buffer, and prefetch buffer make for intensive instruction feeding to the processor execution units. The internal programmable interrupt controller is functionally compatible with the 8259A interrupt controller and services both internal and external interrupts.

The branch prediction system was enhanced in Pentium MMX. In the Pentium processor, two 32-byte prefetch buffers are used for branch predictions: one for fetching instructions in a linear fashion, the other for fetching them according to the contents of the BTB. Subsequently, the needed instruction almost always is prefetched before it is needed for execution. In the Pentium MMX, the number of instruction prefetch buffers is increased to four (each holding

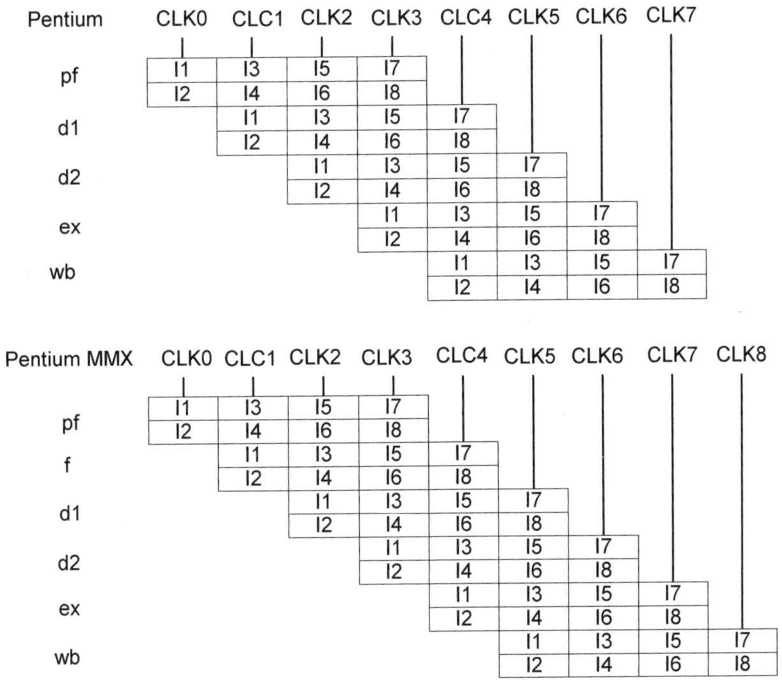

Pentium

	CLK0	CLC1	CLK2	CLK3	CLC4	CLK5	CLK6	CLK7
pf	I1 / I2	I3 / I4	I5 / I6	I7 / I8				
d1		I1 / I2	I3 / I4	I5 / I6	I7 / I8			
d2			I1 / I2	I3 / I4	I5 / I6	I7 / I8		
ex				I1 / I2	I3 / I4	I5 / I6	I7 / I8	
wb					I1 / I2	I3 / I4	I5 / I6	I7 / I8

Pentium MMX

	CLK0	CLC1	CLK2	CLK3	CLC4	CLK5	CLK6	CLK7	CLK8
pf	I1 / I2	I3 / I4	I5 / I6	I7 / I8					
f		I1 / I2	I3 / I4	I5 / I6	I7 / I8				
d1			I1 / I2	I3 / I4	I5 / I6	I7 / I8			
d2				I1 / I2	I3 / I4	I5 / I6	I7 / I8		
ex					I1 / I2	I3 / I4	I5 / I6	I7 / I8	
wb						I1 / I2	I3 / I4	I5 / I6	I7 / I8

FIGURE 2.3 Pentium and Pentium MMX pipeline execution instruction flow.

16 bytes); this makes it possible to prefetch instructions in four different directions and also raises the branch prediction accuracy.

The depth of the execution pipeline was increased to six stages by adding a fetch (F) stage between the prefetch (PF) and decode 1 (D1) stages. The instruction length decoding, which in the previous Pentium models is done at the D1 stage, is done at the F stage in the Pentium MMX processor. The instruction execution process in Pentium and Pentium MMX is shown in Figure 2.3. Increasing the pipeline length allows the average instruction execution time to be shortened, thereby raising the processor performance.

The conventional notations in Figure 2.3 mean the following: I1, I2, ...— instruction sequence; CLK0, CLK1, ...—processor cycle sequence; *pf, f, d1, d2, ex, wb*—instruction execution stages, *prefetch, fetch, instruction decode, address generate, execute,* and *write-back*, respectively.

Along with the enumerated improvements, the main new feature of Pentium MMX is its support of the multimedia instruction set, which is executed in the MMX unit and the floating-point register. The MMX technology introduced the following changes into the Intel microprocessor architecture:

Eight MMX register added (MM0–MM7)—The MMX registers are aliased to the 64-bit floating-point registers and can be used only with MMX data.

Fifty-seven new instructions are added—The instructions are subdivided into the following groups: data exchange, arithmetic, comparison, conversion, logic, shift, and the empty (disable) MMX state instruction (EMMS).

Four new data types are added—Packed bytes (eight bytes in a 64-bit quantity), packed words (four 16-bit words in a 64-bit quantity), packed doublewords (two 32-bit doublewords in a 64-bit quantity), and a quadword (a 64-bit quantity). Arithmetic and logic operations are executed concurrently on each byte of the word or doubleword in the 64-bit MMX register (a SIMD processing type).

The SIMD processing significantly speeds up execution of multimedia algorithms, which are characterized by execution of identical operations on large volumes of same-type data (for example, 16-bit digital sound samplings or 8-bit pixel color codes, among others). Figure 2.4 shows a multiplication with accumulation

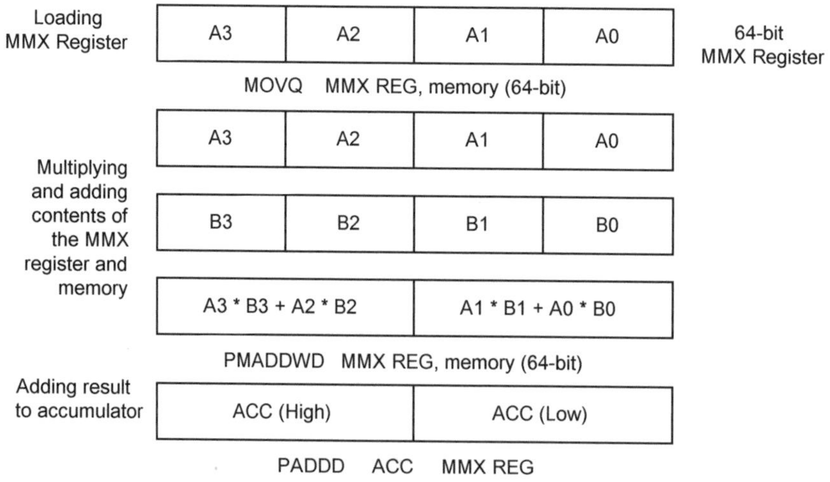

FIGURE 2.4 Performing multiplication accumulation using MMX instructions.

operation of four 16-bit words by another four words using three MMX commands. Execution of the same calculation on the Pentium processor requires the instruction sequence FILD, FMUL, FADD to be executed four times, for the total of 12 instructions.

Using MMX instructions allows the execution speed of multimedia applications to be increased by 60 percent, as compared to the regular Pentium processor, the clock frequencies being identical (data from Intel Media Benchmark).

Because the MMX registers are physically floating-point registers, they are compatible with the previous architecture and, consequently, can be used with previously developed software. However, it takes the processor 50 cycles to switch from the MMX mode into the floating-point mode, thus lowering its efficiency when executing a mixture of MMX and floating-point instructions. An MMX instruction in the U or V pipeline can be executed in parallel with another MMX or integer (but not floating-point) instruction. To the regular execution stages, six more stages are added for MMX commands in the MMX unit.

Expanded pipelining also substantially raises processor performance for non-multimedia applications. Table 2.4 lists comparative data for the Pentium and Pentium MMX processors on SPECint95 and SPECfp95 benchmarks.

TABLE 2.4 Pentium and Pentium MMX SPECint95 and SPECfp95 Performance Indexes

Processor	SPECint95	SPECfp95
Pentium 200	5.47	3.68
Pentium MMX 200	6.41	4.66
Pentium MMX 166	5.59	4.30

2.2.3 Pentium Pro Microprocessor (P6)

Pentium Pro [58] is a sixth-generation processor (P6) from Intel. It is mainly geared for the upper-end workstations and multiprocessor systems. P6, as the previous processors of the family, supports the x86 instruction set. However, the processor's specific architectural features make it efficient when executing 32-bit applications, whereas the execution speed for 16-bit applications can turn out to be substantially lower than in a P5 operating at the same clock frequency.

The processor's package differs from the previous processors. The processor is actually two separate integrated circuits packaged in one ceramic casing. It is different from the Pentium pinout and cannot be used with the old motherboards.

The processor has 5.5 million transistors and dissipates 14 W. The first P6 microprocessors were built on 0.6-micron BiCMOS technology with four metallization layers and operated at 150 MHz clock frequency. With the transition to the 0.35-micron technology, processors operating at clock frequencies up to 200 MHz were produced.

Pentium Pro achieves high performance by using a series of architectural and technological innovations. These are decoupled architecture, dynamic execution, and dual independent bus (DIB).

The dynamic execution technology is based on such performance-increasing methods as speculative instruction execution, instruction reordering, and branch prediction.

The new cache architecture provides for using separate buses for connecting the processor core with the cache and the main system memory. The cache bus operates at the processor's clock frequency, whereas the system bus operates at the system clock speed. This bus separation makes it possible to increase the processor/memory exchange rate by three times.

Owing to the separate cache and main memory buses, there is no longer a need for the external cache. The microprocessor has split 8 KB L1 data and code caches and a unified L2 cache. The L1 data cache has two ports, is non-blocking, and can do a `read` and a `write` operation per cycle. The L2 cache interface operates at the CPU frequency and can transfer 64 bits per cycle. The external bus runs at one-half, one-third, or one-quarter of the CPU clock frequency. The L2 cache is 256 KB with 0.6-micron technology and either 512 KB or 1,024 KB with 0.35-micron technology.

The fact that 16-bit applications do not work faster on Pentium Pro than on a Pentium operating at the same frequency is explained by the performance losses due to the frequent reloading of the long 14-stage pipeline when working with short 16-bit data.

The structure of the Pentium Pro microprocessor is shown in Figure 2.5. It shows the processor's main functional units and their functions.

The 512-item BTB allows the Instruction Fetch Unit (IFU) to fetch cache lines in fewer cycles. The fetch process is pipelined. A new line is fetched every CPU cycle. Three parallel Instruction Decoders (IDs) convert several x86 architecture instruction set instructions into sequences of micro-operations.

The register alias table (RAT) is used to rename registers. The renaming results are stored in the reservation station (RS) and the reorder buffer (ROB).

In the reservation station, the micro-operations with the renamed operands await operands coming independently from several sources. These can be micro-operation execution results, an address from the BTB, or register contents.

Micro-operations are selected from the micro-operation pool and are dynamically executed based on the real data dependencies, as well as the availability of execution units. In general, the order in which they are executed is different from their order in the source program.

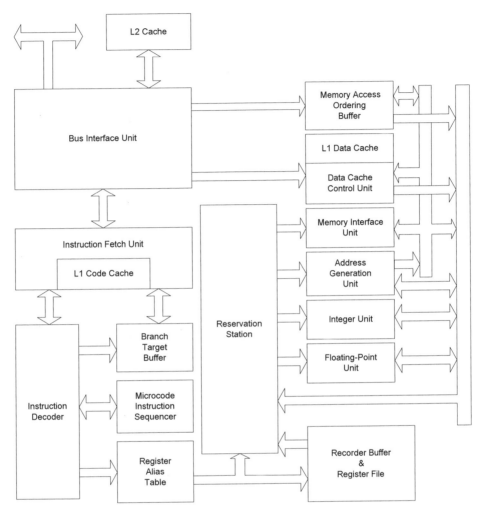

FIGURE 2.5 Pentium Pro block diagram.

Memory references are scheduled using the reservation station, Address Generation Unit (AGU), and Memory Ordering Buffer (MOB).

A micro-operation becomes a potential candidate for retirement as soon as it is executed, the branch address is determined, and the obtained results are sent to the micro-operation that requires them. Status information tags are added to micro-operations when they are entered into the ROB and Retirement Register File (RRF). The retire unit uses these tags to retire micro-operations in the original program order. The retirement process must not only ensure reestablishment of the original program but also do this despite all interrupts, faults, and branch mispredictions. When a micro-operation is retired, its results are transferred from the ROB into the RRF.

Tables 2.5 and 2.6 give main characteristics and performance indexes of the Pentium Pro processors.

TABLE 2.5 Pentium Pro Processor Main Characteristics

CPU Clock Frequency (MHz)	Bus Frequency (MHz)	Technological Process (Microns)	Release Date
150	60	0.6	Nov. 1995
166	66	0.35	1st quarter of 1996
200 (512 KB cache)	66	0.35	4th quarter of 1996
200 (1 MB cache)	66	0.35	Aug. 1997

TABLE 2.6 Microprocessor Performance Comparisons

Microprocessor	SPEC95 int/fp
Pentium Pro (200 MHz), 1 MB cache	8.66/6.8
Pentium (200 MHz)	5.47/3.68
DEC Alpha (300 MHz)	7.3/11.6
Sun UltraSPARC	5.1/7.4

2.2.4 Pentium II Microprocessor

This microprocessor, originally code-named Klamath, combines all the best features of the previous models [59, 60]. It achieves a high performance level by using the three advanced technologies: dynamic code execution, MMX, and dual separate buses.

The processor's architecture in many respects is like the Pentium Pro architecture; however, the L2 cache in the Pentium II is implemented using a less-expensive approach. Pentium Pro was so expensive because of, first of all, the great rejection rate during manufacturing and microassembling in one package both the L2 cache and the processor proper. The quality of the microprocessor could be checked only after it was fully assembled.

A less-expensive L2 cache built from Burst Static RAM (BSRAM) is located on the same board as the Pentium II processor crystal. The L1 split cache size is increased to 32 KB: 16 KB for code and 16 KB for data.

The dual independent buses architecture, first employed in Pentium Pro, increases the memory exchange capabilities of the processor core by three times. As in Pentium Pro, one bus connects the processor to the L2 cache and the other to the main memory. In the earlier Pentium II models, however, the L2 bus clock rate was only half of the CPU clock frequency.

Also as with the Pentium MMX, the Pentium II has a MMX unit.

The processor board with the L2 cache crystal on it is placed into a Single Edge Contact (SEC) cartridge installed into 242-contact Slot 1 motherboard connector (code-named SC242).

Performance of the first Pentium II models, operating at 233 MHz, 266 MHz, and 300 MHz clock frequencies, was higher than Pentium MMX; however, it was lower than Pentium Pro because of the slower L2 cache. At the same time, its price when it first came out was almost half that of the Pentium Pro.

The Pentium II line developed in the following three directions:

- The Pentium II microprocessors for high-performance personal computers and workstations with increased CPU core and system bus frequencies.
- Processors oriented toward use in inexpensive personal computers with low performance requirements. The price of these microprocessors could be lowered by dispensing with the L2 cache (Celeron® processors), while at the same time increasing the CPU core frequency. The Celeron processors were manufactured first in simplified Single Edge Processor Package (SEPP) and then in Plastic Pin Grid Array (PPGA) casing. It must be noted that totally abandoning the L2 cache substantially lowered the processor performance; therefore, later Celerons, manufactured on the 0.25-micron process, had 128 KB L2 on-chip cache.
- Processors with increased clock frequency, increased cache size operating at the CPU core frequency, and multiprocessor support. (Pentium II Xeon processors support eight-processor configuration.) These microprocessors are intended for use in high-end workstations and servers. The Pentium II Xeon processors are also produced in SEC cartridges but of a much larger size. These processors are installed into 330-contact Slot 2 (SC330).

There also are mobile versions of the first and second processor categories. Intel has continued this processor category breakdown in its subsequent microprocessor families.

Table 2.7 lists the chronology of the Pentium II processors and their main characteristics.

TABLE 2.7 General Characteristics of Pentium II Family Processors

Name (Clock Frequency)	Introduction Date	Performance on SPECint95/SPECfp95	Application Area	L2 Cache	Number of Transistors in Million (Technological Process-Microns)	System Bus Clock, MHz
Pentium II (300, 266, 233 MHz)	7.5.97	12.7/8.82	High-performance professional PCs, work-stations, and servers	512 KB	7.5 (0.35)	66
Pentium II (333 MHz)	1.26.98	12.8/9.14	Professional and home PCs, servers, and work-stations	512 KB	7.5(0.25)	66
Mobile Pentium II (233, 266 MHz)	4.2.98	12.3/–	Mobile PCs	512 KB	7.5 (0.25)	66
Pentium II (350, 400 MHz)	4.15.98	16.9/12.0	Professional and home PCs, servers, and work-stations	512 KB	7.5(0.25)	100
Intel Celeron (266 MHz)	4.15.98	8.4/6.6	Low-budget PCs	No	7.5 (0.25)	66
Intel Celeron (300 MHz)	6.8.98	8.8/7.0	Low-budget PCs	No	7.5 (0.25)	66
Pentium II Xeon (400 MHz)	6.29.98	16.3/132	Workstations and servers	512 KB or 1 MB. Works at the CPU clock rate	7.5	100

TABLE 2.7 General Characteristics of Pentium II Family Processors *(Continued)*

Name (Clock Frequency)	Introduction Date	Performance on SPECint95/SPECf p95	Application Area	L2 Cache	Number of Transistors in Million (Technological Process-Microns)	System Bus Clock, MHz
Pentium II (450 MHz)	8.24.98	18.5/13.1	Professional and home PCs, servers, and workstations	512 KB	7.5 (0.25)	100
Intel Celeron (333 MHz)	8.24.98	12.9/10.4	Low-budget PCs	128 KB on-die	19 (0.25)	66
Intel Celeron (300A MHz)	8.24.98	11.9/9.2	Low-budget PCs	128 KB on-die	19 (0.25)	66
Mobile Pentium II (300 MHz)	9.9.98	12.6/—	Mobile PCs	512 KB	7.5 (0.25)	66
Pentium II Xeon (450 MHz)	10.6.98	19.7/15.0	Servers and workstations	512 KB. Works at the CPU clock rate	7.5	100
Intel Celeron (400, 366 MHz)	1.4.99	14.9/11.8	Low-budget PCs	128 KB on-die	19 (0.25)	66
Pentium II Xeon (450 MHz)	1.5.99	21.2/15.3	Servers and workstations	1MB or 2 MB. Works at the CPU clock rate	7.5	100
Mobile Pentium II (266, 300, 333, 366 MHz)	1.25.99	13.2/—	Mobile PCs	256 KB on-die	27.4 (0.25)	66

TABLE 2.7 General Characteristics of Pentium II Family Processors *(Continued)*

Name, Clock Frequency	Introduction Date	Performance on SPECint95/SPECfp95	Application Area	L2 Cache	Number of Transistors in Million (Technological Process-Microns)	System Bus Clock (MHz)
Mobile Intel Celeron (266, 300 MHz)	1.25.99	12.3/–	Low-budget mobile PCs	256 KB on-die	18.9 (0.25)	66
Intel Celeron (433 MHz)	3.22.99	15.8/12.1	PCs	256 KB on-die	19(0.25)	66
Mobile Intel Celeron (333 MHz)	4.5.99	12.6/–	Low-budget mobile PCs	256 KB on-die	18.9 (0.25)	66
Intel Celeron (466 MHz)	4.26.99	17.1/12.7	PCs	256 KB on-die	19 (0.25)	66
Mobile Intel Celeron (366 MHz)	5.17.99	12.8/–	Low-budget mobile PC	256 KB on-die	18.9 (0.25)	66
Mobile Intel Celeron (400 MHz)	6.14.99	13.4/–	Low-budget mobile PCs	256 KB on-die	18.9 (0.25)	66
Mobile Pentium II (400 MHz)	6.14.99	17.1/–	Mobile PCs	256 KB on-die	27.4 (0.18)	66

2.2.5 Pentium III Microprocessor

PENTIUM III MAIN TECHNICAL CHARACTERISTICS

The Intel III (Katmai) processor [61], introduced at the beginning of 1999, inherited the best features of the P6 architecture: dynamic command execution, multiple transaction system bus, and the Intel MMX technology for processing multimedia data. Moreover, the Pentium III implements new streaming SIMD extension: 70 new instructions providing enhanced processing of video, 3D graphics, streaming video and audio data, as well as speech recognition.

The main technical characteristics of the processor follow:

- Clock frequency from 450 MHz to 1 GHz
- 0.25-micron technological process
- An instruction set, including, in addition to the MMX extension instructions, 70 new SIMD instructions that enhance execution of 3D graphics, streaming audio and video, and speech recognition applications
- Improved performance (The Pentium III processor operating at 500 MHz exceeds by more than 90 percent that of the Pentium II processor operating at 450 MHz in 3D graphics processing (as measured by the standard 3D Win-Bench 99–3D Lighting and Transformations benchmark). It is also 42 percent higher when working with multimedia applications (as measured by the standard MultimediaMark 99 benchmark).)
- Dual independent bus (DIB) architecture
- Processor serial-number function implemented as a part of the PC security system proposed by Intel
- Non-blocked on-die 32 KB split L1 cache (16 KB code and 16 KB data) and on-die non-blocked 512 MB L2 cache
- Memory cacheable up to 4 GB of addressable memory
- Scalable dual-processor systems with up to 64 GB of physical memory can be built on the Pentium III basis

EXTENDED INSTRUCTION SET

Performance increases provided by the MMX extension instructions in the Pentium MMX processor did not live up to the expectations of the users working with graphic applications. The performance increase, which amounted to about 10 percent, was mostly due to the enlarged size of the internal cache and not to the vector instructions. SIMD instructions of the MMX extension are not usually used in office applications. Also, it is more important to be able to efficiently process floating-point data than vector commands when performing geometric conversions typical of graphic applications. This cannot be implemented within the framework of the added SIMD integer instructions.

Intel tried to partially solve the problem of raising processor performance in graphics tasks by introducing the graphics processor I740, which was oriented at working in the Pentium II–based systems equipped with AGP. This processor made graphics applications execute faster, due to its floating-point instruction parallel-execution technology.

In its efforts to eliminate the MMX technology shortcomings (lack of vector floating-point instructions and blocking of the floating-point unit when executing MMX instructions), Intel decided to make the necessary additions to the architecture of the sixth-generation processors.

New SIMD instructions and a Streaming SIMD Extensions (SSE) were introduced in Pentium III. The SSE allows SIMD processing of floating-point multimedia data. These innovations made it possible to increase application performance in the following areas:

- 3D graphics
- Signal processing and process simulation with a wide range of parameters change
- Video signal-block coding and decoding algorithms
- Various stream data digital filter algorithms

The new Pentium III instructions are divided into the following four categories:

- Single precision floating-point (SPFP) SIMD instructions
- Additional integer SIMD instructions
- Cache control instructions
- Processor state management instructions

SPFP INSTRUCTIONS

One SIMD floating-point instruction can simultaneously process four 32-bit single-precision floating-point numbers (SPFP in further discussions). SPFP instructions use a new data type: a 128-bit entity containing four packed 32-bit floating-point numbers.

SIMD instructions to operate on SPFP data use eight new 128-bit XMM registers named sequentially from XMM0 through XMM7. Integer registers are used to address memory. Unlike the MMX registers, which are physically 80-bit floating-point registers standard for Intel architecture, the 128-bit XMM registers are new processor components.

Like Intel processors of the previous generations, the Pentium III processor represents floating-point data in the 80-bit extended-precision format; however, packed floating-point data is represented in the 32-bit single-precision format. Therefore, floating-point execution results of the x87 instructions and those of the new SPFP instructions can differ because of the rounding.

The format in which SPFP data is represented is shown in Figure 2.6. According to the IEEE-754 standard, each 32-bit floating-point number has one sign bit, eight exponent bits, and 23 significand bits.

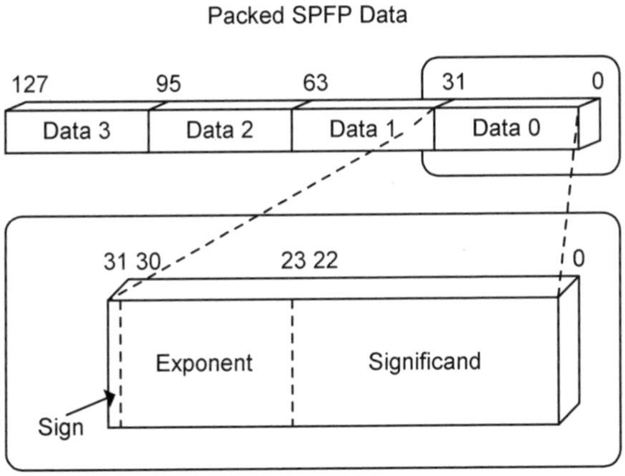

FIGURE 2.6 SPFP data representation format.

Most SPFP instructions have two operands. After an instruction is executed, the first operand data, as a rule, is replaced with the execution results while the second operand data remain unchanged.

The SPFP instructions support two types of operation on packed floating-point data: parallel and scalar.

The parallel operations are performed simultaneously on all four 32-bit data elements of each 128-bit operand. Parallel operation instructions have the suffix ps in their names. For example, the addps instruction sums four pairs of data elements and writes the four obtained sums into the corresponding elements of the first operand.

The scalar operations are performed on the lower data elements of the two operands (bits 0–31). The remaining three data elements do not change. (One exception is the scalar copy instruction movss.)

Scalar operation instructions end in ss (for example, addss).

Table 2.8 gives some examples of floating-point instructions.

The new instruction set of the Pentium III processor includes additional SIMD instructions to operate on integer data. These new commands expand the capabilities of the old MMX instruction set.

The new integer SIMD instructions perform SIMD operations on several integer values packed into 64-bit groups (Figure 2.7), and load and save packed data in the 64-bit MMX registers.

TABLE 2.8 Examples of Floating-Point SIMD Instructions

Operation	Parallel Instruction Mnemonics	Scalar Instruction Mnemonics	Description
Arithmetic Commands			
Addition	addps xmm1, xmm2	addss xmm1, xmm2	The first operand of the instruction is in an XMM register; the second operand is either in an XMM register or in the memory. The result is stored in the first register.
Subtraction	subps xmm1, xmm2	subss xmm1, xmm2	
Multiplication	mulps xmm1, xmm2	mulss xmm1, xmm2	
Division	divps xmm1, xmm2	divss xmm1, xmm2	
Square Root Compute Instructions			
Square Root Compute	sqrtps xmm1, xmm2	sqrtss xmm1, xmm2	The argument is in the second operand; the result is stored in the first operand.
Approximate Reciprocal Compute			
Reciprocal of Value	rcpps xmm1, xmm2	rcpss xmm1, xmm2	Reciprocals of the operand values are calculated approximately by using internal tables. The relative accuracy of the result satisfies the inequality $|a - b| / |b| < 1.5 \times 2^{-12}$, where a is the exact answer and b is the approximate answer. The argument is either in an XMM register or in the memory; the result is stored in the XMM register. Arithmetic exceptions are not generated.
Reciprocal of Square Root of Value	rsqrtps xmm1, xmm2	rsqrtss xmm1, xmm2	

TABLE 2.8 Examples of Floating-Point SIMD Instructions *(Continued)*

Operation	Parallel Instruction Mnemonics	Scalar Instruction Mnemonics	Description
	Determining Maximum and Minimum Value Instructions		
Find Maximum Value	`maxps xmm1, xmm2`	`maxss xmm1, xmm2`	After comparing pairs of operand values, the maximum (minimum) value is stored in the first operand. The second operand can be located in either an XMM register or in the memory. The first operand is located in an XMM register.
Find Minimum Value	`minps xmm1, xmm2`	`minss xmm1, xmm2`	
	Value Shuffle Instructions		
Shuffle Operand Values	`shufps xmm1, xmm2, 9ch`	`not supported`	Two selected values of the destination operand are copied into the lower values of the destination operand, whereas two selected values of the source operand are copied into the upper values of the destination operand. Values are selected using a mask that is defined by the 8-bit immediate operand: the two lower two-bit groups select the values to be moved into the two lower values of the destination operand and the two higher two-bit groups selects the two values to be moved from the source operand into the two higher values of the destination operand.

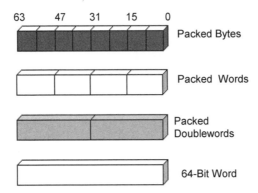

FIGURE 2.7 Data formats in integer SIMD instructions.

Mnemonics of the new integer SIMD instructions have prefixes and suffixes indicating the nature of the performed operation and the data type used. These follow:

Prefix p—Indicates that the instruction performs parallel operations on several data values.

Suffixes b, w, d, and q—Indicate the data type used (packed bytes, words, doublewords, and quadwords, respectively)

Suffixes u and s—Indicate signed or unsigned data, respectively.

For example, multiplication instruction `pmulhuw` multiplies packed, unsigned 16-bit words.

Table 2.9 lists examples of the SIMD integer instructions added to Pentium III.

TABLE 2.9 SIMD Integer Instructions

Mnemonics	Operation	Example	Description
		Word Extraction	
Pextrw	Extracting a word from an MMX register and moving it to an integer register	`pextrw eax, mm1, 1h`	The instruction extracts one of the 16-bit words packed in an MMX register and moves it into the lower word of a 32-bit integer register. The 16 high-order bits or the integer register are zeroed out.
			Which of the four words in the MMX register is to be extracted is determined by the value of the two lower bits of the 8-bit immediate operand.

TABLE 2.9 SIMD Integer Instructions *(Continued)*

Mnemonics	Operation	Example	Description
		Word Insertion	
pinsrw	Inserting a word into an MMX register	pinsrw mm1, eax, 2h	The instruction copies the lower 16-bit integer value from a 32-bit integer register (or from a 16-bit memory location) into one of the words of an MMX register.
			Into which of the four words of the MMX register the insertion is to be made is determined by the value of the two lower bits of the 8-bit immediate operand.
		Maximum (Minimum) Search	
pmaxsw	Finding the largest of four words	pmaxsw mm1, mm2	These instructions compare packed integer values of the source and destination operands and place the greatest or smallest value of each pair into the destination register. The instructions can be used only to compare unsigned bytes or signed words.
			The second operand for these instructions can be located in either an MMX register or in the memory. The first operand must be located in an MMX register.
pminsw	Finding the minimum of four packed words	pminsw mm1, mm2	
pmaxub	Finding the maximum of eight packed bytes	pmaxub mm1, mm2	
pminub	Finding the minimum of eight packed bytes	pminub mm1, mm2	
		Byte Sign Mask Creation	
pmovmskb	Creating a byte mask made up of the sign bits of each of an MMX register bytes	pmovmskb eax, mm1	The instruction copies the high (sign) bits of all eight bytes of the second operand located in an MMX register and creates an 8-bit mask in the lower bits of a 32-bit integer register. The remaining higher 24 bits of the integer register are zeroed out.

TABLE 2.9 SIMD Integer Instructions *(Continued)*

Mnemonics	Operation	Example	Description
		Unsigned Word Multiplication	
pmulhuw	Multiplying four unsigned words	pmulhuw mm1, mm2	The instruction multiplies pairs of four 16-bit unsigned words of the source and destination operands. The higher 16 bits of the 32-bit products are stored into the 16-bit words of the destination operand; the lower 16 bits of the products are discarded.
			The source operand of the pmulhuw instruction can be located in either an MMX register or in the memory. The destination operand must be located in an MMX register.
		Integer Value Shuffle	
pshufw	Shuffling 16-bit integers	pshufw mm1, mm2, 9ch	This integer instruction selects four 16-bit words (not necessarily all different) for the source operand and copies them to the destination operand in the specified order. The order in which the words are written is defined by two-bit fields of the 8-bit immediate operand.
			The source operand can be either an MMX register of a memory location. The destination operand must be an MMX register.

CACHEABILITY CONTROL INSTRUCTIONS

New type of instructions were introduced in Pentium III that provide the following capabilities:

■ Controlling data cacheability to raise the efficiency of cache and lower the number of main memory accesses

■ Preemptive data caching to organize computation pipeline parallel operation and memory exchange

Table 2.10 gives some examples of cacheability control instructions.

TABLE 2.10 Cacheability Control Instructions

Mnemonics	Operation	Example	Memo
	Non-caching memory write instructions		
movntps	Writing data from an MMX register directly to the memory by-passing cache	movntps [ecx], xmm1	The instruction copies four single-precision floating-point values from an XMM register to the memory.
movntq	Writing data from an MMX to the memory by-passing cache	movntq	The instruction copies a quadword from an MMX register to the memory.
	Writing Masked Bytes to the Memory		
maskmovq	Writing data from an XMM register to the memory bypassing the cache	maskmovq mm1, mm2	The instruction selectively writes bytes from an MMX register directly to the memory, bypassing the cache. Bytes are selected using an 8-bit mask consisting of the high bits of the second operand bytes. The mask operand, the same as the source operand, is an MMX register. Whether or not a byte is written to the memory is determined by the most significant bit of the corresponding mask byte: If it is 1, the byte is written; if it is 0 the byte is not written. The memory address to which writing is done is specified by edi register.
	Prefetching Data into Cache		
prefetcht0	Prefetching data from the memory to the L1 or L2 caches	prefetcht0 [esi]	Cache prefetch instructions load data from the memory to specified locations in the cache hierarchy in advance, thereby minimizing memory access delays. The location in the cache hierarchy is specified by a suffix 0, 1, 2, or in the prefetch instruction name.

TABLE 2.10 Cacheability Control Instructions *(Continued)*

Mnemonics	Operation	Example	Memo
prefetcht1, prefetcht2	Prefetching data from the memory to the L2 cache	prefetcht1 [esi]	
	Cache Prefetch		
prefetchta	Prefetching data from the memory to the L1 cache bypassing the L2 cache	prefetchta [esi]	
	Forced Write		
sfence	Ordering of a sequence of memory references and synchronizing it with the cache contents	sfence	The instruction directs to copy to the main memory all data of the preceding write instructions that are stored in the write buffer and the cache.

INSTRUCTIONS TO SAVE AND RESTORE STATE

The Pentium III microprocessor has new XMM registers and a status and control register MXCSR. These registers require support from both the processor and the operating system. To allow application software and the operating system to save and restore the state of these new processor components, several control instructions were introduced.

The new status and control register MXCSR is used to do the following:

- Setting arithmetic exception detection flags
- Setting arithmetic exception masking flags
- Setting the rounding mode
- Setting the flush-to-zero mode

The structure of the MXCSR register is shown in Figure 2.8.

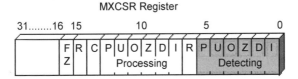

FIGURE 2.8 MXCSR register structure.

Bits 0–5 (exception detection field) contain six flags that indicate arithmetic exception detection. When a flag is set to 1, the corresponding exception has been detected; otherwise, there is no exception.

Bit six is not used.

Bits 7–12 (exception mask field) specify how the detected exceptions are to be processed. If the flag is set, then the corresponding exception is masked and is processed by the processor; if the flag is cleared, a software handler for this exception is called.

Bits 13–14 (RC field) specify the data-rounding mode.

Bit 15 (FZ field) sets the flush-to-zero mode.

Bits 16–31 are not used.

Table 2.11 lists some examples of the state save/restore instructions.

TABLE 2.11 State Save/Restore Instructions

Mnemonics	Operation	Example	Description
	Register Save and Restore Instructions MXCSR		
stmxcsr	Saving contents of the MXCSR register to the memory	stmxcsr	The instruction writes contents of the MXCSR register to a 32-bit memory location.
ldmxcsr	Restoring contents of the MXCSR register from the memory	ldmxcsr	The instruction loads the contents of the stored MXCR register from the memory into the register.

TABLE 2.11 State Save/Restore Instructions *(Continued)*

Mnemonics	Operation	Example	Description
	State Save and Restore Instructions		
fxsave	Saving the processor environment to the memory when switching processor context	fxsave [ecx]	The instruction saves contents of the floating-point data registers, MMX register, and the new Pentium III registers into a 512-byte memory area.
ldmxcsr	Restoring the previously saved processor environment	ldmxcsr [ecx]	The memory area in which the processor environment is saved must be aligned on the word boundary.

PROCESSOR SERIAL NUMBER

The serial number is a processor's unique identifier [62]; it can also be used to identify a computer (user) in a network or used by application programs. The processor serial number is to be used in applications in which it is essential to reliably identify the user or the system. These applications follow:

- Enhanced security-level applications, such as controlled access to the Internet services, and electronic document exchange
- Control applications, such as resource control, remote system booting, and configuring
- Information resource control, such as technical support services and back-up data protection

BUILT-IN PROCESSOR SELF-TEST AND PERFORMANCE-CONTROL FEATURES

The Pentium III processor has the following hardware features to perform self-testing and performance control:

Built-in Self Test (BIST) mechanism—Controls hangings and errors in microcode and large logic matrixes; it also tests code and data caches, TLBs, and ROM segments.

Testing and peripheral scanning standard IEEE 1149.1 port mechanism—Makes it possible to test communications channels between the Pentium III processor and the system using standard interface for this.

Built-in counter units—Monitor the performance indicators and keep track of the events.

Built-in crystal diode monitors—The temperature sensor located on the motherboard controls the temperature of the Pentium III crystal, which is necessary to control the temperature regime.

2.2.6 Pentium 4 Microprocessor

Intel introduced the Pentium 4 processor [63] in June 2000. On one hand, this processor continues the x86 architecture processor line; on the other hand, it undoubtedly has been a major advance in this architecture. As with the Pentium II and Pentium III, Intel produced a whole family of new processors: for personal computers, servers, workstations, and mobile computers.

The new architecture introduced in Pentium 4 is called Intel NetBurst microarchitecture. One of its most important features is the increased operating frequencies of the system bus and the processor ALU. The latter runs at twice the processor frequency. The later models of the family also introduced multithreading or, to use the Intel terminology, hyperthreading.

Architecture of the Pentium 4 processor is oriented toward efficient work with Internet applications.

As in other sixth-generation processors, Pentium 4 uses the following features:

- Harvard architecture internal memory with separate data and code streams
- Superscalar architecture making possible concurrent execution of several instructions in parallel execution units
- Dynamic instruction reordering
- Pipelined instruction execution
- Branch prediction

The microprocessor structure is shown in Figure 2.9.

The following features are the main innovations introduced in Pentium 4 architecture:

- 400MHz system bus effective operating clock frequency
- Doubled ALU operating frequency with respect to the processor clock

- Substantial increase of the execution pipeline depth
- 144 new SIMD SSE2 instructions
- Implementing L1 code cache as trace cache
- On-die integrated L2 cache

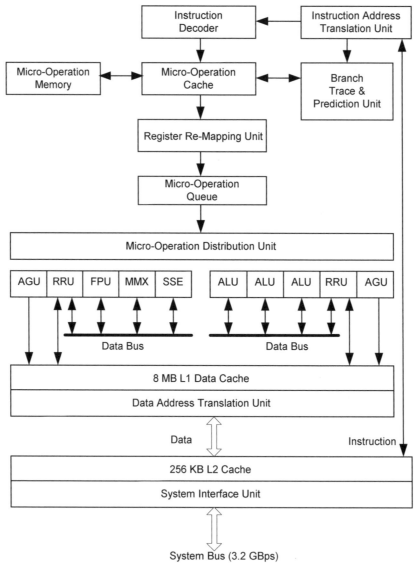

FIGURE 2.9 Pentium 4 microprocessor structure.

The L2 cache is of the eight-way set-associative write-back type. Separate instruction and data streams enter it through the system memory and the external interface unit. Cache size is 256 KB. The cache throughput is twice that of the processors of the previous generation.

Data exchange between the processor and the system bus is done via the external interface unit. The system bus consists of a 64-bit bidirectional data bus and a 41-bit address bus. Thirty-three of the address bus lines are address lines proper (A35-2); the remaining eight lines (A41-36, A2-0) are byte enable lines (BE7–0#). The processor can address up to 64 GB of external memory.

Quadruple data-rate transfer at 100 MHz produces effective 400 MHz bus operation and 3.2 GBps bandwidth.

The instruction decoder, along with the micro-program memory converts, x86 instructions into a sequence of micro-operations, which are then placed into the micro-operation cache, which can hold 12,000 micro-operations. In the cache, micro-operations are ordered according to their execution order, taking into account predicted branches.

The register allocation unit assigns to each logic register micro-operations of one of the 128 physical registers from the register replacement unit (RRU), thereby eliminating data dependencies between individual instructions. The generated micro-operation sequence is placed into the queue, where micro-operations for 126 x86 instructions can be stored (three times more than in Pentium III).

The micro-operation sequencing unit fetches micro-operations from the queue, not in the same order in which they were placed into it but depending on the corresponding operands and execution units becoming available. Here, up to six micro-operations can be executed simultaneously out of order in execution units working in parallel. The micro-operation execution results are sent to the main memory following the same order as the instructions in the program. In the Pentium 4, this is done by the reorder buffer, the same as in the other Intel sixth-generation microprocessors.

The instruction and data translation units generate physical memory addresses; base addresses of the most frequently used memory pages are stored in the internal buffer memory.

The addresses of the operands fetched from the memory are calculated by the Address Generation Unit (AGU), which provides an interface with the 8KB L1 write-through data cache.

The AGU generates 48 addresses to load operands from the memory into a RRU register and 24 addresses for writing from the register to the memory for instructions that have not been fetched for execution yet. When the memory is referenced, the AGU generates two addresses: one to load the operand into the specified RRU register, the other to move the result from the RRU register to the memory.

The cache has separate ports for `read` and `write` operations. The cache access time has been reduced by 1.5 times, compared to the Pentium III processor.

The processor superscalar core contains pipelined execution units: ALU, FPU, and MMX. The integer ALUs operates at double the processor core frequency.

The Pentium 4 employs hyperpipelined instruction-execution technology: the number of pipeline stages reaches 20. Thanks to breaking the instruction-execution cycle into smaller stages, each of which can be executed faster, the processor operating frequency can be raised.

At the same time, increasing the pipeline depth increases the losses associated with branch instruction executions. When the execution results of one branch direction need to be discarded, the pipeline flushed and the instructions from the other branch direction loaded. These losses can be lowered by predicting branches, which is done by the branch-prediction unit. The branch target buffer is a part of the branch-prediction unit and holds 4,092 addresses of the previously performed branches with a prehistory of their execution. The Pentium 4 block prediction unit has a 0.9 branch-prediction success rate.

The Pentium 4 processor introduced the SSE2 unit, which substantially expands the SIMD capability, as compared with the SSE unit in the Pentium III models. The SSE2 unit adds 144 new SSE instructions that can simultaneously operate on several operands, which can be located in the memory or in the 128-bit registers XMM0–XMM7.

Two double-precision floating-point numbers (64 bits) or four single-precision floating-point numbers (32 bits) can be stored and simultaneously operated on in the registers. This unit can also simultaneously process integer operands: 16 8-bit, eight 16-bit, four 32-bit, or two 64-bit integers. As a result, Pentium 4 performance when executing this type of operation is twice that of Pentium III. The SSE2 unit is downward compatible with the SSE unit of the Pentium III processor.

The microprocessor has 42 million transistors. Its Willamette core version is manufactured using CMOS 0.18-micron process technology for clock frequencies from 1.4 GHz to 1.7 GHz. The Northwood core version is manufactured using 0.13-micron process, has six metallization layers, and the clock frequency ranges from 2 GHz to 3.06 GHz.

The Pentium 4 representatives of the server family of the Intel processors are the Willamette core-based Foster and the Northwood core-based Prestonia microprocessors. The former has been produced since 2000, the latter since 2002. These microprocessors perpetuate the Xeon line of the previous generations (Pentium II and Pentium III) and make multiprocessor configuration operation available. Moreover, the Prestonia supports the hyperthreading technology.

The essence of the hyperthreading technology follows. Additional execution units were placed on the processor crystal. These units enable a single physical processor (with one pipeline and common data and code L2 cache) to be recognized by the system as two logic processors, each of which can be loaded with its own task. Most of the processor units are shared, but some are doubled and can perform different tasks.

The hyperthreading technology helps to reduce the processor idle periods by utilizing resources that are not being used in one task to execute instructions of the other task. This can take place in situations such as the following:

- Delays when referencing memory
- Executing a sequence of interdependent instructions
- Branch prediction errors
- Simultaneous integer and floating-point computations

As a result, the performance of the processor's main resources rises, and the total execution time of the two tasks is reduced.

The effectiveness of two parallel tasks sharing the processor resources in the hyperthreading technology substantially depends on the nature of the software: the worse an application is optimized for this micro-architecture, the greater a gain of using hyperthreading may be.

SPECint2000 and SPECfp200 performance benchmarks of Pentium 4 processor running at 1.5 GHz are 536 and 561, respectively.

2.2.7 Pentium M Microprocessor

One of the latest of Intel's achievements, supposed to provide users with new mobile work capabilities, is the development of the Centrino technology. This technology presupposes using new Pentium M microprocessors (which was code-named Banias while in the development stage), new Intel 855 chipset, and a means of access to wireless data networks of the 802.11 standard family.

The main features of systems built on Centrino technology follow. First of all, its energy consumption is low-energy, thanks to the smart microprocessor core frequency and voltage control system, Enhanced SpeedStep. Second, its low weight and size characteristics are achieved by implementing most of the system functions in a high-performance chipset. Finally, its expanded communication capabilities are provided by the built-in radio-Ethernet controller.

The Pentium M microprocessor is the main element of the Centrino technology. Its following new solutions set it apart from the mobile versions of the Pentium III and Pentium 4 microprocessors:

Advanced branch prediction—Three different branch-predicting algorithms are used in Pentium M to analyze conditional and unconditional branches and cycles, as well as the prehistory of program execution. The decision is made on the basis of the most accurate prediction.

Micro-operation combining—The microprocessor combines several micro-operations obtained after decoding a CISC instruction for concurrent execution in different functional units. Parallel execution of several microoperations substantially raises the performance/power consumption ratio.

Advanced stack control—Stack control is implemented on the micro-operation level, which makes this process more energy efficient.

Advanced power-management technology/Enhanced SpeedStep—Unlike the previous version of this technology, which supports only two frequency/power supply voltage ratios, provisions are made in Pentium M for a large number of ratios that make the application performance needed for execution available at the minimum power consumption. Of note is the energy conservation in system bus operations, achieved by the chipset turning on the processor data-reading amplifiers only for the duration of actual data receiving. Energy is also conserved during cache operations by activating only that cache fragment that is currently being accessed.

The microprocessor contains a SSE2 vector operation unit, split 32 KB L1 instruction and data caches, and a unified 1 MB L2 cache. The effective frequency of the processor bus is 400 MHz; the processor core operating frequency ranges from 0.9 GHz to 1.6 MHz. At 1.6 GHz, the processor power consumption is 24.5 W.

The processor is built using the 0.13-micron process technology; it has 77 million transistors.

Performance of 1.7 GHz Pentium M is comparable to that of 2.5 GHz Pentium 4. The average power consumption is from 1 W to 7 W; the maximum power consumption does not exceed 25 W.

2.2.8 IA-64 Architecture Microprocessor

ITANIUM MICROPROCESSOR

The Itanium [64] microprocessor is the first representative of 64-bit microprocessors from Intel. It was first manufactured in 2001. First, microprocessors were built using 0.18-micron process technology for frequencies of 660 MHz, 733 MHz, and 800 MHz. At 800 MHz, the processor performance is 45 SPECint95 and 70 SPECfp95.

The architecture of the Itanium microprocessor (IA-64) implements the explicitly parallel instruction computing (EPIC) concept and is significantly different from the architecture of the previous, 32-bit, microprocessors from Intel.

The main features of EPIC follow:

- A large number of registers
- Scalability with respect to the number of functional units: the number of functional devices can be increased in the following microprocessor models
- Specifying parallelism explicitly in the machine code
- Predicate instruction execution
- Employing hints to prefetch data

Table 2.12 lists the main differences between the IA-64 and x86 processor architectures.

TABLE 2.12 Differences between IA-64 and x86 Processor Architectures

x86 Processors	IA-64 Processors
Use variable length complex instructions processed one at a time.	Use simple same length instructions grouped by threes.
Instructions reordered and optimized dynamically during the execution.	Instructions reordered and optimized during the compilation.
Attempts to predict branches.	Several command sequences are executed without attempting to predict branches.
Data loaded from the memory as needed.	Data prefetch: data are fetched before it is actually needed.

The IA-64 architecture combines the best features of the superscalar processors and VLIW processors. In Itanium, as in other VLIW architecture processors, instruction sequence and the functional unit loading schedule is generated during compilation. The instruction sequence does not change during the program execution. The processor employs dynamic branch-prediction, speculative (upon a hint) instruction execution, hardware support of pipelined cycle execution, postponed memory writes, and prefetching data to L2 cache.

The structure of the Itanium microprocessor is shown in Figure 2.10. It comprises the following:

- Four integer units
- Four multimedia data-processing units

- Two single-precision and two extended-precision floating-point execution units
- Two loading/saving units
- Three branch units

All functional units of the microprocessor are pipelined. Instruction execution takes 10 cycles; up to six commands can be executed in the microprocessor simultaneously. To raise the performance of the processor execution units, instruction execution results are input directly into another instruction, without being stored to registers first. Temporary-result storage memory is used for this purpose.

The advanced floating-point execution unit delivers performance of up to 6 GFLOPS in single-precision operations and up to 3 GFLOPS in extended-precision operations.

The high-speed processor/memory interface has 2.1 GBps bandwidth and provides intensive delivery of data and instructions to the microprocessor.

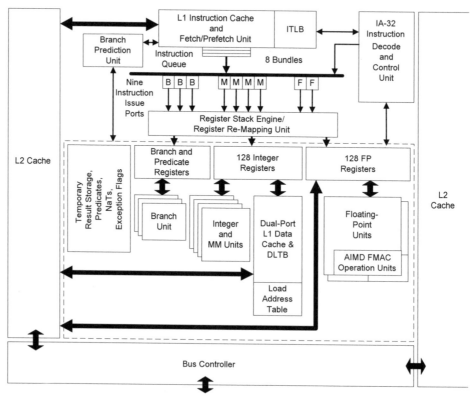

FIGURE 2.10 Itanium microprocessor structure block diagram.

The processor can directly address up to 18 GB of system memory.

Itanium uses a three-level cache hierarchy with the following characteristics:

- L1 is an integrated split instruction and data caches, 16 MB each
- L2 is an integrated unified instruction and data 32 KB cache
- L3 is a unified instruction and data 4 MB cache external to the processor die but located in the same cartridge

The compiler combines instructions into 128-bit bundles. A bundle contains three instructions and a template showing which instructions in the bundle can be executed concurrently. Instructions in bundles can be located in an order different from the original program order and can be both control and data interdependent or not.

The format of instruction bundle follows:

T I Instruction 1 I Instruction 2 I Instruction 3 I

Here *Instruction 1, 2, 3* is a microprocessor instruction; *T* is an 8-bit mask indicating parallel instruction execution capability.

Here are some instruction bundle combinations:

```
i1 || i2 || i3 — All instructions execute concurrently
i1 &  i2 || i3 — First i1 is executed, then i2 and i3 are executed
                    concurrently
i1 || i2 &  i3 — i1 and i2 are executed concurrently, then i3 is executed
i1 &  i2 &  i3 — i1, i2, i3 are executed sequentially
```

When instructions are allocated to the functional units, Itanium analyzes two sequential instruction bundles simultaneously. Mutually independent instructions are selected according to a template and are assigned to functional units. When all instructions in the bundle pair are executed, the next pair of bundles is examined. If some instructions in the current bundle have not been executed, they are assigned to functional units along with allocation of the instructions of the next bundle.

Microprocessor commands have fixed 40-bit length. Their format follows:

Opcode field I PR field I GPR field I GPR field I GPR field I,

Here *Opcode* field is a 13-bit operation code field, *PR* field is a 6-bit predicate field, *GPR* field is a 7-bit common register identifier.

The format of assembler instructions for IA-64 architecture follows:

```
[(qp)] mnemonic [.comp1] [.comp2] dest= src
```

Here:

(qp)—Predicate register. If its value is 1, the instruction is executed.
mnemonic—Specifies a name that uniquely identifies an Itanium instruction.
[.comp1], [.comp2]—Completers (base instruction modifiers). Some instructions can have one or more completers.
dest, src—Represent the destination and source operands. Most of the IA-64 instructions have two or more source operands and one destination operand.

Some examples of microprocessor instructions follow:

A simple instruction—add r1 = r2, r3.
A predicate instruction–(p4) add r1 = r2, r3.
A short word instruction—add r1 = r2, r3, 1.
An instruction with a completer—cmp.eq p3 = r2, r4.

The IA-64 supports the following data types:

Integer—1-, 2-, 4-, and 8-byte.

Floating-point—Single-, double-, and extended double-precision.

Pointers—8-byte.

The Itanium microprocessor has the record number of registers to date. A list of them and their functions are given in Table 2.13.

The microprocessor functional unit load efficiency substantially depends on the instruction bundle sequence generated by the compiler. Therefore, the IA-64 compiler takes into account the specifics of the Itanium architecture to a greater degree than the existing compilers for CISC and RISC microprocessors.

Table 2.13 Itanium Registers and Their Functions

Name	Number	Width in bits	Function
GR0–GR127—general purpose integer registers	128	64 + 1	The GR0 register is read only and its value is always 0.
			Registers GR1 through GR32 are for global data.
			Registers GR33 through GR127 are for stack operations.

Table 2.13 Itanium Registers and Their Functions *(Continued)*

Name	Number	Width in bits	Function
			Each register has 64 bits that are used to hold values. The additional bit NaT (Not a Thing) to indicate validity of data in the register. If the register data is determined to be not needed (for example, if the register contains the result of a speculative instruction execution), the NaT bit is set to 0 and its contents are not used.
FR0–FR127–floating-point registers	128	82	The floating-point unit contains: two read-only registers: FR0 and FR1; their values are always +0.0 and +1.0, respectively.
			32 global and 96 rotating registers.
			Each register has a 64-bit significand, a 17-bit exponent, and one sign bit.
PR0–PR63–predicate registers	64	1	Controlling predicate instruction execution and branching.
			There are 16 static and 48 rotating registers. Predicate registers are used to make a decision as to an instruction execution. Instruction is executed only if the value of its corresponding predicate register is 1.
			The PR0 register is assigned to instructions that are not implicitly assigned predicate registers. Its value is always 1.
BR0–BR7–branch registers	8	64	Specify branch target addresses for indirect branches.
AR0–AR127–application registers	128	64	Used to implement auxiliary functions
IP–instruction pointer	1	64	Holds the address of the next instruction.

The compiler for IA-64 employs tagged-branch technology, which consists of the following. Instructions pertaining to different branches of the algorithm are tagged with different values in the instruction predicate field. Instruction bundles are formed without taking the predicate value into consideration. This makes it possible to execute instructions from different branches concurrently.

During the program execution, after the true branch direction of the computation process has been determined, the results of only those instructions whose predicate value corresponds to the true branch are preserved.

The tagged-instruction technology substantially decreases the negative effect of machine-level branching. It must be noted that not all branches are tagged by the compiler for parallel execution. As a rule, this is done for short, alternate branches. When a branch is not tagged by the compiler, it is predicted by the microprocessor in the course of execution.

Another IA-64 feature is prefetching data into the cache. In essence, this consists in fetching of data, with the instructions that use it separated in time.

While analyzing the source code, the compiler adds to the object code instructions to load the necessary data. Directly before the instructions that use data, the compiler adds instructions to check data availability. This way the delays caused by having to wait for the necessary data to load are significantly reduced.

To provide compatibility with earlier 32-bit applications, the Itanium microprocessor supports two instruction-decoding modes: VLIW and CISC. Switching between the modes is done by software. Additional instructions are added to the x86 instruction set for this purpose.

Itanium has hardware loop execution support. A loop is divided into three execution stages: prolog, kernel, and epilog. All of these stages have corresponding hardware support provided. In particular, the rotating register mechanism is used, along with special instructions and special application registers: loop count (LC) and epilog count (EC) registers.

The microprocessor does not employ dynamic register renaming. To provide pipelined execution of different iterations of the loop, the microprocessor rotates the registers being used. The set of the registers used in the loop is considered as a frame, which is mapped onto the register file starting from some position—frame base. When the next iteration is executed, the frame base is updated by the number of registers in the frame.

THE ITANIUM 2 MICROPROCESSOR

The Itanium 2 microprocessor [65], code-named McKinley, is the second implementation of the 64-bit Intel Itanium architecture. Itanium 2 is aimed at use in high-performance servers and workstations.

The microprocessor is built using the 0.18-micron process technology with clock frequencies 1 GHz and 0.9 GHz. The system bus frequency is 400 MHz, which combined with the 128-bit bus provides interface bandwidth of 6.4 GBps.

Itanium 2 has large caches: L3 cache is 1.5 MB or 3 MB, L2 cache is 256 KB, and L1 cache is 32 KB. Its crystal area being 464 mm^2, Itanium 2 is one of the largest integrated circuits and has 221 million transistors.

Intel is developing next versions of a server processor. A mass production of a renovated version of Madison core based Itanium 2 started in 2003. This microprocessor is built using the 0.13-micron process technology, has 6 MB L3 cache, and a clock frequency of 1.5 GHz. In the same year, production of the 0.09-micron process technology version of the processor based on the Deerfield core started. This processor is intended for use in low-power-consumption systems.

Itanium 2 holds the performance record to date. Table 2.14 lists the benchmark results for the fastest modern processors.

TABLE 2.14 Processor Performance Comparisons

Processor	Frequency	Cache	SPECint2000	SPECfp2000
Intel Itanium 2	1 GHz	Three-level 3 MB	760	1,350
Intel Itanium	800 MHz	Three-level 4 MB	400	701
Compaq Alpha 21264C	1 GHz	Three-level level 8 MB	679	960
AMD Athlon XP 2100	1.677 GHz	Two-level level 256 KB	759	642
IBM RD64 IV	750 MHz	Two-level level 8 MB	458	410
IMB Power 4	1.3 GHz	Two-level level 16 MB	839	1,266
Sun UltraSPARC III	1.05 GHz	Two-level level 8 MB	610	827

2.2.9 Fifth-Generation Processors from NexGen

When considering microprocessors produced by Intel rivals, the first ones that need to be considered, in chronological order, are the developments of NexGen company, presently absorbed by AMD [66, 67]. The original architectural approaches first used in the processors from this company were later used by many other designers. NexGen managed to beat AMD and Cyrix to building a fifth-generation processor, which was produced in 1995.

In the Nx586 microprocessor, a new architecture was implemented for the first time. This architecture, on one hand, possessed all the advantages of the RISC architecture, while on the other hand provided instruction compatibility with the x86 family microprocessors. Analogous solutions were later used by Intel in Pentium Pro and by AMD in K5.

The main idea was to convert complex x86 CISC instructions into RISC-like operations to be executed by the RISC-type processor core. This approach has turned out to be quite promising for the processors perpetuating the x86 line.

The Nx586 microprocessor had three execution units and a three-way super-scalar core; however, it could decode only one x86 instruction during one clock cycle. Another distinctive feature of the Nx586 processors was the absence of an integrated floating-point coprocessor. The designers assumed that this micro-processor would be used in personal computers, in which floating-point opera-tions are performed rarely and their execution speed is not critical.

Processors from NexGen were named Nx586-P100, Nx586-P90, Nx586-P80, and Nx586-P75. By controlling the internal frequency using a phase locked-loop circuit, the Nx586 microprocessors could work at practically any clock frequency (limited by their architecture, of course). Thus, the Nx586-P100 worked at 93 MHz, the Nx586-P90 worked at 84 MHz, the Nx586-P80 worked at 75 MHz, and the Nx586-P75 worked at 70 MHz. The operating frequency of the system inter-face bus was half of the processor frequency.

NexGen was planning to further develop the main ideas of the Nx586 in the sixth-generation microprocessor Nx686. However, because of financial problems (the low-volume microprocessor sales was caused mainly by having to use non-standard motherboards with the Nx586), NexGen had to agree to a merger with AMD. Further work on the Nx686 project, now named K6, was continued as a part of AMD.

2.2.10 Microprocessors from AMD

Products from AMD give Intel microprocessors a good run for their money. Mi-croprocessors from this company have taken lead positions in some specification figures. Some of the interesting architectural and technical approaches first used in the AMD processors were later used by other manufactures in their products, including Intel microprocessors.

K5 MICROPROCESSOR

For many years AMD lagged behind Intel at least one processor generation. They relied mostly on licensed technology and introduced only minor constructive changes into their microprocessors. Appearance of the Pentium microprocessor threatened to push AMD out of the market. This stimulated the company to intensify its work on developing a new family of x86-compatible microprocessors. A goal was formulated to design a processor that would be superior to the Intel Pentium and change the position of following the leader that the AMD found themselves in. Work on the K5 started when the details about the Pentium processor were not yet available. AMD engineers had to develop their own microarchitecture, while at the same time ensure compatibility with the existing software for the x86 processors.

At first AMD planned to start deliveries of its 100–120 MHz microprocessor in 1995; however, only a few thousand of these processors were produced, and their clock frequency was only 75 MHz. Main deliveries of K5 started in the first quarter of 1996, after the company had moved to the 0.35-micron process technology developed jointly with Hewlett-Packard. This made it possible to bring the number of transistors up to 4.2 million on a crystal area of 167 mm^2.

K5 [68] is the first processor where, with the exception of the microcode, AMD did not use any of Intel's intellectual property. At the same time, its performance characteristics are superior to those of the Intel microprocessors. Many applications, such as Microsoft® Excel and Word and CorelDRAW®, ran on the K5 series processors 30 percent faster than on the same frequency Pentiums. This performance was attained mainly by the increased cache and more progressive superscalar architecture. The RISC86 architecture used in the AMD microprocessors (first proposed by NexGen) is shown in Figure 2.11.

x86 instructions are of variable length and have complex structure, which makes their decoding and analysis of the existing data dependencies in instructions difficult. In the architecture proposed by AMD, the decoder is the most complex unit of the microprocessor. It decomposes long CISC instructions into short RISC-like components, named RISC operations (ROP).

ROPs are similar to x86 microprocessor microcode instructions. The first x86 architecture microprocessors executed their complex instructions by fetching microcode from their internal ROM. In the latest x86 microprocessors, microcode use has been minimized by using simple, hardware-implemented instructions.

K5 takes another approach: here most of the ROPs are generated dynamically not by microcode but by the decoder. However, the microcode is still used to process complex and rarely used x86 instructions, such as string and complex-number operations. Nevertheless, even in these cases, the end result is a stream of ROPs. K5 can convert an x86 instruction in 1 to 4 R-operations. These operations are scheduled for execution in the processor core, which is adopted from the RISC architecture. Dynamic register renaming, branch prediction, instruction reordering, and other dynamic execution methods are employed. Consequently, K5 implements hybrid CISC/RISC architecture.

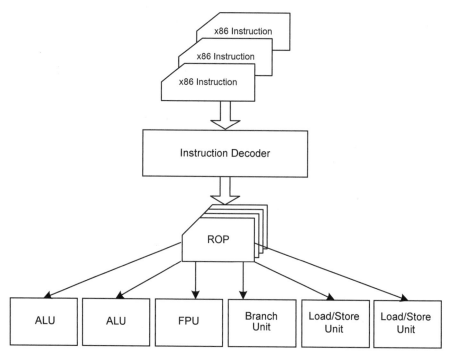

FIGURE 2.11 AMD RISC86 architecture.

Unlike the Pentium, instead of two pipelines for performing parallel operations on integers, K5 has six units functioning in parallel. Floating-point, load/write-back, or branch instructions can be executed simultaneously with integer operations. The load/write-back unit can fetch two instructions in one cycle. Another difference from Pentium is that K5 can reorder instructions.

The FPU meets the x86 standards; however, its performance is somewhat lower than that of the Pentium processor FPU.

The combined use of CISC and RISC principles in the K5 architecture made it possible to overcome the x86 instruction-set limitations. At the price of increased processor complexity, AMD managed to raise its performance, while at the same time preserving the x86 instruction set compatibility. The latter is quite important, taking into account that software for this microprocessor architecture is widespread.

The main features of the K5 microprocessor follow:

- Superscalar architecture with four-way instruction issuance
- Five-stage execution pipeline
- Six parallel functional units: two ALUs, an FPU, a branch unit, and two load/write-back units

- 16 KB code cache and 8 KB data cache
- Instruction reordering, branch prediction, improved microcode
- 30 percent higher performance than the same clock frequency Pentium
- 75–200 MHz clock frequency
- Bus frequency multipliers: 1, 1.5, 2, and 3
- 4.1 million transistors
- Process technology: first 0.5-micron, then 0.35-micron with three metallization layers
- 3.3 V power supply
- Pentium-compatible instruction set and electrical signals

K6 MICROPROCESSOR

The K6 microprocessor is a product of joint efforts of AMD and engineers from the former NexGen. In fact, K6 is the realization of the Nx686 project that had been started by NexGen but not completed because of financial problems.

The K6 microprocessor [69] was introduced in 1997. It was produced on CMOS 0.35-micron process technology, has five metallization layers, contains 8.8 million transistors on 162 mm^2 crystal area, operates at 166 MHz, 200 MHz, and 233 MHz clock frequency, and is installed into Socket 7.

As in K5, K6 implements superscalar RISC86 architecture with decoupled instruction decoding and execution, which provides compatibility with the x86 instructions set and high performance inherent to the sixth-generation microprocessors. K6 is equipped with a multimedia instruction set extension: MMX. K6 surpasses in performance Pentium MMX running at the same clock frequency and is comparable to Pentium Pro. Unlike Pentium Pro, K6 equally successfully works with both 32-bit and 16-bit applications.

Processor's high performance is achieved by the following new architectural and technological approaches:

- x86 instructions are pre-decoded when fetched to cache; each instruction placed into the L1 cache is annotated with pre-decode bits, which show the offset of the next instruction in the cache (from 1 to 15 bytes)
- On-chip split L1 code and data cache, 32 KB each
- High-performance floating-point unit
- MMX standard high-performance multimedia operation unit
- Decoding multiple x86 instructions per cycle into simple RSIC operations
- Parallel decoders, centralized operation scheduler, and seven superscalar execution units in a six-stage pipeline
- Speculative instruction execution with instruction reordering, data prefetch, register renaming

■ The processor supports branch prediction logic using 8,192-entry branch history table, BTB, and the return address stack. This brings the prediction success rate up to 0.95. The processor employs a two-level branch-prediction system. The branch history table does not store predictions of the target addresses; they are calculated by the address ALU during instruction decoding. The BTB augments predicted-branch performance by reducing the time losses during memory references and supplies the decoders with the first 16 bytes of instructions fetched according to the branch prediction.

Figure 2.12 shows a block diagram of the K6 microprocessor structure.

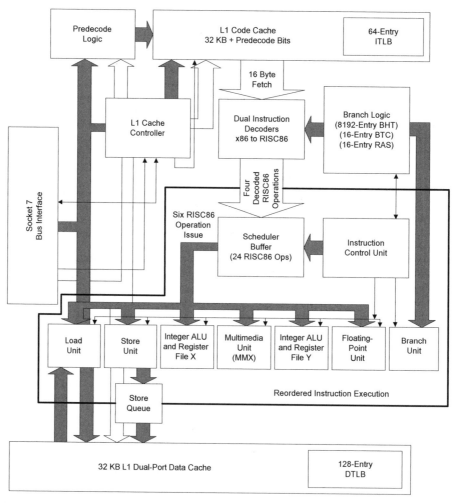

FIGURE 2.12 AMD K6 microprocessor structure block diagram.

The microprocessor RISC core can execute up to six ROPs in one cycle. The core uses 48 physical registers to execute ROPs: 24 general-purpose registers and 24 registers for renaming.

The K6 scheduler can buffer up to 24 RISC86 operations at a time. It can simultaneously issue up to six RISC86 operations for execution to the following seven execution units: load, store, branch, two ALU, MMX, and FPU units. The buffer allows the scheduler to dynamically analyze and change the selected instruction sequence, providing maximum workload to the execution units.

The MMX unit executes all MMX extension instructions. The integer unit X executes arithmetic and logic operations, including multiplication, division, shift, and cyclic shift. The integer unit Y executes main one- and two-word arithmetic-logic operations: ADD, AND, CMP, OR, SUB, XOR, zero-extend, and sign-extend. The FPU performs computations with real numbers. The branch unit updates the processor environment (register, flags) after a predicted conditional branch is confirmed, thereby providing a speculative-branch instruction execution capability up to seven branches deep.

At the beginning of 1998, new processor versions were built. These utilized the 0.25-micron process technology with five metallization layers and ran at 266 MHz and 300 MHz.

K6-II MICROPROCESSOR

At the end of May 1998, AMD announced the K6-II microprocessor. This microprocessor employs superscalar processing of MMX instructions. It is also equipped with 3DNow! technology with 21 new instructions used to accelerate processing of 3D graphics, stereo sound, and video [70]. This type of technology was implemented in Intel microprocessors significantly later, when the Pentium III (Katmai) microprocessor was introduced.

The 3DNow! technology speeds up multimedia application execution by processing multimedia data in SIMD mode. The 3DNow! instructions are analogous to MMX instructions but process floating-point data. The AMD K6-II microprocessor executes up to two 3DNow! instructions in one cycle in each of the two pipelines. Consequently, up to four floating-point instructions (addition, subtraction, or multiplication) can be executed in one cycle.

The multimedia unit of the AMD-K6-II microprocessor merges the existing MMX instructions with the 3DNow! instructions. This merger makes it possible to execute integer MMX instructions and floating-point 3DNow! instructions without wasting time on switching between the execution units. The 3DNow! instructions use the same registers as the MMX instructions, have similar coding, and can be executed in parallel with the MMX instructions.

The difference between the MMX and 3DNow! technologies is that the former is intended to accelerate integer computations and improve integer operations in such applications as video editing and playing. The latter complements and enhances the graphic accelerator capabilities, speeding up floating-point computations in the early stages of the graphics pipeline.

The K6-II microprocessor has 9.3 million transistors and was initially produced for 266 MHz, 300 MHz, and 333 MHz clock frequencies. Later processors for 350 MHz, 366 MHz, 380 MHz, and 400 MHz frequencies appeared. The K6-II microprocessor was the first Socket 7 microprocessor to operate with 100 MHz system bus frequency.

The applications that derive the largest performance benefits from the 3DNow! follow:

- 3D games
- Multimedia encyclopedias
- Multimedia Internet resources
- Presentations, word processors, and electronic spreadsheets
- CAD/CAM tools
- Spatial sound processing
- Voice-recognition applications
- Image transmission over the Internet
- High-quality reproduction of DVD movies and MPEG video

At the end of February 1999, AMD introduced its next microprocessor: K6-III with 400 MHz and 450 MHz operating frequencies. The main distinctive feature of this microprocessor is the integrated on-die 256 KB L2 cache. K6-III can also work with external L3 cache ranging from 512 KB to 2 MB. As compared to K6-II, its performance increased by 15 to 20 percent.

K7 MICROPROCESSOR

In June 1999, AMD introduced its next generation microprocessor: K7 (code-named Athlon) [71]. K7 has more than 22 million transistors on the crystal 184 mm^2 in area. At first it was produced using the 0.25-micron process technology, had six metallization layers, and operating frequencies of 500 MHz, 550 MHz, 600 MHz, and 650 MHz. Later, the 0.18-micron process technology was used and the operating frequency raised to 1 GHz and higher. The processor power supply voltage is 1.6 V.

The processor die is packaged into a cartridge that is mounted into the motherboard via Slot A connector developed by AMD. Athlon and Slot A use Digital Alpha EV6 bus protocol, which has several advantages over the GTL+ protocol used by Intel. Thus, EV6 allows point-to-point topology to be used for multiprocessor systems. Moreover, EV6 utilizes both the leading and the falling edges of the clock signal, which at 100 MHz bus frequency gives effective data transfer frequency of 200 MHz and 1.6 GBps interface bandwidth. In the following processor models, the operating bus frequency reached 133 MHz and then 200 MHz, the effective data transfer frequencies being 266 MHz and 400 MHz, respectively.

A block diagram of the K7 microprocessor is shown in Figure 2.13.

The architecture implemented in Athlon is named QuantiSpeed™. It defines superscalar, super-pipelined instruction execution, pipelined FPU, hardware data prefetch into the cache, and enhanced branch-prediction technology.

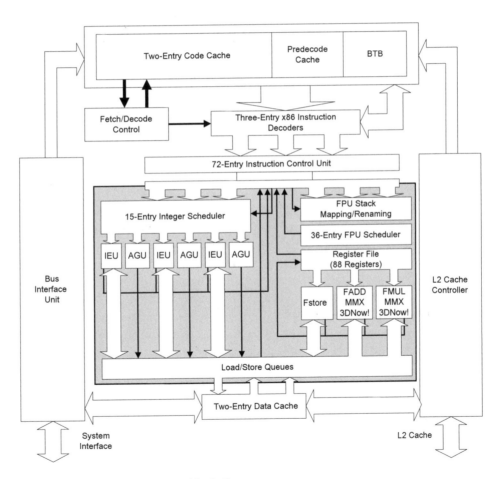

FIGURE 2.13 K7 microprocessor block diagram.

Athlon has nine execution units: three integer execution units (IEU), three address generation units (AGU), and three floating-point and multimedia units (one for loading and storing floating-point data (FSTORE), and pipelined units for executing FP/MMX/3DNow! instructions). The integer pipeline is 10 stages deep; the floating-point pipeline is 15 stages deep. (FPUs in the previous models were not pipelined.) One Athlon FPU operation allows two floating-point instructions to be executed simultaneously.

Athlon can decode three x86 instructions into six ROPs. After decoding, ROPs are placed into the scheduler buffer where they await their turn to be executed in one of the processor functional units. The scheduler buffer in K7 holds 72 operations (three times more than K6) and issues nine ROPs to nine execution units.

The branch history table has been increased to hold 2,048 entries (twice more than Pentium III), and the branch prediction algorithm has been enhanced. This makes for a 95 percent branch-prediction success rate as compared to only 90 percent in Pentium III.

The SIMD instruction set of 3DNow! has been supplemented by 24 instructions for a total of 45 instructions; it is called Enhanced 3DNow!.

Athlon has a 128 KB split L1 cache: 64 KB for data and 64 KB for code. Interactions with the L2 cache are carried over a special bus, similar to the Intel P6 architecture. L2 cache is located on a die separate from the processor core but packaged in the same cartridge as the processor die. It operates at half the processor core frequency.

Using 0.18-micron process technology with copper interconnects allowed a 256-KB L2 cache to be integrated on the processor die and operate at the processor core frequency. The new processor core was code-named Thunderbird. A microprocessor with the Thunderbird core has 37 million transistors on 120-mm^2 crystal area and operates at frequencies ranging from 750 MHz to 1400 MHz.

The next processor with K7 architecture built on the Thunderbird core was Duron, a budget version of the microprocessor oriented toward inexpensive PCs. Its main feature is the reduced 64 KB L2 cache. Duron has 25 million transistors on 100-mm^2 crystal area and operates at frequencies from 600 MHz to 1200 MHz.

The integrated L2 cache allowed dispensing with the cartridge and going back to mounting the processor in a 462-pin Socket A instead of a slot. Cache in Duron and Athlon processors operates using an algorithm that provides cache data exclusivity—data in L1 and L2 caches are not doubled, thus increasing the effective volume of cached data.

Owing to the new architectural and technical approaches utilized in K7, the AMD microprocessors achieve 7 to 10 percent better performance than the Pentium III processors operating at the same clock frequencies.

Further enhancements of the K7 family microprocessor architecture and manufacturing technology gave rise to two new Athlon versions: Athlon XP and Athlon MP.

The Athlon XP processor is based on the new Palomino core, which has the following advantages over the Thunderbird core:

- The 3DNow! SIMD instruction set has been extended and is fully compatible with the previous versions of 3DNow! and also supports Intel SSE instructions. (The new 107-instruction SIMD instruction set is named 3DNow! Professional.)
- Owing to the enhanced core, Athlon XP power consumption was reduced by 20 percent as compared to the same-frequency Thunderbird core processors.
- The branch prediction algorithm has been enhanced.
- The mechanism to prefetch instructions from the main memory to cache has been improved.
- The TLB has been modified.
- Crystal temperature can be monitored using the on-die integrated thermo sensor.

The new microprocessor is oriented toward use in high-performance PCs. The designers believe that it should be especially effective when running under Windows XP, hence the XP in the processor's name.

The processor packs 54.3 million transistors on the 101-mm^2 crystal area. Athlon XP has been manufactured first using the 0.18-micron process technology and then using the 0.13-micron technology with copper interconnects.

Athlon MP is intended for use in servers and high-end workstations and can be used in two-processor configuration systems.

The main difference between Athlon MP and Athlon XP is the use of the Smart MP technology in the former. This is a combination of a high-speed dual-system bus and cache-coherence protocol MOESI controlling the memory bandwidth, which is necessary for optimal processor operation in multiprocessor systems. Bus bandwidth is 2.1 GBps per processor.

The processor is produced with operating frequencies from 1 GHz (using 0.18-micron technology) and up to 2.133 GHz (using 0.13-micron technology, Thunderbird core).

HAMMER FAMILY MICROPROCESSORS

The new family of microprocessors from AMD is called Hammer [72]. Its members are the microprocessor for PCs ClawHammer and for servers Sledge-Hammer. Both processors are manufactured using the 0.13-micron silicon-oninsulator (SOI) technology with copper interconnects. Microprocessors of this family accommodate 64-bit processing. However, unlike the Intel IA-64 architecture, their architecture is compatible with the x86 instruction system and, consequently, is called x86-64 architecture.

The main advantages of the x86-64 architecture follow:

■ Downward compatibility with the x86 instructions
■ 64-bit general-purpose registers (GPR)
■ Eight new GPRs available only to 64-bit applications
■ SSE and SSE2 support; eight new SSE2 registers
■ Increased addressable memory
■ High performance executing 32-bit applications; 64-bit application support

The differences between the x86 and x86-64 architectures are shown in Figure 2.14.

A simplified block diagram of the Hammer structure is shown in Figure 2.15.

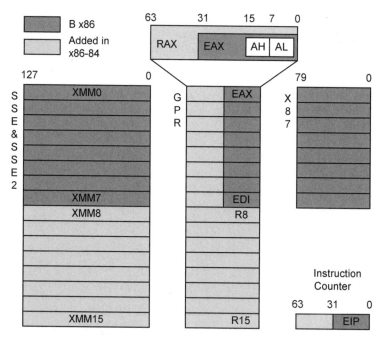

FIGURE 2.14 Registers of x86-64 architecture.

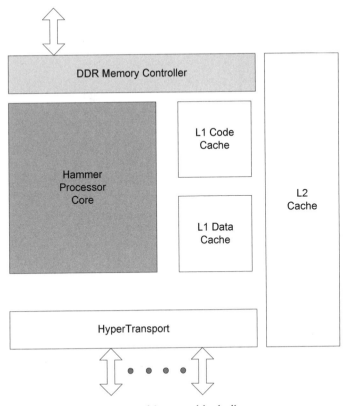

FIGURE 2.15 Hammer architecture block diagram.

When developing the new architecture, AMD designers strived to increase microprocessor performance not only by increasing the clock frequency and the number of functional units but also by increasing their workload levels. With this aim, the following changes were made in the microprocessor:

- The depth of the integer pipeline was increased to 12 stages; the depth of the floating-point pipeline was increased to 17 stages
- A memory controller was integrated
- The branch-prediction unit was enhanced
- A coherent interface HyperTransport was built in

It is obvious that the workload of the microprocessor functional units cannot be increased without intensifying data supply to the microprocessor core. By integrating the memory controller on the crystal, AMD was trying to raise the performance of the main memory exchange subsystem, which is the data bottleneck

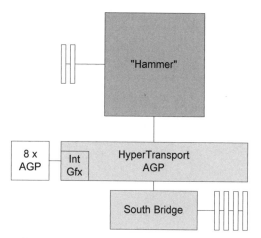

FIGURE 2.16 Structure of a computer system based on Hammer MP.

in the modern microprocessors. Optimized for the given processor and maximally proximate to it, the memory controller makes it possible to organize effective memory exchange at the processor core frequency, bypassing the system bus.

Hammer processors have either a 64-bit or 128-bit DDR SDRAM controller; in the future, DDR II memory will be used. The structure of a computer system based on the Hammer microprocessor is shown in Figure 2.16. With the exception of the AGP controller, functions of the north bridge are provided by the processor.

The built-in coherent HyperTransport (HT) interface allows Hammer to be used in a multiprocessor configuration.

The enhancement to the branch-prediction unit is the capability to detect and take into consideration information about the branch type: whether the branch is static, that is, the branch address does not change, or dynamic.

The difference between processors for one- and multi-processor configurations is mostly in the size of the L2 cache and the number of HT interface links. Thus, the Athlon ClawHammer-DT (DT stands for desktop) microprocessor of the following configuration is intended for use in one-processor systems: one 72-bit DDR SDRAM DDR200/266/333 (PC1600/PC2100/PC2700) channel, one 16-bit HT link with 3.2 GBps bandwidth, 256 KB or 1 MB L2 cache. For two-processor systems (Figure 2.17), the Athlon ClawHammer-DP (DP means dual processing) is intended, of the following configuration: one 72-bit DDR SDRAM DDR200/266/333 (PC1600/PC2100/PC2700) channel, two 8-bit HT links with 3.2 GBps bandwidth each, and 512 KB or 1 MB L2 cache. The microprocessor Opteron SledgeHammer-MP (MP stands for multi-processing) of the following configuration is intended for use in multiprocessor systems of up to

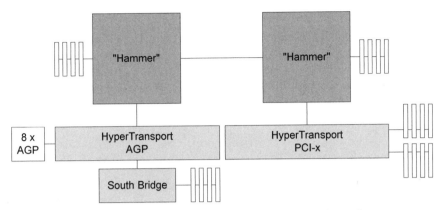

FIGURE 2.17 Hammer microprocessor in two-processor configuration.

eight processors: two 72-bit DDR SDRAM DDR200/266/333 (PC1600/PC2100/PC2700) channels, three 16-bit HT links with 3.2 GBps bandwidth each, and 1 MB L2 cache.

The first two microprocessors are packaged into a 754-contact mPGA casing; the latter is packaged into a new 940-contact mPGA casing.

2.2.11 Cyrix Microprocessor

Microprocessors from Cyrix hold the third place after Intel and AMD in the x86 architecture microprocessor market. Not laying a claim to being performance leaders, nevertheless, these microprocessors hold a steady position in the lower-priced PC market segment owing to the balanced combination of such characteristics as performance, power consumption, and cost.

CYRIX 5X86 (M1SC) MICROPROCESSOR

Cyrix developed the 5x86 microprocessor as an alternative to Pentium and to the transitional technology from the fourth to the fifth microprocessor generation proposed by Intel: ChipUp. To make it easier for the users to switch to the new processor, the company made its pinout compatible with the x846 microprocessor.

Such architectural and structural approaches characteristic of the fifth-generation processors as 64-bit internal architecture, branch prediction, data prefetch, executing several instructions in one cycle (made possible by decoupling load and store units) are implemented in 5x86 [73]. An 80-bit FPU and a unified 16 KB cache are integrated into the processor crystal.

A distinctive feature of the microprocessor is its highly efficient power-management system for the processor and periphery. The power management system can disconnect the FPU and other internal units if they are not being used. Low energy consumption by the processor (at 100 MHz and 3.3 V power supply voltage, the 5x86 consumes less than 3.5 W) makes it preferable for use in mobile computers, where power consumption and dissipation are important considerations.

The 5x86 architecture is a compromise between the performance and simplicity of implementation, thanks to which the number of transistors on crystal and the power consumption could be reduced.

The processor has a branch-prediction unit equipped with a BTB, a unified 16 KB, write-back code and data cache, an FPU, an instruction fetch unit, an instruction decode unit, a memory control unit equipped with a 32-entry TLB, loading and storing units functioning in parallel, and an address generation unit. The functional units are interconnected by two 32-bit buses providing nonblocked data exchange. The 128-bit instruction fetch bus supplies 16 bytes of instructions in one clock cycle to the three-level buffer of the decode unit.

The integer module fetches, decodes, and executes instructions in the six-stage pipeline. In the first stage, *fetch*, a continuous high-speed instruction stream is fetched from the internal cache. Up to 128 bits of code are read in one clock. In the second stage, *decode*, the instruction stream is analyzed and the number of bytes in each instruction and its type are determined. In the third stage, *address calculation*, addresses are calculated in the pipeline that includes AC1 and AC2 stages. The AC1 stage is used to calculate linear address if the instruction references a memory operand. In the AC2 stage, memory, cache, and register access operations are performed. If a floating-point instruction is detected, the AC2 hands it over to the FPU. In the *execution stage* instructions are executed under microcode control. In the *write-back* stage the result is saved in the register file by the integer unit; saving in the memory is carried out by the load/store unit.

The FPU is connected to the cache and integer unit by a 64-bit interface and supports the x87 instruction set, including the extended 80-bit format.

The memory control unit contains the load/store unit, the TLB, and the address calculation unit. The address calculation unit calculates addresses, sets the program counter, and initiates load/store operations. The load/store unit can execute operations in parallel and reorder three load and four store operations.

The internal 64-bit processor bus changes to the 32-bit external system bus. The external system bus operates at 33 MHz to 50 MHz; the multiplier raises the internal frequency to 100 MHz–120 MHz.

In terms of performance, Cyrix 5x86 is comparable with the lower-end Pentium models (75 MHz, 90 MHz) but is substantially less expensive.

CYRIX 6X86 MICROPROCESSOR

Cyrix presented the first units of the 6x86 microprocessor [74, 75], named M1, in October 1995. Operating at 100 MHz, the processor performance surpassed Pentium by 30 percent. However, because of the large crystal size, the processor generated lots of heat, which negatively affected its reliability. Consequently, the processor in the first implementation did not gain wide use, and its mass production began only when the new version came out in the first quarter of 1996. Using the same 0.6-micron process technology, Cyrix switched over to using five metallization layers instead of the previous three. This allowed the crystal size to be reduced from 394 mm^2 to 225 mm^2, with a corresponding drop in the heat generation. The new processor version was produced with 100 MHz, 120 MHz, 133 MHz, and 150 MHz operating frequencies. This corresponds to the Pentium ratings used in marking Cyrix processors of 120+, 150+, 166+, and 200+. The processor power supply is 3.3 V; a low-voltage version was also produced: the 6x86L, which was intended for use in mobile computers. The 6x86L processor core was powered by 2.8 V and the I/O subsystem by 3.3 V.

The 6x86 microprocessor implemented such progressive architectural performance increasing methods as instruction reordering, dynamic elimination of instruction interdependencies, register renaming, speculative execution, and branch prediction.

The block diagram of the microprocessor structure is shown in Figure 2.18.

The Cyrix 6x86 microprocessor has two independent seven-stage pipelines (X and Y), which can execute several instructions in one cycle. The processor has two caches: a unified 16 KB code and data cache, and an additional 256-byte direct-mapped code cache. The dedicated code cache makes it possible to avoid frequent conflicts when referencing data and code in the unified cache.

Like the 5x86, the 6x86 can execute integer, floating-point, and postponed and reordered load/store instructions in parallel.

The operation cycle of the integer ALU consists of the following pipeline execution stages:

- Instruction Fetch (IF)
- Instruction Decode 1 (ID1)
- Instruction Decode 2 (ID2)
- Address Calculate 1 (AC1)
- Address Calculate 2 (AC2)
- Execution (EX)
- Write-back (WB)

Thus, the instruction decode and address calculate stages are also pipelined.

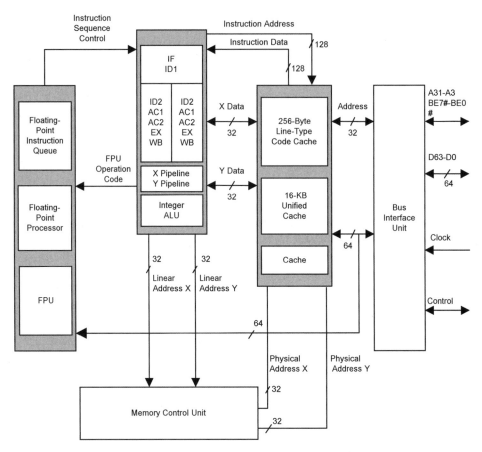

FIGURE 2.18 Cyrix 6x68 structure block diagram.

In the IF stage, common for the X and Y pipelines, 16 bytes of code are fetched from the cache in one clock cycle. In the same clock cycle, the instruction stream is checked for the presence of branch instructions. If a branch instruction is detected, the branch-prediction logic generates the branch address, and instructions in the predicted-branch direction are fetched.

The super-pipelined instruction decode stage has two substages. The ID1 is a common stage for both pipelines. In this stage the instruction size in bytes is determined. Then instructions for each pipeline are passed to the ID2 stage. In this stage, instructions are decoded and passed for execution to either the X or Y pipeline. The specific pipeline is selected, depending on which of the instructions currently being executed will complete first.

Addresses are calculated in two stages analogous to those already considered for 5x86.

In the execution stage, data obtained in the previous address-calculation stage are used. The execution results are written to a register directly in the integer ALU or are passed to the cache control unit if they are to be stored in the memory.

What makes the 6x86 processor different from the Pentium and 5x86 processors is its instruction-reordering capability. If an instruction will execute faster than the preceding one in the other pipeline, they will be reordered. Up to the execution stage, all instructions are processed in order, but in the EX and WB stages the order can be broken. Hardware-blocking devices ensure the required execution and write-back order of instructions with data interdependencies.

To eliminate data interdependencies, the processor employs the following techniques:

- Register renaming
- Data prefetch
- Data passing

The 6x86 processor has 32 physical GPRs. Each of the registers can be temporarily associated with an x86 architecture register (EAX, EBX, ECX, EDX, ESI, EDI, EBP, or ESP). For each `write` operation to the register, a new physical register is chosen to preserve the previous data. Register renaming effectively eliminates WAW and WAR dependencies.

Renaming registers is not sufficient to eliminate RAW dependencies. For this purpose, 6x86 also employs data and result prefetch.

To reduce performance losses caused by the data and result prefetch when eliminating RAW dependencies, the 6x68 employs the data-passing technique. This technique is used when two instructions manipulate the same data: one of the instructions writes the data to the memory while the other reads it from there. The data-passing mechanism passes this data directly from the first instruction to the second, saving a memory read cycle. Data passing is employed only for the cached areas of the memory.

The 6x86 processor uses a 256-position, four-way set associative BTB for predicting branches. A correctly predicted branch instruction takes one processor cycle to execute.

CYRIX 6X86 MX MICROPROCESSOR

The next development from Cyrix was a processor with a multimedia instruction set extension: 6x86MX, also known as M2 [76–78]. Cyrix proposed this processor as an inexpensive alternative to Pentium II. Like the Pentium Pro, M2 is optimized to work with 32-bit software; however, unlike the latter, 6x86MX could also run 16-bit software and was installed into a standard Socket 7 connector.

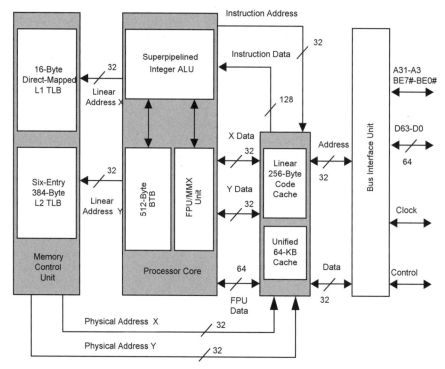

FIGURE 2.19 Cyrix 6x86MX structure block diagram.

The processor was manufactured using CMOS 0.35-micron and 0.25-micron technology with five metallization layers and operated from dual voltages: 2.9 V for the processor core and 3.3 V for the I/O subsystem. The processor could work at 66 MHz, 75 MHz, and 83 MHz system bus frequencies. Its Pentium-rate performance was from PR166 to PR433.

The microprocessor structure block diagram is shown in Figure 2.19.

The 6x86 microprocessor employs branch-predicting and speculative execution. The speculative execution can be up to four branches deep. Speculative instruction execution results are not cached or stored in the main memory until the correction of the predicted branch has been confirmed. Instructions continue to execute speculatively until one of the following events happens:

- The depth of speculative branching has exceeded four levels
- An exception arose or a wrong branch prediction discovered
- Overflowing of the speculatively executed instruction result buffer
- An attempt to modify resources not saved at a control point (segment registers or system flags, for example)

The main differences between the 6x86MX and 6x86 processors are shown in Table 2.15.

The 6x86MX microprocessor has two caches: a 64 KB four-entry multi-associative write-back unified cache and a 256 KB high-speed direct-mapped code cache. A part of the cache can be used as line-type main memory to support multimedia operations; in this case, it is not used for caching.

The 6x86MX has two TLBs: a smaller TLB1 and a larger TLB2. The direct-mapped TLB1 has 16 positions; the six-entry multi-associative TLB L2 has 384 positions.

TABLE 2.15 Differences between Cyrix 6x86 and 6x86MX

Parameter	Cyrix 6x86MX	Cyrix 6 86	
		6x86	**6x86L**
Power Supply Voltage			
Core	2.9 V	3.3 V or 3.52 V	2.8 V
I/O subsystem	3.3 V	3.3 V	3.3 V
L1 Cache	64 KB	16 KB	
TLB	L1: 16 bytes	L1: 128 bytes	
	L2: 384 bytes		
Branch Prediction	512-byte BTB	256-byte BTB	
	1024-byte BHT	512-byte BHT	
MMX	Yes	No	
Using L1 cache for line type main memory	Yes	No	
Frequency multiplier	2x; 2.5x; 3x; 3.5x	2x; 3x	

The FPU has a 64-bit interface, a 4-entry instruction input queue, and a 6-entry result write-back queue. The FPU has an MMX extension and can execute floating-point and MMX instructions simultaneously with integer instructions.

The 6x86MX has an effective power-management system for potential mobile applications.

Table 2.16 compares benchmark results, including for graphic and multimedia applications, for the Cyrix and Pentium processors.

TABLE 2.16 Cyrix and Pentium Processor Performance Comparisons

Microprocessor	ZD Winstone 97 for Windows 95	ZD Winstone 97 for Windows NT
PR233 Cyrix 6x86MX	49.4	63.5
Pentium II–233 MHz	49.9	65.7
PR200 Cyrix 6x86MX	45.9	59.1
Pentium MMX–200 MHz	43.3	57.7
PR166 Cyrix 6x86MX	43.8	56.2
Pentium MMX–166 MHz	40.8	54.6
PR200+ Cyrix 6x86	41.5	50.6
Pentium–200 MHz	40.5	51.0

IDT WINCHIP AND VIA CYRIX MICROPROCESSORS

At the end of the 1990s, VIA, a well-known chipset manufacturer, entered the microprocessor market. Antecedent to this move was the company's purchase of Cyrix and Centaur Technology, the latter a microprocessor division of IDT. IDT is known as the developer of the x86 architecture WinChip microprocessors. Consequently, new microprocessors from VIA combined the best features of the Cyrix and WinChip microprocessor architectures.

WINCHIP MICROPROCESSORS

A few words must be said about the IDT WinChip microprocessors, which appeared on the market in 1997 and were oriented at the budget PC sector. These microprocessors were functional analogs of Pentium MMX. Although they were not superachievers performance-wise, they nevertheless possessed several advantages, such as low power consumption, low price (almost half that of Pentium MMX at the same performance level), and connector (Socket 5) and voltage (3.3–3.5 V) compatibility with the old motherboards designed for Pentium processors. Consequently, switching to WinChip was an alternative to replacing the motherboard when upgrading from Pentium to Pentium MMX or from AMD K5 to K6.

The WinChip C6 is a rather simple microprocessor. It has only one execution pipeline, no branch-prediction capabilities, no speculative execution, or any other techniques aimed at raising performance. The WinChip C6 made the bet on the effective implementation of the most frequently executed instructions as well as on using a large cache (64 KB) integrated on-die and operating at the processor core frequency. The microprocessor supports MMX instructions; however, its FPU, unlike the ALU, is not pipelined, and its multimedia and floating-point instruction performance was quite low.

The WinChip C6 microprocessor was produced using the 0.35-micron technology with four metallization layers for operating frequencies from 180 MHz to 240 MHz. It could work with 60 MHz, 66 MHz, and 75 MHz buses, consumed about 10 W, and contained 5.4 million transistors on the 88 mm^2 crystal area.

WINCHIP2 MICROPROCESSOR

In the fall of 1998, IDT introduced another microprocessor, WinChip 2 [80], which deserves to be given a closer look. The main innovations of the microprocessors were a pipelined FPU, another MMX pipeline, and a 3DNow! unit.

The microprocessor dynamically translates x86 instructions into internal microcode. x86 instructions are fetched and decoded asynchronously with respect to the execution pipeline. The processor does not employ instruction reordering. If there is no data in the cache, the execution pipeline halts until it is swapped in from the main memory. One integer or floating-point instruction can be fetched and executed per cycle in the original program order. MMX and 3DNow! instructions can be fetched and executed simultaneously.

Although its architecture is quite simple, the microprocessor attains a high performance level by efficiently utilizing frequently used instructions, pipelined FPU and MMX units, a large cache, and a TLB integrated on-die.

The internal structure of the microprocessor is shown in Figure 2.20. The discussion of the main functional units and their operation follows.

The X86-to-Microcode Translator fetches up to 16 bytes from the I-Cache in one cycle, places them into the X86 Inst Buffer, and converts the x86 instructions into internal instructions. Two MMX or two 3DNow! instructions can be converted in one cycle. The stream of microinstructions (x86 Inst Parms) and immediate operands (X86 Immed Data) from the translator output queue (Queue) or internal instructions (Internal Insts) from the microinstruction memory (ROM) enter the execution block.

The branch-prediction unit, which also contains the return address stack (BR Prediction & Ret Stk), predicts branch directions with a 90 percent success rate. This information is used to control the fetch unit (X86 Fetch) when loading program code into the code cache.

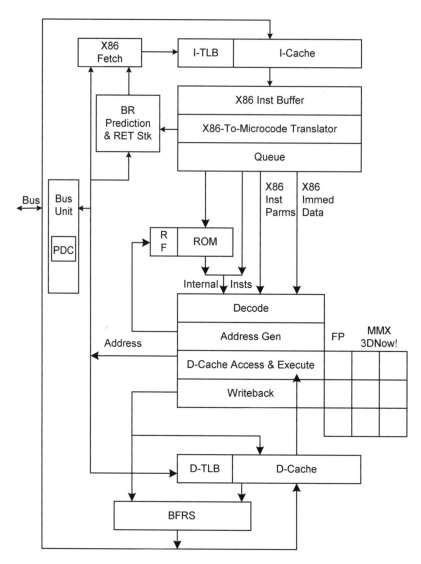

FIGURE 2.20 WinChip 2 microprocessor structure.

The microprocessor has split L1 data and code caches. The code cache (I-Cache) is 32 KB, two-way set associative, with the 32-byte line size. The data cache (D-Cache) is also 32 KB and four-way set associative. The processor has a provision for using external L2 cache from 512 KB to 2 MB in size.

Microinstructions are executed in the pipeline in four stages: decoding (Decode), address generation (Address Gen), fetching data from the cache and execution (D-Cache Access & Execute), and storing the result (Write-back). The processor can process multimedia and floating-point data simultaneously with integer calculations.

The 80-bit FPU is fully pipelined and can perform one addition or multiplication operation per cycle. Each of the two MMX units has an adder and a logic unit; additionally, one of them has an adder/multiplier whereas the other has a data shift and format conversion unit.

The MMX units share hardware resources with the 3DNow! units, whereas the FPU is independent. The FPU and MMX registers are architecturally identical; however, physically they are different and are clocked at the hardware level.

The microprocessor Bus Unit provides interface with the system bus operating at 60 MHz, 66 MHz, 75 MHz, 83 MHz, 95 MHz, and 100 MHz.

WinChip 2 was produced using the 0.25-micron technology with operating frequencies from 200 MHz to 300 MHz and the core voltage 3.3 V or 3.5 V.

Despite the substantial performance increase as compared to the previous IDT processor, especially concerning multimedia and floating-point data processing, WinChip 2 was not able to catch up with the leaders in the field: microprocessors from Intel. At the same time, this microprocessor surpassed Intel's closest rivals, AMD and Cyrix, in the performance for the dollar area. WinChip also was the only processor whose upgrading did not necessitate replacing the motherboard.

The next (and the last) independent design of Centaur Technology was the WinChip 3 microprocessor, whose main feature was the integrated 128 KB L2 cache. However, because the company was acquired by VIA, this microprocessor has never been put into production. Its next microprocessors Centaur Technology was developing as a part of VIA jointly with Cyrix designers.

VIA CYRIX III MICROPROCESSOR

The VIA Cyrix III [81] microprocessor was introduced in two versions: one with the Joshua core developed by Cyrix and the other with the Samuel core, previously known as the WinChip 4, developed by Centaur Technology.

Cyrix had been developing this microprocessor under the code names of Jedi and Gobi. After the company was acquired by National Semiconductor, it was renamed Jalapeno, and then Joshua when Cyrix was acquired by VIA.

The VIA Cyrix III Joshua was manufactured using the six-layer 0.18-micron technology. It has a 64 KB L1 cache and a 256 KB L2 cache, 2.2 V power supply voltage, system bus frequencies 66 MHz, 100 MHz, and 133 MHz, and the following bus frequency multipliers: 2.5, 3, 3.5, 4, 4.5, 5, 5.5, 6, 6.5, and 7. The processor was designed to be installed into the Socket 370 connector, thereby providing compatibility with the motherboards designed for Pentium III and Celeron.

Joshua uses two pipelines to process instructions. It implements register renaming to eliminate instruction data interdependencies and supports MMX and 3DNow! SIMD instruction sets. Its MMX and FPU have two pipelines each.

The VIA Cyrix III Joshua microprocessor was introduced in February 2000; however, because of its substantially lower-than-expected performance, it has never been put into serial production.

The VIA Cyrix III Samuel was announced at the beginning of June 2000. This microprocessor had the following characteristics:

- 0.18-micron technology with six aluminum interconnect layers
- 11.2 million transistors on 76 mm^2 crystal area
- Socket 370 connector
- 66/100/133 MHz system bus
- On-chip split L1 code and data caches, 64 KB each
- No L2 cache
- 1.9 V power supply voltage
- Power consumption under 10 W
- 3DNow! and MMX SIMD instruction set
- 500 MHz, 533 MHz, 600 MHz, and 667 MHz operating frequencies

The microprocessor has a 12-stage execution pipeline and four functional units: one ALU, two MMX/3DNow!, and one FPU.

The absence of L2 cache substantially affected processor's performance and the Cyrix III with the Samuel 2 core had 64 KB integrated L2 cache. Besides the added L2 cache, this processor also has a modified FPU and is manufactured using the 0.15-micron technology, one of the most advanced today. The crystal size has shrunk to 52 mm^2. The processor operating frequency ranges from 700 MHz to 900 MHz. A Samuel 2 core-based processor operating at 700 MHz with the core power supply voltage of 1.5 V consumes only 5 W. The low power consumption makes it possible to dispense with the processor cooler.

Benchmark results show that in both integer and floating-point calculations Cyrix III lags behind Celeron and Duron processors; moreover, the floating-point lag in quite substantial.

VIA's further plans are to produce Ezra processors based on the new Nehemiah 0.13-micron technology core. The processor has an integrated 256 KB L2 cache, operates at frequencies from 800 MHz to 1200 MHz with the system bus frequency of 266 MHz. The system bus operates in the DDR mode. In addition to the MMX and 3Dnow!, the processor will support SSE instruction set.

Striving to make computer systems based on its processors even less expensive, VIA has begun producing microprocessors with the integrated north bridge. This microprocessor is called Mark, is based on the Nehemiah core, and runs at 1 GHz.

2.2.12 Microprocessors from Transmeta

Transmeta presented its first microprocessors at the beginning of 2000. The microprocessor family, called Crusoe, has a distinctive architecture based on several innovations patented by Transmeta [82, 83]. The main architectural distinctive feature of these microprocessors is the dynamic conversion of the binary program code into internal VLIW core code performed by a software/hardware component called Code Morphing by the designers. The software part of the converter is loaded from flash ROM into the SDRAM when the processor is initialized and can be modified to support a new input instruction set or a new processor core.

The existing implementations of the microprocessor support x86 instruction set; however, there are no principle restrictions on adjusting the processor to work with other instruction sets.

The designers were able to efficiently emulate execution of x86 instructions in the VLIW core by saving the converted code of each instruction in the special cache to be used later. They also took into account the specifics of the executed programs when generating the VLIW instruction sequence: the code-morphing software analyzes the most frequently executed program fragments, branches, subroutine calls, and so on with the aim of implementing them more efficiently in the internal VLIW code.

The Crusoe's core comprises five different type units: two integer units, one FPU, one memory operation unit, and one branch unit.

Accordingly, every VLIW instructions (called a molecule in the Transmeta terminology and from 64 to 128 bits long) can consist of four RISC-like operations of these types (atoms in the Transmeta parlance). All atoms are executed in parallel, each by its corresponding unit (Figure 2.21).

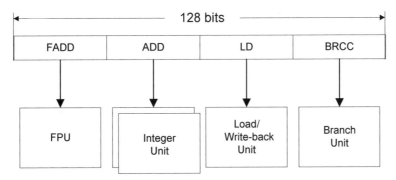

FIGURE 2.21 Executing a VLIW instruction in the Crusoe microprocessor.

Potential instruction-level parallelisms are detected by software during the code-conversion stage, not dynamically during the execution stage as most modern superscalar microprocessors do. This characteristic makes it possible to use fewer transistors, decrease the crystal size, and, consequently, lower the processor cost and power consumption. Other architectural and technical approaches are directed at lowering power consumption in Crusoe microprocessors, in particular the ACPI-compliant power management LongRun™ technology that allows the clock frequency and the core voltage to be changed during a program execution.

Low power consumption coupled with high performance has established Crusoe family microprocessors in the mobile device market. To date, TM3120 (later designated as TM3200), 5400, TM5600, TM5900, and other microprocessors belong to this family.

TM3120 MICROPROCESSOR

TM3120 is the entry-level processor of the family. It is produced for 333 MHz, 366 MHz, and 400 MHz clock frequencies.

TM3120 is a highly integrated device, intended for building on-chip systems. In addition to the processor proper, the crystal integrates an on-chip 96 KB split L1 cache (64 for code and 32 KB for data), a 66 MHz to 133 MHz SDRAM controller, a PCI 2.1 controller, and the controller for flash memory containing the code-morphing software.

The microprocessor has 1.5 V power supply and consumes only 0.015 W in the deep sleep state. Its power consumption in the active computing state is an order of magnitude lower than in other x86 architecture processors.

This microprocessor was manufactured using the 0.22-micron technology. At the time of introduction, it cost from $65 to $89, depending on the clock frequency.

TM5400 MICROPROCESSOR

The next member of the Crusoe family is TM5400. This processor has a higher performance level and operates at the frequencies ranging from 500 MHz to 700 MHz. It had a 256 KB L2 cache added, and the L1 data cache increased to 64 KB. The power management unit regulates the core voltage from 1.2 V to 1.6 V, depending on the intensity of the calculations. The processor power consumption is less than 1.8 W.

The internal structure of the processor is shown in the block diagram in Figure 2.22.

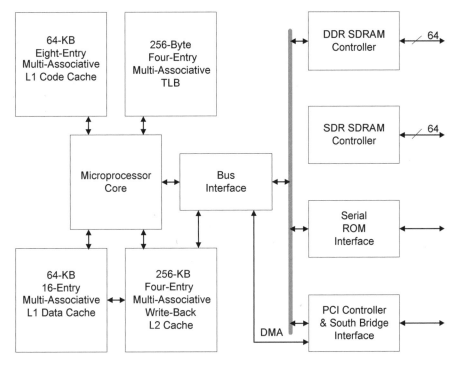

FIGURE 2.22 TM 5400 microprocessor structure.

An additional item of the system integration is the on-chip DDR SDRAM controller. The microprocessor is manufactured using the 0.18-micron technology and cost from \$119 to \$329 at the time it was introduced.

The performance of a 667 MHz TM5400 microprocessor is about that of a 500 MHz Pentium III.

FUTURE DEVELOPMENT PROSPECTS FOR CRUSOE MICROPROCESSORS

The development of the Crusoe family microprocessors is going in the direction of raising the clock frequency, lowering power consumption, enhancing the code-morphing software, and expanding the range of integrated devices. Thus, the TM5800 microprocessor has hardware support to implement cryptographic algorithms complying with the advanced encryption standard (AES).

In August 2003, first second-generation processor, TM8000 Astra, was renamed Efficeon, and the first batch of processors was produced in September 2003.

The TM8000 microprocessor has integrated RAM and AGP controllers, is equipped with a high-speed HyperTransport system bus and 4x AGP bus, and supports DDR400 memory. The length of the microprocessor core VLIW instructions has increased from 128 to 256 bits, which allows eight 32-bit instructions to be processed in one cycle.

Astra is built using the 0.13-micron technology and operates at frequencies over 1 GHz. Its mass production was to start in the third quarter of 2003.

2.3 ALPHA ARCHITECTURE MICROPROCESSORS

The Alpha microprocessors from DEC held the performance leadership for a number of years [84–89]. The first 64-bit microprocessors were developed within the framework of the Speed Daemon concept; afterwards, starting from the model 21264, the designers started to use approaches typical of the Brainiac concept.

2.3.1 Alpha 2106x Microprocessors

The Alpha 21064 microprocessors are a good example of the Speed Daemon concept: obtaining high performance levels by raising the clock frequency while keeping the functional logic relatively simple.

The architecture of the Alpha microprocessor was first introduced in February 1992 at an ISSCC conference, and already by February 1993, the first microprocessor, Alpha 21064, was produced. It was built using CMOS 0.75-micron process technology, had four metallization layers, and ran at 200 MHz. The processor had 1.68 million transistors; its crystal area was 238 mm^2.

Thanks to its high clock frequency and deep pipeline (up to 10 stages per operation), this microprocessor held the leading performance position for a long time. The main application area of the processor was high-end workstations and servers.

Next, Alpha processors were introduced in September 1993. These were the 166 MHz Alpha 21066 and the 66 MHz Alpha 21068. These processors were oriented at less-expensive systems: entry- and middle-level workstations. The core architecture of these processors is analogous to that of Alpha 21064. They both have DMA, PCI, and video controllers integrated on crystal. This made it possible to lower their cost and to make constructing systems based on them easier.

Introduction of the Alpha 21064A microprocessors with 225 MHz and 275 MHz operating frequencies strengthened the performance leader position of DEC microprocessors even more. The Alpha 21064 microprocessor has 64-bit superscalar architecture with two execution pipelines. The microprocessor executes up to two integer, floating-point, or branch instructions in one cycle. The branch prediction block has an 80 percent prediction success rate.

The microprocessor has separate L1 code and data caches, each 16 KB, integrated on crystal. It has 32 floating-point and 32 integer registers. The L2 is external, ranging in size from 128 KB to 16 MB. The system bus clock frequency can be from 2.5 to 10 times lower than the microprocessor operating frequency. The external address and data bus are 43 and 128 bits wide, respectively.

The Alpha 21064 processors did not utilize complex instruction-reordering and register-renaming logic. The responsibility for loading functional units efficiently was entrusted to the compiler.

Main characteristics of the Alpha 2106x microprocessors are listed in Table 2.17.

TABLE 2.17 Alpha 2106x Processor Main Characteristics

Model	Clock Frequency (MHz)	Technology (microns)	Power Consumption (Watts)	External Cache (MB)	Performance	
					SPECint92	SPECfp92
21064	200	0.75	27	up to 16	130	184
21064A	275	0.75	33	up to 16	190	290
21066	166	0.68	21	0.064—2	70	105
21066A	233	0.68	23	0.064—2	100	112
21068	66	0.68	9	0.064—2	30	50

2.3.2 Alpha 21164 Microprocessor

Next, the Alpha architecture microprocessor Alpha 21164 was introduced in September 1994, and had operating frequencies of 266 MHz and 300 MHz. This was the first microprocessor to have performance of over 1 billion instructions per second (1.2 BIPS). Using the then new 0.5-micron technology allowed 9.3 million transistors to be placed on the 298 mm^2 area crystal.

The microprocessor structure is shown in Figure 2.23.

The architecture of this processor is oriented at achieving the maximum high performance by increasing the operating clock frequency. Compared to the previous generations, the number or functional units in Alpha 21164 doubled while the pipeline depth decreased.

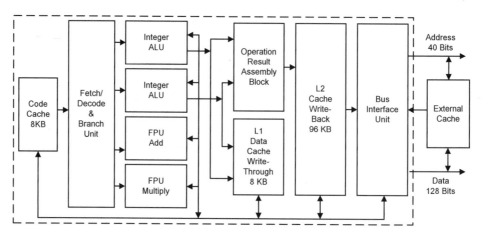

FIGURE 2.23 Alpha 21164 microprocessor structure.

Four instructions are fetched simultaneously from the 8 KB code cache and are placed into one of the two code buffers, each of which can store four instructions. Instructions are fetched from the buffers in the order determined by the software. One buffer must be emptied completely before the other one can be used. This limits the instruction fetch throughput, but makes the processor much easier to control.

When instruction operands are ready (are in the register file or are available in passing between an execution unit and the register file), the instruction is passed for execution into an appropriate functional unit. To assure proper execution of interrupts, the processor cannot change the instruction execution order: register values are modified in the order specified by the program.

The processor can execute up to four instructions in one cycle: two floating-point and two integer instructions. The two integer units are not identical. One of them can execute store, shift, and integer multiply instructions; the other, in addition to the general integer instructions, executes conditional branches. The FPUs also differ. The FPU+ executes pipelined additions, simple additions, divisions, and floating-point branches. The FPUx executes pipelined and simple multiply instructions. Alpha 21164 has 32 registers for floating-point and for integer operations each.

Like earlier Alpha microprocessors, Alpha 21164 makes intensive use of pipelining. The first four stages of instruction execution are common for all instructions and are executed in the instruction module. These are instruction prefetch, branching, decoding, and check/issue operations. ALUs add three more stages to the instruction processing: execution 1, execution 2, and write-back. FPUs add five more stages: floating-point register access, floating-point instruction executions 1, 2, and 3, and write-back.

Unlike the previous models, Alpha 21164 updates registers directly. No complex instruction-tracking mechanism that updates registers in the necessary order is needed. To avoid performance losses due to waiting for instructions to complete, Alpha 21164 employs roundabout ways that make operands available prior to instruction completion. This is similar to the look-ahead data-feed techniques employed in other microprocessors.

In addition to the split write-through L1 code and data caches, the crystal integrates a unified 96 KB set associative write-back L2 cache. The crystal also integrates an L3 cache controller, making both synchronous and asynchronous operating modes available.

A branch history table associated with the code cache is used to predict branches. Each cached instruction has an entry in the table containing its branch history. Table entries use two-bit counters. Predictions are made only one level deep. If a new branch instruction is encountered in the stream of the predicted instructions, it is not predicted, but the execution process is halted until the true branch direction of the previous branch instruction has been determined.

The L1 cache can support a certain number of unprocessed misses. The processor has a miss address file (MAF), each of whose six entries contains the address and register for loading in event of a miss. If the miss addresses belong to one cache line, it is considered as one element in the MAF. This makes it possible to accumulate a large number of misses in the MAF.

Using several levels of cache made it possible to increase the intensity of the instruction and data caching. The external 128-bit processor data bus can operate at the clock frequency 1 to 15 times lower than the core frequency.

In November 1995, a 333 MHz processor was introduced. In March 1996, the 336 MHz and 400 MHz Alpha 21164A processors, manufactured using the 0.35-micron technology, were introduced.

The further development of the microprocessor family is characterized by increasing the clock frequency (the 500 MHz version of Alpha 21164 produced in July 1996, the 600 MHz version in March of 1997), steep price reductions for the processors (almost by half), as well as a search for ways to adapt the Alpha architecture for use with the Windows NT systems.

The Alpha 21164PC microprocessor was introduced in March 1997. It was supposed to aid the penetration of DEC microprocessors into the desktop system market segment, thereby competing with the Intel microprocessors. The x86 instruction set compatibility is provided by the software binary code translator DIGITAL FX!32. The distinctive features of Alpha 21164PC are the increased to 16 KB L2 cache and the Motion Video Instruction unit, which provides real-time data coding/decoding in compliance with the MPEG-2 standard.

The Alpha 21164PC microprocessor was produced with 400 MHz, 466 MHz, and 533 MHz operating frequencies and has SPECint95 and PSECfp95 benchmark indexes of 11/13, 12/15, and 14/17, respectively.

At the end of 1997, a 600 MHz version was produced; its SPECint95 and SPECfp95 performance measures at 18.0 and 27.0, respectively.

2.3.3 Alpha 21264 Microprocessor

In May 1997, DEC introduced a new-generation microprocessor, Alpha 21264, whose performance substantially surpassed the previous family processors. At 500 MHz clock frequency, its SPECint95 and SPECfp95 performance benchmark indexes are 30 and 50, respectively. The microprocessor has 15.2 million transistors and was manufactured on a crystal 310 mm^2 in area using CMOS technology with six metallization layers.

Unlike the previous microprocessors of the family, along with the high clock frequency, a complex dynamic instruction execution mechanism was employed in Alpha 21264. It consisted of dynamic scheduling with instruction reordering, register renaming, and speculative instruction execution.

Instruction execution consists of the following sequence of stages:

■ Instruction fetch accounting for any possible predicted branch
■ Sending data for the instruction into the register-renaming unit
■ Renaming registers
■ Selecting instructions for execution from the queue
■ Executing integer or floating-point instructions
■ Write-back

Simultaneously, 80 instructions are examined for dynamic execution—more than in any other processor. After an instruction is decoded, it is placed into either the integer or floating-point unit queue. Instructions that have all their operands available compete for access to the execution devices. Instructions that have spent longer time in the queues have higher resource-access priorities. Up to six instructions can execute simultaneously.

The structure of the Alpha 21264 microprocessor is shown in Figure 2.24.

The processor has two FPUs performing add, multiply, divide, and square root operations; it also has two general-purpose ALUs and two address ALUs. The address ALUs, along with performing simple arithmetic and logic operations, execute all load and store instructions both for integer and floating-point data. The general-purpose ALUs do arithmetic and logic, shift, and branch operations. One of the ALUs also performs multiply operations whereas the other executes the multimedia instructions. Forty-one from the 80 integer registers and 41 out of the 72 floating-point registers are available for dynamic renaming.

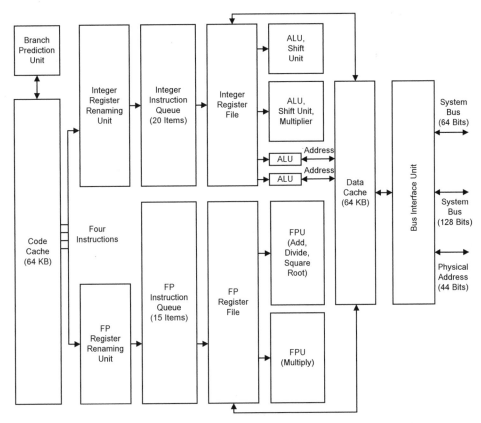

FIGURE 2.24 Alpha 21264 microprocessor structure.

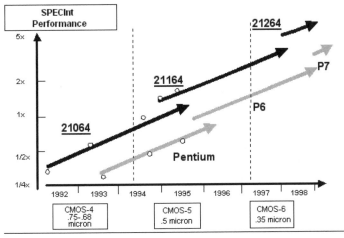

FIGURE 2.25 DEC and Intel microprocessor family performance comparisons.

The microprocessor has no L2 cache; however, its L1 code and data cache sizes are increased to 64 KB each. The crystal integrates the L2 cache controller and the bus interface unit; the former provides 5.3 GBps cache exchange bandwidth, the latter provides 2.6 GBps main memory exchange bandwidth.

DEC was planning to use the new microprocessor not only in UNIX systems but also in 64-bit Windows NT and Windows 95 systems (by using the DIGITAL FX!32 binary code translator).

Figure 2.25 shows the development of the DEC and Intel microprocessor families (Alpha and P5, P6, and P7, respectively).

2.3.4 Alpha 21364 Microprocessor

In 1998, DEC was acquired by Compaq, which in turn merged with Hewlett-Packard. Development of the Alpha microprocessor line was continued in the new settings. The new microprocessor of the family was introduced at the end of 2002.

The microprocessor has the same core as Alpha 21264 but also incorporates several significant additions. Unlike in the previous processor, the new crystal integrates a six-way set associative 1.75 MB L2 cache, an eight-channel Direct Rambus dynamic memory controller, and a network interface. The memory exchange bandwidth is 12.8 GBps. Data exchange between the L1 and L2 caches and the system memory is buffered (16 buffers for each hierarchy level).

A simplified block diagram of the microprocessor structure is shown in Figure 2.26.

Integrating system components on crystal allows systems based on this microprocessor to be made substantially simpler and less expensive. Linking microprocessors into high-performance multiprocessor systems is easier because of the integrated network interface. The network interface supports four point-to-point interprocessor links with 6.4 GBps bandwidth each. The exchange delay in links is 15 ns. The network interface ensures cache coherence in multiprocessor systems and implements asynchronous data exchange with adaptive routing.

An example of a multiprocessor system based on the Alpha 21364 microprocessor is shown in Figure 2.27.

The processor also has a fifth I/O port with 3 GBps bandwidth. This port can be used for connecting a switch [90].

The Alpha 21364 microprocessor was produced on 0.18-micron technology with six copper interconnect layers with a 1.2 GHz operating-clock frequency. Its energy consumption is 125 W. The crystal is 397 mm^2 in area and contains 152 million transistors (138 million out of which are used for cache). The microprocessor SPECint2000 and SPECfp2000 benchmark results are 804 and 1253, respectively.

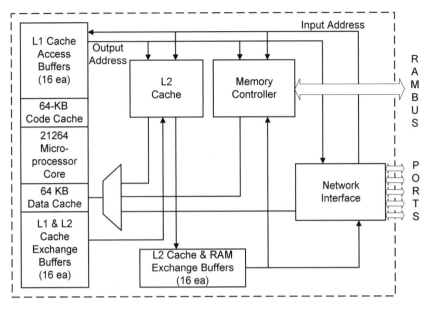

FIGURE 2.26 Alpha 21364 structure block diagram.

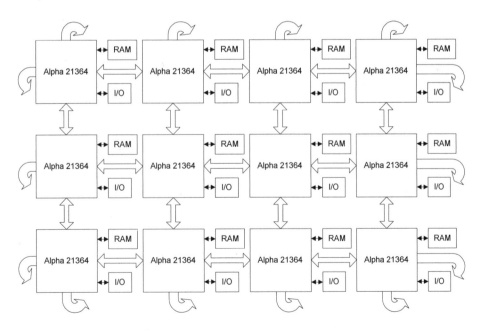

FIGURE 2.27 Block diagram of a multiprocessor system based on Alpha 21364.

Compaq planned to produce a new multithread microprocessor Alpha 21464 on 0.13-micron technology. However, the company subsequently switched to support the development of the IA-64 architecture and announced its plans to terminate the development and use of Alpha processors by 2004. All operations of the company's processor division were to be folded within two years. The key software tools and processors technologies will be turned over to Intel. The 64-bit Alpha (and MIPS) Compaq servers will be adapted to IA-64 microprocessors.

2.4 SPARC ARCHITECTURE MICROPROCESSORS

The SPARC architecture microprocessor family includes 32-bit MicroSPARC, SuperSPARC, HyperSPARC, and 64-bit UltraSPARC microprocessors. The main application areas of SPARC processors are the high-end workstations, servers, and supercomputers.

This section will consider the 32-bit SuperSPARC and the 64-bit UltraSPARC microprocessors.

2.4.1 SPARC Architecture

The SPARC architecture was developed by Sun Microsystems in 1985. It was based on the RISC I and RISC II research conducted at the University of California Berkley from 1980 through 1982. In particular, the concept of register windows proposed at Berkley was included into the SPARC architecture. The register window concept simplifies construction of single-pass compilers and significantly lowers number of memory-access instructions as compared to other implementations of the RISC architecture.

The main features of the SPARC architecture follow:

- Support of 32-bit linear address space
- Use of 32-bit fixed-structure instructions of three base formats
- Implementing memory and I/O access by using load/store instructions
- Small number of addressing methods; addresses are calculated as either *register+register* or *register+immediate operand*
- Use of three-address register instructions; most instructions operate on two operands placing the result into the receiving register
- A large register file with register windows; at any given moment, the program has available eight global integer registers and a 24-register register windows mapped onto the register file—using register windows makes it possible to significantly reduce the overhead associated with environment switching when executing parallel processes

- A separate floating-point register file; software can interpret this file as either a 32 single-precision (32-bit) register set, a 16 double-precision (64-bit) register set, an eight quad-precision (128-bit) register set, or a mix of different precision registers
- Postponed control hand-off; the processor always fetches the instruction following the postponed control hand-off instruction—this instruction can be executed or not, depending on the status of the canceling bit in the control hand-off instruction
- Fast interrupt handlers; interrupt generation causes a new register window to be created in the register file
- Tagged instructions; tagged arithmetic instructions interpret the two lower bits (tags) in operands as the information about the operand type—these instructions set the overflow bit in the status register at arithmetic overflow or when one of the operand tag bits is not a 0, and some instruction versions do not generate interrupts under these conditions
- Interprocessor synchronization instructions; one instruction executes a continuous `read` followed by `write` operation, while another instruction executes a continuous register/memory exchange operation
- Coprocessor support; the architecture defines a simple coprocessor instruction set, which can be used concurrently with the ALU
- Binary compatibility of user software for all SPARC implementations; software executed in the user mode must behave the same on all SPARC systems

A description of the main components of the SPARC v8 microprocessor architecture follows.

2.4.2 SPARC Processor

The SPARC processor [91] comprises the following functional units: Integer Unit (IU), Floating-Point Unit (FPU), and an optional Co-Processor (CP). Each unit has its own register set. Registers of all functional units have a fixed width of 32 bits. The exception from this rule is coprocessor registers; their width is determined by the specific implementation. In general, instruction operands are separate registers, register pairs, or register quads.

The processor can switch between two modes: user and privileged. In the privileged mode, the processor executes all instructions, including the privileged (those that can only be executed in this mode). In the user mode, an attempt to execute a privileged instruction causes an interrupt and control transfer to the special system subroutine. User application software is executed only in the user mode.

The IU has a general-purpose register set and a program counter (PC). The IU executes integer instructions, calculates memory addresses, and also controls the FPU and CP operation.

Depending on the specific hardware implementation, the IU can have from 40 to 520 32-bit general-purpose registers. The IU register file is broken into groups of eight global general-purpose registers plus a circular stack of from 2 to 32 sets of 16 registers each (register windows).

At any given time the application program has available eight global registers and a 24-register register window. The register window registers are broken into three groups: eight input, local, and output registers each. The output registers of one window overlap the input registers of the next register window. The current window is specified by the current window indicator field in the processor status word.

When accessing memory, the address is complemented with the address space identifier (ASI). This identifier contains information about the processor operation mode (user or privileged) as well as about what memory (instruction or data) is being referenced.

The SPARC architecture allows memory (and/or) cache to be broken into data and instruction type (Harvard architecture). If the program modifies instructions, they must issue FLUSH instructions for the addresses to which the modified instructions were written.

The FPU has 32 32-bit registers (*f*-registers) to store *f*-data. Double-precision *f*-data takes two registers, with an even and odd numbers, to store. Quad-precision *f*-data (128 bits) takes a set of four adjacent registers to store. Consequently, the most that the FPU register file can store is 32 single-precision, 16 double-precision, or eight quad-precision floating-point numbers.

Data between *f*-registers and the memory is exchanged using floating-point load/store instructions. Memory addresses are generated by the IU.

The SPARC instruction set allows connecting one coprocessor. The coprocessor has its own set of 32-bit registers, whose exact number is determined by the specific coprocessor implementation. Memory exchange is performed using coprocessor load/store instructions.

DATA FORMATS

The SPARC architecture supports the following three main data types:

- Signed integers: 8, 16, 32, and 64 bits
- Unsigned integers: 8, 16, 32, and 64 bits
- Real: 32, 64, and 128 bits

The widths of the data types follow:

- Byte: 8 bits
- Halfword: 16 bits
- Word (single word): 32 bits
- Tagged word: 32 bits (30 bits for the value and two bits for the tags)
- Doubleword: 64 bits
- Quadword: 128 bits

MEMORY

The SPARC architecture uses a memory model called total score order (TSO). This memory model is applicable to both one- and multi-processor shared memory systems. Using this model guarantees that all write, reset, and non-interrupted read/write instructions are processed in batches by the memory subsystem on any processor in the same order that they are completed by the processor. The architecture also supports another memory model, called the partial store order (PSO). In some cases, using this memory model can raise the system performance. The choice of the memory model is made by the designers of the specific system.

INPUT/OUTPUT

The SPARC architecture allows a whole range of I/O, memory, and cache control devices to be connected.

The SPARC architecture assumes I/O registers are accessed by alternate address space memory read/write instructions, regular read/write instructions, coprocessor instructions, or auxiliary status register read/write instructions. When alternate address space read/write instructions are used, I/O registers can be accessed only in the privileged mode. Accessing I/O registers with coprocessor instructions not in the non-privileged mode is determined by the specific implementation. The contents and addresses of the I/O registers are also determined by the specific implementation.

INTERRUPT PROCESSING

Exceptions caused by interrupt requests (external or software) create *traps*. A trap is a vectored transfer of control to the operating system through a special trap table, each entry of which contains the first four instructions of the specific trap event handler. The table's base address is set by software by writing to the trap base register (TBR) in the IU. The offset of each trap event is determined by the trap type number. A half of the table is reserved for hardware traps; the other half is allocated to the software traps generated by interrupt request instructions.

Prior to executing an instruction, the IU checks for the presence of exceptions and interrupt requests waiting to be serviced. If such exist, the IU selects that which has the highest priority and executes the corresponding trap.

A trap causes the current window pointer to switch to point to the next register window and the numbers of the current and the following instructions to be written by hardware to two registers of the new window. A trap handler can access the saved registers and six other local registers of the new window.

2.4.3 SPARC Architecture Implementation

Being the developer of the SPARC microprocessor architecture, Sun grants licenses to produce microprocessors of this architecture according to its specifications. Several companies produce microprocessors with this architecture, among them are Texas Instruments, Fujitsu, LSI Logic, Philips, Cypress Semiconductor, and others.

The first SPARC processor was manufactured by Fujitsu in 1986. This processor had 16.67 MHz operating frequency and was used to build the first workstation, Sun-4, in 1987. The workstation's performance was 10 MIPS.

Various manufacturers of the following implementations of the SPARC architecture widely used principles of superscalar processing, reduced the crystal-process engineering requirements, increased the operating clock frequency, and so on. All of this made it possible to substantially raise the processor performance.

The most common implementation of the 32-bit SPARC architecture was the SuperSPARC microprocessor from Texas Instruments. It became the foundation for a series of SPARCstation/SPARCserver 10 and 20 workstations and servers. The structure of the SuperSPARC microprocessor is shown in Figure 2.28.

The superscalar microprocessor had separate integer and floating-point arithmetic pipelines, had integrated split L1 cache (20 KB code and 16 KB data), executed up to three instructions per cycle, and had instruction bandwidth of 205 MIPS at 76 MHz.

2.4.4 UltraSPARC Architecture (V9)

The new generation of microprocessors significantly enhanced graphic- and video-processing capabilities of SPARC systems. UltraSPARC is one of the first general-purpose processors that implemented these functions by hardware. It has a special module for processing video data and graphics in RGB or alpha format. The video-processing functions can work with eight image elements simultaneously. Images are processed in the FPU, which usually processes 64-bit floating-point data.

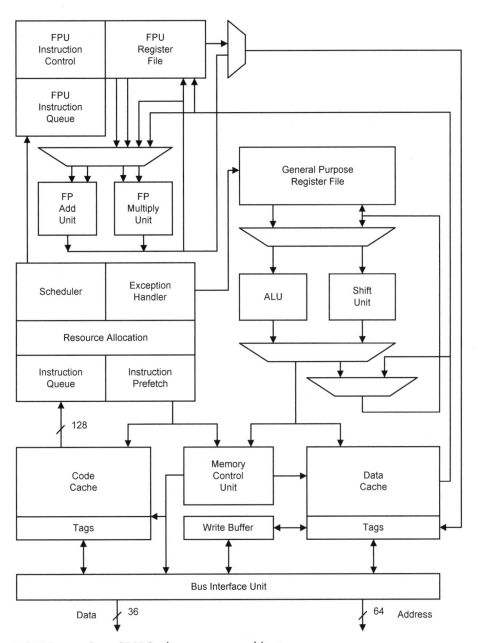

FIGURE 2.28 SuperSPARC microprocessor architecture.

The microprocessor's instruction set includes the visual instruction set (VIS), which allows data to be loaded and processed in 64-bit blocks. The VIS has 30 instructions, allowing it to efficiently process multimedia, graphic, image, and other integer-processing-oriented algorithms. The VIS contains add, multiply, and subtract instructions that allow up to eight integer operations on bytes or halfwords to be carried in one cycle.

In MPEG codec algorithms, analyzing movements and comparing each portion of the current frame with the previous takes the bigger portion of the algorithm's time. The special instructions in UltraSPARC perform these actions as one graphic operation. The special memory subsystem automatically loads image elements into 8-byte blocks. No separate instruction is needed to do this. Because these special instructions are pipelined, the microprocessor can execute one instruction per cycle. Using these specialized instructions increases image-processing speed by 80 times as compared to other SPARC processors.

UltraSPARC also has several hardware memory-operation enhancing means. The biggest change is a new instruction to move a block of data over the processor/main memory bus with the speed of 600 MBps. This allows the system's main processor to work as a video processor exchanging data with the video memory. The block move operations also provide performance gains in other applications, such as network software, for example.

The following 64-bit SPARC architecture microprocessors have been produced: UltraSPARC, UltraSPARC II, and UltraSPARC III.

The structure of the UltraSPARC microprocessor is shown in Figure 2.29.

UltraSPARC is equipped with split L1 code and data caches, 16 KB each. Each cache has its own TLB. UltraSPARC integrates an L2-cache controller. Instruction fetch is closely integrated with the L1 code cache. Instructions are stored in the cache and are pre-decoded to speed up the processing. Every two instructions in the cache are associated with two bits that are used to predict branches. The two bits make it possible to trace four different states that encode the last two branches executed by these instructions. The prefetch mechanism uses these bits for dynamic branch prediction. UltraSPARC branch-prediction success rate is 88 percent in the SPECint92 and 94 percent in the SPECfp92 benchmarks.

The execution pipeline of UltraSPARC has nine stages and can execute up to four instructions per cycle. The first two stages are standard: instruction fetch and decode. In the third stage, all instructions that can be handed over to the execution unit are grouped together.

The microprocessor does not reorder instructions. The functional-unit load scheduling is done statically in the compilation stage. Two integer instructions, two floating-point or graphic instructions, and one load/store or one branch instruction can be issued for execution in each cycle. Thus, although potentially six instructions can be executed simultaneously, only four can actually be executed. In the same stage, information is received from the registers.

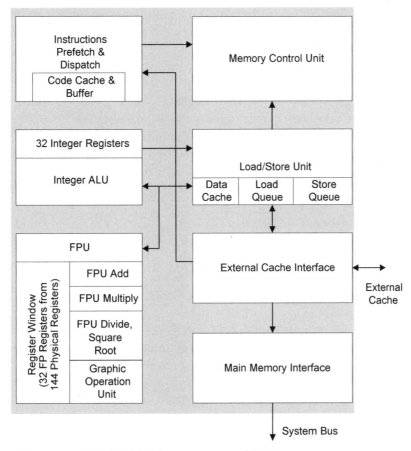

FIGURE 2.29 UltraSPARC microprocessor architecture.

Floating-point instructions are passed into the three-stage pipeline, which does all the processing with the exception of *f*-divide and *f*-square root. These functions are performed by a separate unit. Although the microprocessor issues instructions sequentially, the results are not necessarily presented in the same order.

Basic integer instructions are executed in one cycle. Others, such as integer multiply and divide, can take a varied number of cycles to execute. The remaining stages of the integer pipeline perform load/store operations.

UltraSPARC uses bus hierarchy to provide high data throughput. The 128-bit memory bus operates at the processor core frequency. The SBus is used to conduct exchange with peripheral devices. The interface with this bus is implemented on the hardware level using a bus switch microchip that is a part of the processor chipset. The bus switch chip makes it possible to isolate the memory bus from the I/O bus and to perform memory read operations concurrently with I/O to peripheral devices. As a result, a high bus use rate is obtained producing a 1.3. Gb/s throughput.

The performance of UltraSPARC running at 167 MHz is 5.6 and 9.1 in SPECint95 and SPECfp95, respectively.

Like SuperSPARC, UltraSPARC employs the register window concept. Instead of 32 base registers, these microprocessors make eight overlapping register windows, 24 registers each, available to applications. The current window does not need to be saved when a new procedure or a processing branch is initiated; the new process will use a new register window. An additional mechanism is added to UltraSPARC II that makes a new 8-register register window available at each interrupt.

The performance of a 250 MHz UltraSPARC II in SPECint95 and SPECfp95 is 8.5 and 15, respectively.

UltraSPARC III [93] is a member of the third generation of V9 architecture microprocessors from Sun Microsystems. This microprocessor is intended for use in a wide range of computer systems, from workstations to high-end servers and supercomputers.

Unlike microprocessors of the previous SPARC family generations, which require their architectural specifics taken into account on the operating system level, this microprocessor is compatible with all operating systems and applications developed for SPARC processors.

Like in the previous generations, UltraSPARC III employs branch prediction, static ILP detection at the compilation stage, and register windows. Instructions are executed in a 14-stage pipeline.

Architectural and work registers are decoupled in the microprocessor. Out-of-order instruction execution results are stored in work registers and, if necessary, can be either cancelled (restored from the architectural registers) or stored (written into the architectural registers).

The architecture of the UltraSPARC III microprocessor is shown in Figure 2.30.

The microprocessor has six main functional units.

The Instruction Issue Unit (IIU) predicts branches and selects instructions for execution, taking into account the predicted path. Selected instructions are placed into the queue to two functional units: the ALU and FPU. The IIU is equipped with a four-way set associative 32 KB code cache, a TLB, and a 16 KB branch prediction table.

The Integer Executive Unit performs all integer-processing operations: data load and store, arithmetic and logic operations, and shift and branch operations. The four independent data supply paths allow up to four integer operations to be executed in one cycle.

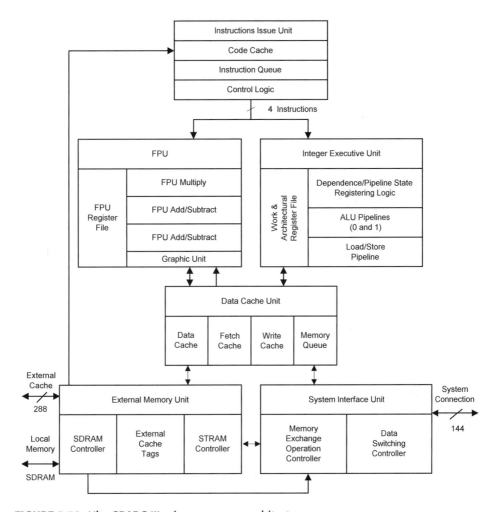

FIGURE 2.30 UltraSPARC III microprocessor architecture.

The FPU processes all floating-point instructions and some integer instructions (graphic instructions from the VIS extension of the SPARC instruction set). The FPU can execute up to three instructions simultaneously.

Together, the IEU and FPU can simultaneously execute up to six instructions per cycle.

The Data Cache Unit (DCU) has a data address translation buffer and three types of cache: a 64 KB L1 data cache, a 2 KB data prefetch cache, and a 2 KB data write cache. Requests for writing data to the main memory are first placed into the DCU queue and then into the data write cache.

The external memory unit controls the external L2 data cache and the system SDRAM. The L2 cache controller has 90 KB memory for the external cache tags. The main memory controller supports up to four memory banks for a total of up to 4 GB.

The system interface unit provides an interface with the external memory, I/O devices, and other processors. The unit can process out of order up 15 transactions awaiting execution. UltraSPARC III provides for memory coherence hardware support when the processor is used in multiprocessor systems.

Due to the use of hierarchic cache and the wide, 128-bit, memory interface, the processor/memory exchange speed is 2.4 GBps.

UltraSPARC III is produced for operating clock frequencies ranging from 600 MHz to 1 GHz. The performance of the 600 MHz microprocessor is 35 SPECint95 and 60 SPECfp95. UltraSPARC III is produced on CMOS 0.25-micron technology with six metal interconnect layers; it has 360 mm^2 crystal area on which 16 million transistors are located (12 million logic, and 4 million memory). In the future, Sun is planning to switch to producing the microprocessor on 0.18-micron technology.

2.5 MAJC ARCHITECTURE MICROPROCESSORS

Microprocessors with MAJC architecture [84, 95], which was developed by Sun, are intended for use in the multimedia data-processing systems and Internet applications.

Sun places big hopes on this architecture and believes that it will determine the direction of the microprocessor development for the nearest 20 years, as the SPARC architecture did in its time.

MAJC 5200, the first processor with this architecture, was introduced in the middle of 2000. The processor consists of two 128-bit VLIW microprocessor cores integrated on one crystal.

The microprocessor implements four levels of parallelism:

- Multiprocessor crystal structure
- Multithread program execution
- Instruction-level VLIW parallelism
- SIMD processing

The instruction set of these RISC processors is oriented toward stream multimedia information processing and hardware support of high-level Java constructs. The name MAJC is an acronym for microprocessor architecture for Java computing.

The MAJC architecture is scalable. One MAJC crystal can contain several identical processors, the exact number depending on the specific implementation. Each of these processors can contain from one to four functional units: RISC processors and code caches. Each of the crystal's processors can execute up to four instructions per cycle.

The structure of the MAJC 5200 microprocessor is shown in Figure 2.31.

In addition to the functional units (FU) and the 16 KB code caches, the crystal contains shared by all RISC processors 16 KB data cache, a Rambus memory controller, a graphic pre-processor, 64-bit high-speed interfaces for interprocessor and graphic subsystem connections, and a switch that interconnects the RISC processors and the other crystal components.

A MAJC instruction word is called a *MAJC packet* and comprises from one to four 32-bit instructions, each of which is intended for one of the four RISC processors. In the case when all four processors cannot be used simultaneously, individual instructions may be missing; therefore, a MAJC packet can be of a variable length. The first two bits of the packet are used to indicate the number of instructions in it: 00—one instruction, 01—two instructions, 10—three instructions, and 11—four instructions. Instructions are assigned to the functional units by their numbers.

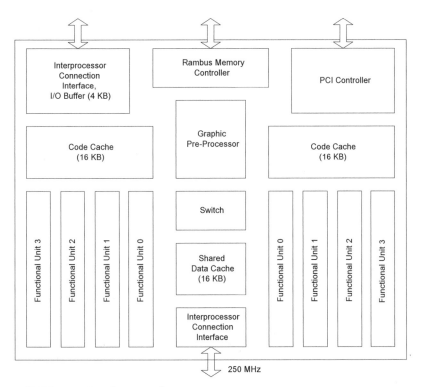

FIGURE 2.31 MAJC 5200 microprocessor structure.

Number 0 functional unit of the RISC processor executes different types of instructions than the other three units do. In addition to arithmetic-logic instructions, it executes load/store and branch instructions. Arithmetic-logic instructions are executed on integers (8-, 16-, 32-, and 64-bit long), 16-bit fixed-point numbers, 32- and 64-bit floating-point numbers, as well as on bit sequences. There also are SIMD instructions to process pairs of 16-bit fixed-point data.

Each RISC processor has a local register file, instruction-decoding control logic, and status registers. There also are global registers available to all RISC processors.

The hardware-implemented multithread program execution support allows up to four instruction streams to be executed concurrently.

For integer operations, the performance of a 500 MHz MAJC microprocessor is 7 BIPS for 32-bit data and 13 BIPS for 16-bit data. For floating-point operations, the performance is 1.5 BIPS for single-precision data and 6.16 BIPS for double-precision data.

MAJC-5200+ is the next modification of the processor. The main differences between these microprocessors are listed in Table 2.18.

TABLE 2.18 Technological Characteristics of MAJC Microprocessors

Microprocessor	*MAJC-5200*	*MAJC-5200+*	
Manufacturing Technology	0.22-micron, six copper interconnect layers	0.18-micron, seven copper interconnect layers	
Clock Frequency	500 MHz	500 MHz	700 MHz
Consumed Power	15 W	10 W	15 W
Power Supply Voltage	1.8 V	1.5 V	1.5 V
Crystal Area	220 mm^2	130 mm^2	130 mm^2

2.6 PA ARCHITECTURE PROCESSORS

Hewlett-Packard (HP) was one of the first companies to enter the RISC microprocessor market. Its 32-bit Precision Architecture PA-RISC microprocessor was introduced back in 1986. Steadily developing the RISC architecture principles, HP introduced the PA-8000 microprocessor in 1996. This microprocessor implemented fully the main dynamic instruction execution principles (intellectual execution in the HP parlance) [96–99].

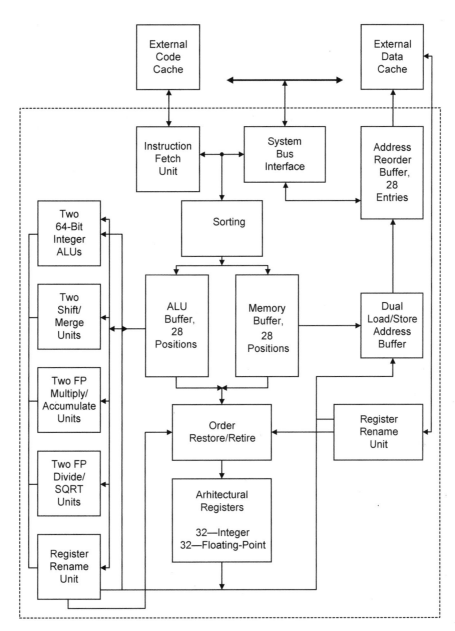

FIGURE 2.32 PA-8000 microprocessor structure.

The processor was manufactured on CMOS 0.5-micron technology; its performance is 11.8 SPECint95 and 20.2 SPECfp96 at 180 MHz clock frequency. The structure of the PA-8000 microprocessor is shown in Figure 2.32.

PA-8000 has 64-bit four-pipeline architecture supporting instruction reordering. The processor has 10 functional units: two integer ALUs, two integer shift/merge units, two floating-point multiplication/accumulation units, two division/square root units, and two load/store units. The multiplication/accumulation units are pipelined and process single-precision data in three cycles. Division operation is not pipelined and takes 17 cycles to perform.

PA-8000 uses instruction reordering buffer (IRB). The buffer examines the next 56 instructions in the instruction stream and determines which of them can be executed in parallel. Actually, the IRB is made up of two parts of 28 instructions each. The ALU part of the buffer holds integer arithmetic instructions; the other part is used for floating-point and load/store instructions.

An instruction placed into the IRB waits for the data—results from the previous instructions—to become available. It is selected for execution as soon as all the necessary data has been received and the needed functional unit is available. Each of the IRB units allows scheduling for execution of two instructions per cycle, for a total of four instructions per cycle. Register renaming is used to eliminate instruction data interdependencies.

The microprocessor implements a branch-prediction mechanism based on the majority history evaluation principle for each branch point. A BHT is used to predict branches. It has 256 three-bit entries, one for each branch point. The branch prediction success rate is 0.8.

HP designed PA-8000 especially for scientific and engineering calculations on potentially great volumes of intensively used data. This necessitates large code and data caches. PA-8000 uses main external types of data and code caches up to 4 MB in size. The address reordering buffer (ARB) that monitors all load/store instructions makes it possible to reduce the delays associated with addressing the external cache. Large latency of the external data cache is compensated by its large size and the efficient cache-exchange control features, such as high-speed cache control lines and instruction and code prefetch from the main memory to cache.

The development of the next microprocessor of the family, PA-8200, pursued the goal of raising the performance while making minimum design changes. The main enhancements follow:

- The increased BHT size (from 256 to 1,024 entries) and taking into account presence of several branch directions in one cycle allowed the branch prediction success rate to be raised
- Increasing the TLB size from 96 to 120 entries made it possible to lower cache miss rate
- The processor operating clock frequency was raised to 220 MHz
- The maximum size of code and data caches was increased to 2 MB
- Use of CMOS 0.35-micron process technology

All introduced changes did not affect the electrical and binary compatibility with the previous microprocessors of the family. The processor performance is 15.5 SPECint95 and 25 SPECfp95, which is 35 percent better than the PA-8000 performance.

The shortcoming of the system's increased cost due to the use of external cache was overcome by HP in the new PA-8500 microprocessor, which was manufactured on 0.25-micron technology. High component density made it possible to integrate a 1 MB data and a 0.5 MB code caches on crystal. PA-8500 has 140 million transistors; its performance measures 26.0 SPECint95 and 28.1 SPECfp95 running at 360 MHz. 360 MHz, 440 MHz, and 500 MHz versions of the processor were also produced.

HP introduced the next microprocessor of the family, PA-8600, at the end of 1999. This microprocessor is a modernized version of PA-8500. The microprocessor's cache operations are enhanced by using the least recently used eviction algorithm. The cache is equipped with an error-detection and correction mechanism to enhance its reliability. The branch-prediction algorithm was also improved. The superscalar FPU produces up to four results per cycle, giving 2.2 GFLOPS at 550 MHz.

In March 2000, the next microprocessor of the family, PA-8700, was introduced. The microprocessor is manufactured on SOI 0.18-micron technology with seven copper interconnect layers, able to integrate 1.5 MB L1 data cache and 750 KB code cache on crystal and to raise the operating clock frequency to 800 MHz and higher. The processor's instruction throughput is 3.2 BIPS.

PA-8700 has 10 functional units. These are two load/store units, two ALUs, two shift units, two FPUs, two multiply/divide units, and two square root extraction units.

The use of 44-bit addressing instead of the 40-bit, as in the previous models, allows the processor to access up to 16 TB of physical memory.

The latest development to data from HP is the PA-8800 microprocessor, code-named Mako. It is a single-crystal dual-processor system based on a modified PA-8700 core. The processor crystal is packed in a cartridge together with a 32 MB L2 cache.

The crystal area is 366 mm^2 and houses 300 million transistors, 25 million of which are used for logic and the remaining implement cache.

Mako's system bus is the same as in Intel's McKinley: 128-bit wide with 6.4 GBps bandwidth at 400 MHz. The processor is produced on 0.13-micron SOI technology with low-k dielectrics and eight copper interconnect layers.

The processor's performance at 1 GHz operating frequency is 900 SPECint2000 and 1,000 SPECfp2000.

HP's plans are to produce another microprocessor, PA-8900, for 1.2 GHz to 1.3 GHz operating frequency and to cooperate with Intel in developing IA-64 architecture microprocessors.

2.7 POWER AND POWERPC ARCHITECTURE MICROPROCESSORS

IBM has been developing the Power architecture since the end of the 1980s. It is based on the principles of the RISC architecture: fixed instruction format, register operations, one-cycle instruction execution, simple addressing techniques, and a large register file. At the same time, the Power architecture has several significant features that make it different from the architectures of other RISC processors. Some of these features follow: an independent register set for each functional unit; inclusion of CISC-like instructions into the register set (a block load/store, bit field manipulation, combined floating-point multiply/divide instructions, for example); absence of the postponed branch mechanism (that is, an advance execution of the instruction following the conditional branch instruction); and an original method of implementing conditional branches (the operation code of each instruction has a conditional execution bit and there are several condition registers).

The main application area of the Power architecture microprocessors is the high-end servers and supercomputers. The desire of the developers to use the architecture they developed in lower-performance, less-expensive systems has led to creation of a version of the Power architecture for personal computers and entry-level work stations. This architecture was named PowerPC [100].

The first microprocessor with this architecture, PowerPC 601, appeared as a product of joint efforts of three companies: the creator of the Power architecture—IBM, one of the personal-computer developer leaders—Apple, and the manufacturer of computers for Apple—Motorola.

Currently, the PowerPC architecture is employed in processors from IBM and Motorola, which are used in controllers, telecommunications equipment, personal computers, servers, and workstations. At the same time, IBM independently continues to develop Power architecture microprocessors for use in high-end multiprocessor systems.

The latest designs from Motorola and IBM combine new architectural approaches with progressive microchip-production technologies, such as copper interconnect and SOI, that make it possible to reduce the crystal size, lower power consumption, and raise the microprocessor operating clock frequency. PowerPC, G3, G4, G5, PowerPC 970, Power 3, and Power 4 are the most interesting recent designs. Introduced by Motorola multimedia data-processing technology, AltiVec also deserves a closer look.

2.7.1 PowerPC 620 Microprocessor

The first 64-bit implementation of the PowerPC architecture, the PowerPC 620 microprocessor, was announced in 1995 [101]. If the previous versions of PowerPC microprocessors were oriented toward personal computers, PowerPC 620 was intended for workstations and high-performance servers.

The processor was manufactured on CMOS 0.5-micron technology with four metal interconnect layers. The crystal has a 311-mm^2 area, on which 7 million transistors are located. PowerPC has 3.3 V power supply and consumes under 30 W power.

The microprocessor's structure is shown in Figure 2.33.

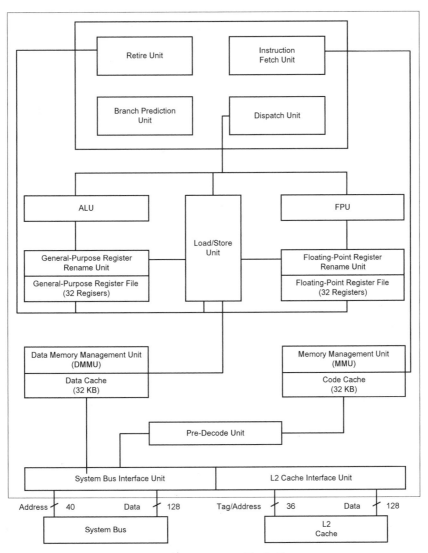

FIGURE 2.33 PowerPC 6210 microprocessor block diagram.

PowerPC 620 has a four-pipeline superscalar architecture with six functional units: three integer ALUs, an FPU, a load/store unit, and a branch unit. The processor can execute up to four instructions per cycle.

To provide efficient loading of the functional units, the processor employs dynamic branch prediction along with speculative instruction execution up to four predicted branches deep. To predict branches, a BHT is used in which execution results of each branch instruction are registered. Branch instructions and addresses are cached in the branch-target address cache (BTAC). The processor's branch prediction success rate is 90 percent.

PowerPC 620 utilizes Harvard architecture with split code and data caches, 32 KB each. Each cache has its own Memory Management Unit (MMU) and functions independently of the other.

A control circuit for the external L2 cache is integrated on crystal. The external cache can be up to 128 MB and work at the processor frequency, or one-half or one-quarter of this frequency, which makes for flexible system memory configuration choice. Data cache supports write-through and write-back modes and the multiprocessor system cache coherence protocol MESI.

Before they are placed into the internal cache, instructions are pre-decoded to take into account their interdependencies during scheduling. Pre-decoded instructions are stored in the code cache until they are fetched by the scheduler/execution unit. The final instruction decoding is combined with the microprocessor pipeline-load-scheduling stage, which allows the number of pipeline stages to be reduced to five (fetch, decode/schedule, execution, completion, and write-back).

The processor is binary compatible with the earlier versions of PowerPC, allowing it to execute 32-bit PowerPC software along with the new specially designed 64-bit software. The processor can work in either the 32- or 64-bit mode: switching between the modes is software-controlled. A unique feature of the PowerPC microprocessors is the software-controlled switching between the Intel/Motorola addressing modes. This allows a workstation based on PowerPC 620 to execute applications for different operating systems and do this at high performance levels.

Running at 133 MHz, the microprocessor's performance is 225 SPECint92 and 300 SPECfp92.

2.7.2 AltiVec Technology

Currently, many manufacturers expand functional capabilities of their microprocessors by adding specialized units to them, such as, for example, multimedia application accelerators. Processors from Intel, AMD, Sun, and DEC have such units (MMX and SSE, 3DNow!, VIS, and MVI, respectively).

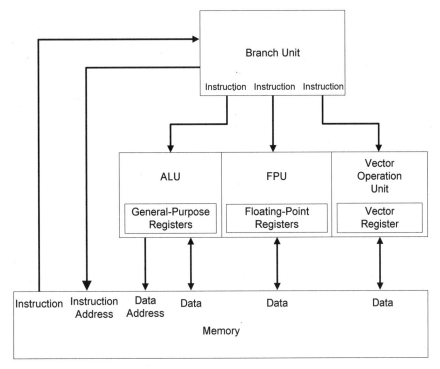

FIGURE 2.34 PowerPC with the vector-processing unit block diagram.

Motorola introduced the AltiVec technology [23, 102] to speed up multimedia data processing in PowerPC microprocessors. The technology envisions extending the PowerPC architecture by adding a 128-bit vector-processing unit functioning in parallel with the already existing ALU and FPU.

The structure of PowerPC with the vector-processing unit is shown in Figure 2.34.

The AltiVec technology allows parallel SIMD processing of vectors 4-, 8-, or 16-elements long.

The vector-processing unit can process up to 16 operations per cycle. With the total vector unit width of 128 bits, the following can be processed in one cycle:

- 16 8-bit signed or unsigned integers or symbols
- Eight 16-bit signed or unsigned integers
- Four 32-bit signed or unsigned integer or floating-point numbers

The AltiVec technology allocates a dedicated register file to the vector-processing unit. It has 32 128-bit registers. These registers are used to store the source data and their execution results in the vector unit. These registers are loaded and their contents are stored to the memory by special vector instructions.

AltiVec vector unit instructions can have up to three source and one destination operands. Each operand is one of the 128-bit vector unit registers. There also are instructions of the *memory/register* and *register/immediate operand* type. The total number of vector unit instructions is 162. The instructions are divided into several types.

LOCAL OPERATION INSTRUCTIONS

Local operations are executed in parallel and independently on the data in the source registers with the results placed into the corresponding fields of the destination register. Figure 2.35 shows an example of a local operation. There are three source operands, vA, vB, and vC, and the destination operand vT. Sixteen operations are performed on 8-bit vectors.

Instructions on signed and unsigned numbers for corresponding units are defined. These are addition, subtraction, multiplication, multiplication with accumulation, min, max, average, and converting between integer and floating-point number formats. In addition, logic instructions AND, OR, NOT, XOR, NAND, compare instructions, shift, and right and left cyclic shift instructions are performed. There also is a multiplexing instruction that sends data from one of the two source registers to the destination register. A sequence of compare and select instructions makes it possible, among other things, to efficiently mask and replace data in 128-bit fields.

GLOBAL OPERATION INSTRUCTIONS

Global operations make it possible to sum products of components of several vectors, as well as to sum components of one vector (Figure 2.36). In particular, these operations can be used to calculate scalar product of vectors.

Moreover, global operations can be used to shift vectors, do packing and unpacking, merging, and other operations characteristic of signal and image compressing and converting.

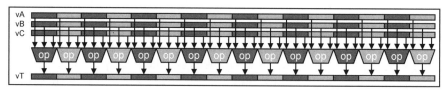

FIGURE 2.35 Executing local operations.

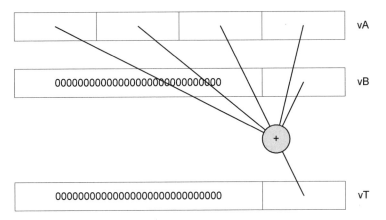

FIGURE 2.36 Executing global operations.

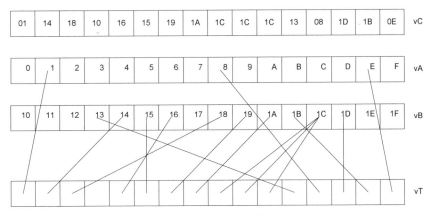

FIGURE 2.37 Executing a transposition instruction.

There also is a transposition instruction that allows selection of bytes from two 16-bit source registers into the 16-byte destination register, as is shown in Figure 2.37.

Such transpositions used to pack/unpack 8- and 16-bit data elements can produce significant time gains when storing packed elements in the memory, unpacking them before calculations, and packing them back after the calculations.

The global instructions can also be used to work with 128-bit data.

2.7.3 PowerPC 750/740 (G3) Microprocessors

The 32-bit PowerPC 750 and PowerPC 740 [103] microprocessors are produced by IBM and Motorola for use in mobile and desktop systems (PowerPC 750) and are imbedded in communications equipment (PowerPC 740).

Both microprocessors are binary and compatible pin-wise with PowerPC 603 and PowerPC 604. The main difference between PowerPC 750 and PowerPC 740 is that the former has a built-in interface and integrated L2 tag cache. L1 cache is split, eight-way set associative, and 32 KB code and data each.

The microprocessors can execute up to three instructions per cycle in six independent functional units: two integer units, two FPUs, a branch unit, a load/store unit, and a system register unit.

A great deal of attention in the microprocessors is given to lowering power consumption. Functional units have separate voltage control. Microprocessors are produced on CMOS 7S 0.22-micron technology.

PowerPC 750 is produced for 200 MHz, 233 MHz, 266 MHz, 276 MHz, 333 MHz, 350 MHz, 366 MHz, and 400 MHz; PowerPC 750 is produced for 200 MHz, 233 MHz, 266 MHz, and 300 MHz. The size of the external L2 cache can be 256 KB, 512 KB, or 1 MB. The external cache operates at half the processor clock frequency.

PowerPC 750 has 6.35 million transistors and consumes 4.1 W running at 400 MHz. Its performance at this frequency is 18.8 SPECint95 and 12.2 SPECfp95. The performance of PowerPC 740 operating at 300 MHz is 13.5 SPECint95 and 9.0 SPECfp95.

2.7.4 PowerPC G4 Microprocessor

The PowerPC microprocessor line continued with the G4 chip [104]. This microprocessor uses base architecture of G3 and contains several important additions.

G4 is the first microprocessor that uses the AltiVec technology proposed by Motorola, which makes it possible to speed up graphic processing and to substantially raise processor performance in digital-signal processing tasks.

Another addition is the capability to cluster up to four microprocessors, with each of them having access to the cache of any other.

The microprocessor can work with L2 cache size ranging from 512 KB to 2 MB. The crystal area is 82 mm^2 with 10.5 million transistors. At 1.8 V power supply voltage G4 consumes 8 W.

The first microprocessors were produced on 0.2-micron technology with copper interconnect layers for 400 MHz. With the switch to 0.18-micron technology, the clock frequency was raised to 1 GHz.

At 400 MHz, G4's performance is 21.4 SPECint95 and 19.5 SPECfp95.

The microprocessor is not compatible with the previous generation in the pinout and power supply voltage, which necessitates using new motherboards.

2.7.5 G5 Microprocessor from IBM

The G5 [105] microprocessor from IBM is a RISC processor whose instruction set is compatible with the ESA/390, IBM S/360, and S/370 architecture generations that IBM has been developing since 1964.

The G5 microprocessor is produced on CMOS 0.25-micron technology with six metal interconnect layers for operating clock frequencies 385 MHz, 417 MHz, and 500 MHz. At a 1.9 V power supply voltage, G5 consumes 25 W.

Physically, the processor crystal is placed on a multichip module, which it shares with the L2 cache, the air-cooling system, and some other auxiliary circuitry. The L2 cache operates at half the processor core frequency. The processor crystal is 214.6 mm^2 in area and contains 25 million transistors: 7 million for logic, 13 million for L1 cache, and 5 million for supplementary functional units.

The G5 microprocessor has an integrated 4-entry multi-associative 256 KB L1 cache. The L1/L2 exchange speed is 4 GBps. The 2,048-entry two-way set associative BTB of the processor employs a two-bit branch prediction algorithm.

Like in the ESA/390 architecture systems, the G5 instruction system is based on microcode concept (IBM uses the term *millicode* for G5), which is a collection of built-in microprograms called to execute high-level CISC instructions of the microprocessor.

Most of the microprocessor's instructions are of the RX format (operands are stored in registers and the memory), which is equivalent to executing two register instructions. Optimizing the 10-stage execution pipeline of the microprocessor has allowed executing two instructions per cycle.

Unlike the ESA/390 architecture, G5 uses an IEEE-754-compatible FPU. The ESA/390 architecture uses Hexadecimal Floating Point data representation format, which is also supported by G5. The microprocessor integration features allow a symmetrical multiprocessor system of up to 12 microprocessors to be assembled. For fault-tolerant applications, microprocessors can be used in a dual mode, which is transparent to the system user. The microprocessor provides hardware parity check and testing of the logic circuit and service array state.

G5 implements hardware means to emulate a two-level virtual machine, which allows several operating systems to be run concurrently.

Due to its high operating clock frequency and a careful microcode design, this microprocessor's performance is almost twice that of the G4 microprocessor.

The application area of the microprocessor is mainframe construction.

2.7.6 Power 3 Microprocessor

The 64-bit Power 3 microprocessor [106] was developed by IBM as an alternative to high-performance Intel and Alpha microprocessors and was intended for use in high-end servers and workstation. The processor's multiprocessor mode support features allow this microprocessor to be used to assemble SMP systems.

Power 3 is a superscalar microprocessor that supports reordering and speculative execution of instructions. It can execute up to eight instructions per cycle: two load/store, two floating-point, two short integer, one long integer, and one branch instructions.

The microprocessor's structure is shown in Figure 2.38.

Power 3 has three execution units: two FPUs, three fixed-point execution units (FXU), and two load/store units (L/ST).

The processor has split L1 cache: 64 KB for code and 32 KB for data. Both caches are 128-entry, set associative, and non-blocking. Power 3 can use from 1 MB to 16 MB L2 cache, which is connected to the dedicated 256-bit bus operating at 200 MHz. The L2 cache bandwidth is 6.4 GBps. Data exchange with the main memory is carried out over the 128-bit system bus operating at 100 MHz. Its bandwidth is 1.6 GBps.

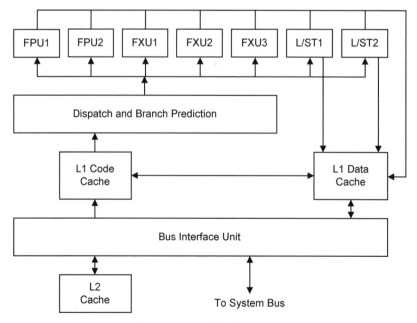

FIGURE 2.38 Power 3 microprocessor block diagram.

The processor's distinctive feature is the minimal time it takes to issue instructions for execution in the pipeline—only three cycles. This lowers pipeline delays caused by incorrectly predicted branches.

The processor was produced on 0.25-micron technology. Its crystal area is 270 mm^2 and houses 15 million transistors. The processor operates at 200 MHz. In the future, IBM is planning to manufacture the processor using 0.2-micron technology with copper interconnect layers and then go over to 0.18-micron technology and raise the operating frequency to 500 to 600 MHz.

At 200 MHz, Power 3 performance is 13.2 SPECint95 and 30.1 SPECfp95; at 600 MHz, it is 30 and 70, respectively.

2.7.7 Power4 Microprocessor

The Power4 microprocessor was introduced by IBM in 2001. It is a single-crystal dual-processor system (Figure 2.39). Power4 also has a large integrated cache, a high-speed external memory port, and links for interconnecting microprocessors into distributed shared memory (DSM) systems.

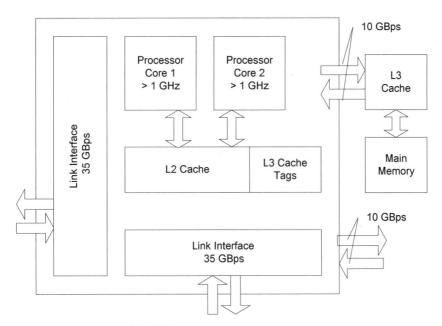

FIGURE 2.39 Power4 block diagram.

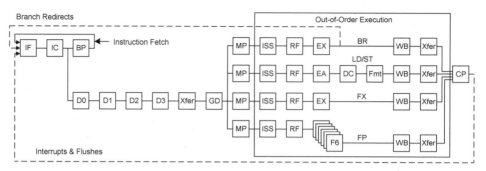

FIGURE 2.40 Power4 execution pipeline.

Along with ILP, the processor employs thread-level parallelism (TLP). Detecting parallelism dynamically raises the efficiency of the processor resource use by preventing functional unit downtime caused by cache misses or non-removable instruction data interdependencies.

Each of the on-chip processors has superscalar architecture and out-of-order and speculative instruction execution capabilities. Up to eight instructions can be fetched at one time, more than 200 instructions examined for execution, and up to eight instructions executed in eight functional units. The processor's functional units follow: two FPUs, two FXUs, two load/store units (LD/ST), one branch unit (BU), and one logic operation unit (CR). On average, each processor executes five instructions per cycle.

The processor's execution pipeline is shown in Figure 2.40.

The *instruction fetch* (IF), *instruction caching* (IC), and *branch prediction* (BP) cycles correspond to fetching a new instruction. Cycles D0 through GD are *instruction decode* and *group formation* cycles. At the *Mapper* (MP) cycle, instruction interdependencies are determined, resources are allocated, and instructions are placed into queues to the appropriate functional units. Further processing is carried out in four independent pipelines.

Instruction execution starts from extracting instructions from the queue in the ISS cycle and finishes with writing the execution results to registers in the WB cycle. In the Xfer and CP cycles, a check for execution completion of all previous instructions (according to their program order) is performed and the results are furnished.

The processors have split L1 data and code caches, 64 KB each. The crystal also integrates a shared, eight-way set associative, 1.5 MB L2 cache, and the controller and the tag memory of the external, up to 32 MB, L3 cache.

The external L3 cache is connected using the bidirectional 16-bit port operating at one-third the processor frequency, which provides a bandwidth of 10 GBps.

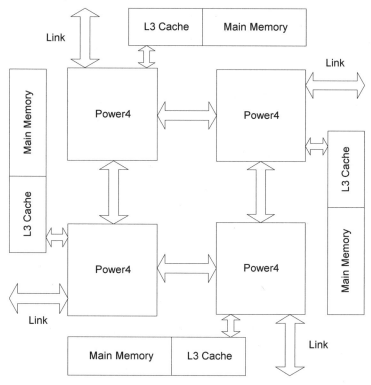

FIGURE 2.41 A Power4-based four-processor cluster.

The processor has three links to form multiprocessor configuration with the total bandwidth of 35 GBps. Figure 2.41 shows a four-processor system based on the Power4 microprocessor.

The microprocessor supports the IBM ISA instruction set, which is implemented in RS/6000 and AS/400, and is fully compatible with the PowerPC instruction set.

Power4 is produced on SOI 0.18-micron technology with five copper interconnect layers for 1.1 GHz and 1.3 GHz frequencies. The crystal is 400 mm^2 in area and contains 170 million transistors.

The performance of a Power4-based system is 140 GFLOPS.

At the end of 2001, IBM introduced a new 64-bit microprocessor, PowerPC 970, oriented toward use in Apple computers.

PowerPC 970 is a simplified version of the Power4 server processor. The processor can work both with 32- and 64-bit applications and has a special instruction set for efficient graphics processing.

Processor's performance at 1.8 GHz operating frequency is 7.2 BIPS.

2.8 MICROPROCESSORS FROM MIPS (SILICON GRAPHICS)

The application area of MIPS architecture ranges from game stations and pocket computers to high-performance servers and workstations.

MIPS microprocessors are a good example of the Brainiac concept implemented in processor architecture.

The 64-bit R10000 microprocessor developed by MIPS Technology, Inc., is based on the previous-generation processors (R2000, R3000, R4000, and R5000).

This microprocessor is based on the fifth-generation superscalar RISC technology that was earlier implemented in the R8000 microprocessor, which was oriented toward use in supercomputers. However, unlike the multicrystal R8000 optimized for high-performance scientific calculations, R10000 is a single-crystal, general-purpose processor for use in desktop PCs, workstations, and servers. It provides a better balance between integer and floating-point operations than R8000, making it more suitable for a wide range of applications. R10000 was designed in such a way so that it could be used equally well in home PCs running Windows NT, in UNIX workstations, or in multiprocessor database servers.

The processor was produced on CMOS 0.35 technology with four metal interconnect layers. Its area is 298 mm^2, and it holds 6.8 million transistors, 4.4 million of which are used for L1 cache. The processor is powered by 3.3 V and consumes 20 to 30 W. Its internal operating clock frequency can be programmed to 200 MHz, 133 MHz, 100 MHz, 80 MHz, 67 MHz, 57 MHz, or 50 MHz. At 200 MHz, processor's performance is 10.7 SPECint95 and 19.0 SPECfp95.

The processor's structure is shown in Figure 2.42.

The main features of the R10000 processor's architecture are the high level of microoperation execution parallelism, efficient prediction of branches, and functional unit load scheduling. R10000 has five functional units: two FPUs, two ALUs, and one load/store unit.

The branch prediction mechanism uses a 512-entry BHT. The prediction success rate is 87 percent. When performing a predicted branch, the processor saves its state in a four-position branch stack. The alternate branch address, the entire integer and floating-point register-mapping table, and various control bits are pushed onto it. This information is needed for restoring the processor state in case a wrong branch prediction was made. While the stack is being filled, instructions are continued to be decoded until a new branch instruction arrives, after which event it is ceased until a stack position becomes available. In the case of a wrong branch prediction, all instructions fetched after it are discarded. These instructions are identified by a four-bit mask associated with a specific position in the stack.

During instruction prefetch stage, four 32-bit instructions are cached and pre-decoded. Each cached instruction is supplied with a four-bit tag, which is used for further decoding and classification of the instruction. For branch instructions, the result address is calculated.

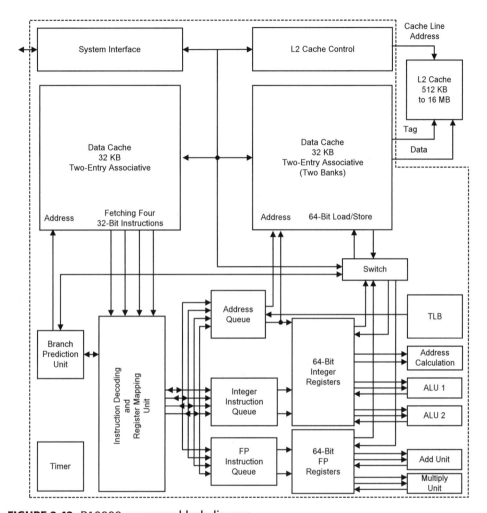

FIGURE 2.42 R10000 processor block diagram.

In the second decoding stage, instruction data interdependencies are eliminated (if this is possible). A register-renaming mechanism is used for this purpose. Memory dependencies are determined by the contents of the memory access queue. Independent instructions can be executed out of the order specified in the program. All out-of-order instruction execution results are temporary and can be cancelled in case of a wrong branch prediction. R10000 has 64 integer and 64 floating-point registers, which are dynamically mapped onto 33 integer and 32 floating-point registers.

After decoding, instructions are divided into three 16-entry queues where they await operands and functional blocks to become available. The queues are serviced as functional units become available. Instructions are executed in one of the functional units' pipelines. Integer and floating-point functional units are independent; they have separate register files and data paths. The length of the integer pipeline is one cycle, that of the load/store unit is two cycles, and the FPU pipeline is three cycles long. With the potential to execute five instructions per cycle, R10000 allows only four instructions to be fetched, leaving elbow room for scheduling loading of the functional units. This allows the R10000 to execute four instructions per cycle most of the time, correspondingly producing up to four results per cycle.

Each of the integer ALUs can perform addition and logic operations. Additionally, one of the ALUs performs shift operations and branch predictions while the other carries out integer multiply and divide. One of the two double-precision FPUs performs addition while the other does multiply/divide and square root operations. The latter operation is broken down into two suboperations that are executed in parallel.

R10000 has a high-speed (up to 1.6 GBps) intercrystal processor/bus interface, which allows up to four processors to be linked into a multiprocessor configuration without using additional interface circuits.

The architecture of the next microprocessor, R12000, is slightly different from R10000. The differences follow:

- A four times larger than in R10000 branch history table: 2,048 lines
- A new, 32-line BTB
- Forty-eight as compared to 32 maximum number of instructions that can be executed out of order
- Enhanced L2 cache operations
- Increased depths of the pipelines

The internal structure of the R12000 microprocessor is shown in Figure 2.43.

R12000 operates at 300 MHz and 400 MHz. Its SPEC performance indexes are shown in Table 2.19.

TABLE 2.19 R12000 Microprocessor Performance

Microprocessor	Frequency (MHz)	SPEC int95	SPEC fp95	SPEC int95	SPEC fp95
R12000	400	23.8	37.1	320	319

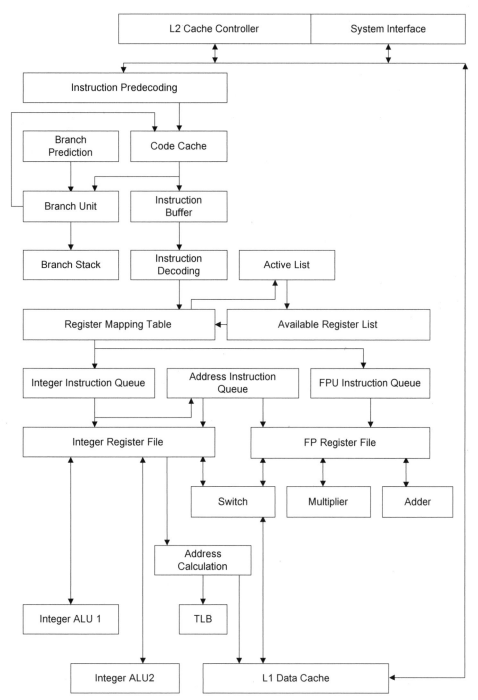

FIGURE 2.43 R12000 microprocessor block diagram.

In 2001, Silicon Graphics introduced the next microprocessor of the family—R14000. This microprocessor is produced on 0.13-micron copper technology and operates at 500 MHz. Its architecture has not been significantly changed from the R12000.

The next processor in the R1x000 line is the R16000, which currently operates at 700 MHz, with a slight rise planned during its lifetime. It is built on advanced 0.11-micron technology and has the theoretical peak performance of 1.4 GFLOPS. Recently, it has become clear that SGI will quickly fold its MIPS platform entirely and use IA-64 architecture processors from Intel and HP in its servers and workstations. Therefore, it is in doubt whether the SGI's plans for the successors of the R16000—the R18000 and R20000—which were projected for the middle of 2004 and on, will ever be implemented.

2.9 OTHER MICROPROCESSORS

To give a full processor review, some processors from nontraditional manufacturers will be considered, namely those designed and manufactured in Russia.

Currently, the overwhelming majority of Russian microprocessor system designers and manufacturers use foreign-made components. This is due to the large market selection of relatively inexpensive microprocessors. Mass-produced microprocessors from various manufacturers are available at the consumer-electronics level on the market. In most cases, to design a microprocessor system it is enough to simply give a most general description of the microprocessor without going into details of its implementation. In other words, consider it as a black box, all of whose activity is reflected externally and, moreover, can be perceived using the freely distributed software. However, there are some applications for which the importance of predictability of the device functioning comes into play. In this case, it is not enough to simply observe the external activity of the processor; a clear idea of not only what it does but also of how it does it is necessary. This kind of predictability can be provided only by a full-scale development of all system components, including the microprocessor.

Several Russian manufacturers have designed microprocessors with both original and cloned architectures. To the former pertain, for example, microprocessors of the Kvant series, described in Chapter 4. Examples of the latter are designs from the Systems Research Institute of the Russian Academy of Science (SRI of RAS) and the Moscow SPARC technology center (MSTC).

2.9.1 MSTC Architecture Microprocessors

During the past several years, MSTC has been developing SPARC-compatible microprocessors of the MSTC family. The microprocessors have 32-bit architecture complying with the SPARC v8 specification described in Section 2.4.

To date, the MSTC-R150 [110] microprocessor has been introduced. It is produced on 0.35-micron technology with four metallization layers, has 2.8 million transistors, and operates at 150 MHz. Work is being conducted to produce a prototype of the 500 MHz MSTC-R500 and to develop the 1 GHz MSTC-R1000 [110]. The latter two microprocessors will be produced on 0.13-micron technology with eight metallization layers and will have 4.2 million transistors.

The performance of the microprocessor follows:

MSTC-R150—140 MIPS, 64 MFLOPS

MSTCR500—400 MIPS, 170 MFLOPS

2.9.2 MIPS Architecture Microprocessors

A version of the popular MIPS-I architecture was implemented in the 1B812 microprocessor developed at SRI of RAS. This microprocessor is an architectural analog of the MIPS R3000 microprocessor [111].

The MIPS R3000 was developed by MIPS Technology, Inc., in 1988. Microprocessors with this architecture were widely used in the computer industry and were employed in workstations and servers, telecommunications, and office equipment.

The processors work at 33 MHz or 40 MHz giving performance of 20 SPECint92 and 23 SPECfp92.

The R3000 consists of two closely linked processors on a single crystal: a central RISC processor and a system control coprocessor. The R3000 is a 32-bit microprocessor: all its registers are 32 bits wide. The central processor has 32 general-purpose registers, a program counter, and two registers for storing the high and low words of the floating-point multiply/divide instruction results. GP register 0 is hardware zero: reading it always produces 0 and data written to it is lost. GP register 31 is used to store the return address in jump with return instructions.

Doubleword (64 bits), word (32 bits), halfword (16 bits), and byte (8 bits) are the formats supported by the processors.

The processor has an integrated memory management unit, which processes memory requests from the central processor. This unit converts virtual address into physical ones, protects memory, controls cache, and performs bus arbitration. Split data and code caches are connected to the processor. Each cache can be from

4 KB to 256 KB. The on-chip cache control logic provides cache coherence in multiprocessor systems. The cache is direct-mapped of the write-through type.

The specification of the R3000 architecture provides for using up to three external coprocessors (CP1–CP3). The CP1 is a floating-point coprocessor; the CP2 and CP3 are to be used for system expansion. Coprocessors' memory exchange is serviced by the central processor; it generates addresses and controls memory interface.

Central processor instructions are 32 bits wide and have three formats: register (I-type), branch (J-type), and direct operand (R-type) instructions. Instructions are executed in the five-stage pipeline. Results of the next instruction are produced every cycle.

In terms of their functions, the instructions are divided into the following groups:

- Load/Store (I-type) instructions
- Calculation (arithmetic, logic, shift operations) (R- or I-type) instructions
- Unconditional (R- or J-type) and conditional (I-type) branch instructions
- Coprocessor instructions (instruction format depends on the coprocessor used)
- Coprocessor 0 instructions for manipulating CP0 registers
- Special instructions (R-type) to handle system calls, halts, exchanges between the special and GP registers, and so on

The system control coprocessor CP0 supports virtual memory, handles exceptions, controls switches between the *core* and *user* modes, provides diagnostic and error recovery means, controls cache, and so on.

The CP0 has 10 special registers, four of which are used by the virtual memory system; the other six are used in processing exceptions.

The R3000 address space is 4 GB and is divided into two parts: 2 GB for the user (low addresses) and 2 GB for the core. Because the physical address space is smaller than the virtual, the MMU converts virtual addresses into physical (using the on-chip TLB) and also caches memory pages. Fully associative TLB has 64 entries, each of which corresponds to a 4 KB page.

The CP0 usually works in the user mode, switching to the core mode to process exceptions. The following exceptions are considered: arithmetic overflows, I/O interrupts, system calls, cache misses, and the like.

The FPU coprocessor R3010 operates on 64-bit numbers. The FPU has 16 64-bit registers, which can hold single- and double-precision numbers.

The FPU pipeline is six stages deep and works in parallel with the central processor pipeline. Like the integer instructions, FPU instructions are executed on average one per cycle.

R3000 ARCHITECTURE IMPLEMENTATION

Many semiconductor manufacturers have purchased licenses to produce MIPS architecture processors. Among them are such as Broadcom, IDT, LSI Logic, NEC, NKK, Philips, Toshiba, and others.

The SRI of RAS purchased license for the R3000 microprocessor and the FPU coprocessor R3010 [112] from MIPS Technology, Inc. They have developed their own version of the MIPS architecture.

The 1B812 [113] microprocessor developed at the SRI of RAS is a functional analog of the R3000 microprocessor with the R3010 coprocessor. It contains the following functional units:

- A fixed-point numbers processor (including the system coprocessor CP0), which is a functional analog of the R3000 microprocessor
- A FPU coprocessor, which is an architectural analog of the R3010 coprocessor
- An 8 KB code cache
- An 8 KB data cache
- A system bus controller
- A write buffer, which allows delays during the write cycle to be avoided
- A read buffer, which support block and single reads

The processor has the following characteristics:

- Operating clock frequency: 33 MHz
- System bus interface operates at the processor frequency of half its rate
- The performance is 24.5 MIPS, 8.7 MFLOPS
- A programmable energy-saving mode; energy consumption in the operating mode does not exceed 1 W, and in the energy saving mode, it is less than 0.1 W
- The processor is built as a MSIC CMOS chip on 0.5-micron technology with about 1.6 million transistors on the crystal; the case has 108 output pins

The microprocessor structure is shown in Figure 2.44.

The 1B812 uses the same integer instruction set as the R3000A.

The processor core (CPU) is a 32-bit RISC processor. The SPU has a five-stage pipeline and 32 orthogonal 32-bit registers.

Pipelined execution allows up to one instruction per cycle executed. Up to five instructions in various execution stages can be processed in the processor pipeline simultaneously.

The 1B812 has an on-chip arithmetic coprocessor (CP1), a R3010 analog, which works in parallel with the CPU. The CP1 is a high-performance coprocessor performing single- and double-precision add, multiply, and divide floating-point operations.

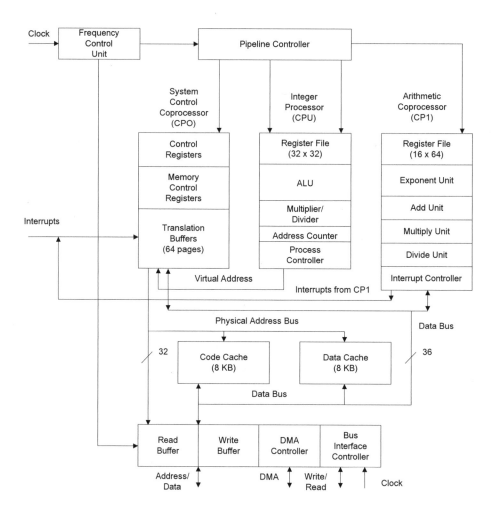

FIGURE 2.44 1B812 microprocessor block diagram.

The CP1 has 16 orthogonal 64-bit registers. For some read/write operations executed by the CPU, these registers look like 32 32-bit registers. The structure of the CP1 is shown in Figure 2.45.

The processor has on-chip split data and code caches 8 KB each.

The system interface of the 1B812 uses a 32-bit multiplexed address and data bus and provides bus arbitration for implementing DMA mode. A four-word buffer is used for external memory exchange, which can be conducted by either single words or in four-word bursts.

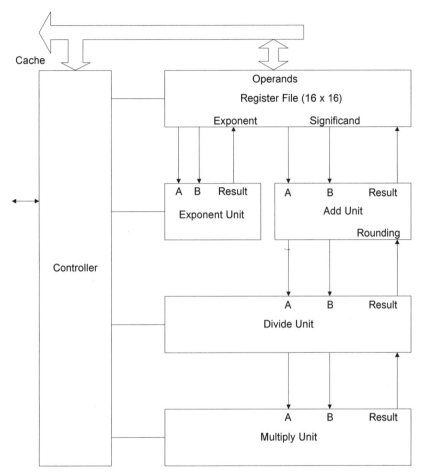

FIGURE 2.45 Structure of the arithmetic coprocessor CP1.

Microprocessors developed by the SRI or RAS are produced at the Angstrem production plant in Russia [112].

2.10 CURRENT STATE AND FUTURE DEVELOPMENT PROSPECTS OF UNIVERSAL PROCESSORS

2.10.1 Universal Microprocessor Main Development Trends

Review of processor families from different developers confirms their common development tendencies: striving for interfamily binary compatibility, raising the clock frequency, increasing the size and bandwidth of the memory subsystem, and increasing the number of parallel functional units.

In the 20th century, it was impossible to implement in one microprocessor a record of parameter figures for all these tendencies because of technological limitations and economic restrictions on the costs per processor, in particular, and on the microelectronic production, in general. Faced with limited hardware resources, each microprocessor developer had to choose some such architectural and structural techniques, concentrating on whose development would make its processor superior to others. Therefore, every specific microprocessor type is a result of many compromises made by its developers.

The architectures of modern microprocessors from different developers have much in common, which even makes one think about a unified architecture. Being able to place a large number of transistors on chip makes it possible to use all known performance-enhancing techniques in one microprocessor, only taking their compatibility into account.

2.10.2 Software Compatibility

Backward software compatibility is of a great importance for developing and accumulating software products. Actually, it is impossible to imagine computer technology development without preserving backward compatibility.

Of course, providing backward software compatibility may be at odds with the striving to raise microprocessor performance. An especially good example of this is the development of the x86 architecture microprocessors. The demand potential of these microprocessor is so great that it has exerted and continues to exert a stimulating influence on the manufacturers, forcing them to search for technological and price niches to advance their products.

Research directed at optimizing the architecture gave rise to the conception of RISC processors. During this research, it came out that the complex instructions of the x86 architecture get in the way of raising processor performance. This discovery, however, has not led to abandoning the architecture. Intel, the main producer of the x86 architecture microprocessors, continued to follow the traditional approaches. However, creative minds in a more flexible small company endeavoring to get a larger market share suggested a solution that allowed building RISC processors while preserving the x86 architecture. They achieved it by hardware translation of x86 instructions into RISC instructions. Adding a RISC core allowed x86 architecture processors to bridge their performance gap with the pure RISC processors. Moreover, this approach makes it possible to move to processors with multiple instruction sets, each of which is translated by its on-chip hardware translator into a set of the processor core RISC operations. This approach has been subsequently accepted by all x86 architecture microprocessor manufacturers, including Intel.

Of course, this approach requires additional hardware expenditures, but these expenditures are fully justified by the obtained performance gains.

Transmeta attempted to develop a processor more efficient in terms of performance, power consumption, and hardware amount while preserving the x86 architecture by using the VLIW architecture.

Other manufacturers aim at offering processors advantageously oriented toward computers in specific price ranges.

Moreover, developers of microprocessors with architectures others than the x86 try to equip their products with means allowing them to emulate the x86 architecture. Thus, at one time DEC solved the portability of the existing x86 software problem by using the optimizing FX!32 translator. The translator converts the x86 object code into optimized DEC Alpha object code. Actually, this approach can be employed to provide compatibility with other architectures. It is also possible to equip processors with programmable logic analogous to PLIC that can be configured to the needed instruction system.

Therefore, preserving backward software compatibility is not so much of a hindrance to microprocessor development as a factor that stimulates this development.

2.10.3 Increasing Clock Frequency

Increasing clock frequency is achieved by using more advanced, lower-figure technological processes, increasing metallization layers, using more advanced circuit design with fewer cascades and more advanced transistors, as well as packing functional units on the chip more densely.

Thus, all manufacturers have moved to the CMOS technology, although Intel, for example, used BiCMOS for its first Pentiums. Exponential Technologies made an even more exotic attempt by trying to develop bipolar circuitry to build PowerPC and x86 processors. However, at 466 MHz, a 150 mm^2 crystal dissipated about 80 W, which created a serious heat-removal problem, because without an efficient heat sink the crystal turns into a glowing-hot metal plate. The company tried to raise the clock frequency to the promised level but could not do it, in its words, because of financial problems.

Generally speaking, it is known that at high frequencies both bipolar and CMOS circuitry dissipate about the same amount of heat, but CMOS circuitry is more technological, which was what determined its predominance in microprocessors.

Reducing transistor size accompanied by lowering power supply voltage raises the operating speed and lowers the dissipated heat energy. All microprocessor manufacturers have moved from the 0.25-micron process technology to the 0.18-micron and are now steadily moving toward 0.13–0.09 micron.

The problem of cutting down on the crystal interconnect lengths is solved by increasing the number of metallization layers. Thus, by increasing metal layers from three to five, Cyrix, while preserving the 0.6 CMOS technology, decreased the chip size by 40 percent and reduced the dissipated heat, taking care of the earlier chip-overheating problem.

Reducing the lengths of chip interconnections is important for raising the operating frequency, because a significant portion of the cycle period is taken by signal propagation over the conductors inside the chip. Thus, special measures to cluster processing aimed at localizing interacting processor elements were undertaken in the Alpha 21264 processors.

2.10.4 Increasing Memory Subsystem Bandwidth

The memory subsystem supplies processor functional units with work. There are several approaches to increasing its bandwidth. One of these approaches is building one- or multilevel caches. Another is increasing the processor/cache interface bandwidth and the processor/main memory interface. Interfaces can be enhanced by increasing bus bandwidth (by raising bus operating frequency and/or its width). They can also be enhanced by adding additional buses, which eliminate conflicts between the processor, cache, and the main memory. In the latter case, one bus operates at the processor frequency and provides an interface with the cache. The other bus provides an interface with the main memory and operates at its working frequency. With this approach, working frequencies of the second bus in Pentium Pro-200, PowerPC 604E-225, and Alpha 21164-500 are 66 MHz, 66 MHz, and 166 MHz, respectively. (The microprocessors themselves operate at 300 MHz, 225 MHz, and 500 MHz, respectively.) The buses being 64, 64, and 128 bits wide, at these frequencies their main memory interface bandwidths are 512 MBps, 512 MBps, and 2,560 MBps, respectively.

The general tendency toward increasing cache size is implemented in different ways. Some of them follow:

- The HP PA-8000 is equipped with external data and code cache from 256 KB to 2 MB in size and two-cycle access time.
- The Pentium Pro microprocessor is equipped with a separate L2 cache crystal packaged in the same casing with the microprocessor.
- The Alpha 21164 processor uses on-chip split L1 8 KB data and code caches and an on-chip unified 96 KB L2 cache.

The most-often-used approach is to employ on-chip split L1 data and code caches and possibly off-chip L2 cache. For example, Pentium II uses on-chip split data and code caches of 16 KB each, working at the processor clock frequency; it also uses an off-chip L2 cache operating at half the processor clock frequency.

2.10.5 Raising the Internal Parallelism Level

Every next generation of each microprocessor family has the number of functional units increased. Moreover, functional unit characteristics are enhanced in both temporal (reduced number of pipeline stages and decreased duration of each stage) and functional (introducing MMX extension instruction set, for example) respects.

Currently, processors can execute up to six operations per cycle. However, the number of floating-point operations is limited to two in R10000 and Alpha 21164 and to four in PA-8x00.

To achieve efficient functional unit loading, register renaming and branch-prediction techniques are used, which eliminate instruction data and control interdependencies. There are practically no established approaches in this area, because each microprocessor demonstrates the inventiveness of its designers in creating hardware/compiler symbiosis for static and dynamic elimination of instruction interdependencies.

Currently, VLIW and multithread architectures are in the practical approbation testing stage.

REVIEW QUESTIONS TO CHAPTER 2

1. List the main architecture types of universal microprocessors and their typical application areas.

2. Name the main representatives of the x86 architecture microprocessors.

3. What is a multimedia instruction set extension? Give examples of multimedia instruction set extension implementations in RISC and CISC architecture microprocessors.

4. What is the purpose of using separate processor buses for working with cache and the main memory?

5. List the constituents of energy-saving technology.

6. State the specific cache organization features of the sixth-generation Intel microprocessors.

7. How do RISC and CISC processor concepts combine within the framework of the x86 architecture?

8. What are the distinctive features of the x86 architecture implementations in the competing products from Intel, AMD, and Cyrix? Compare main structural and functional characteristics of these families: instruction fetch unit, branch control, cache operation, and instruction dispatch and execution implementation.

9. List the main features of Transmeta microprocessors.

10. Enumerate the architectural features of the Alpha 21064, 21164, and 21264 microprocessors. Comment upon the development path of the Alpha architecture microprocessor family. Name the processors' main units and their functions.

11. What are the typical features of SPARC architecture? Name the main units and their functions.

12. In your opinion, what was the reason for the emergence of MAJC family microprocessors?

13. Comment on the development of PA architecture microprocessor family from Hewlett-Packard. Describe the structure and organization of their cache. Name the main functional units. Describe the branch prediction mechanism.

14. List the features of the PowerPC and R-1xxxx family microprocessors from the standpoint of their being classic decoupled architecture RISC processors.

15. Compare the approaches taken by different manufactures to microprocessor cache organization. State the development tendencies in cache organization.

16. Name the number of instructions that can be executed simultaneously by different microprocessors. What stands in the way of increasing this number? State the pipeline depth for each microprocessor.

17. Compare the implementations of branch-prediction mechanisms in different microprocessors. Name the prevalent development tendencies.

18. Compare the instruction dispatch and execution methods in different microprocessors.

19. Point out the mechanisms of creating multiprocessor configurations embedded in different microprocessors.

20. Comment on the non-Western microprocessor development.

ENDNOTES

54. IBM's Cell Completes Design Phase. *IEEE Micro*, 2002, Vol. 22, No. 5.

55. *Pentium Processor Family. Developer's Manual*, Intel Corporation, 1997.

56. *Intel Architecture. Optimization Manual.* (Order Number: 242816-003), Intel Corporation, 1997.

57. *Pentium Processor with MMX Technology,* (Order Number 243185-001), Intel Corporation, January 1997.

58. *Pentium Pro Family Developer's Manual.* Volume 1: Specifications. (Order Number 242690), Intel Corporation, 1996.

59. *Intel Delivers the Next Level of Computing with the New Pentium II Processor,* Intel Corporation, May 1997.

60. *Pentium II Processor at 233 MHz, 266 MHz and 300 MHz,* (Order Number: 243335-001), Intel Corporation, April 1997.

61. *Introduction to Streaming SIMD Extensions,* Intel Corporation, 1999.

62. Fischer, S., Mi, J., Teng, A. Pentium® III Processor Serial Number Feature and Applications, *Intel Technology Journal,* 1999, Q2.

63. *IA-32 Intel® Architecture Software Developer's Manual Volume 1: Basic Architecture.* Intel Corp., 2003.

64. Sharangpani, H., Arora, K., Itanium Processor Microarchitecture. *IEEE Micro,* 2000, Sept.–Oct.

65. *Intel® Itanium® 2 Processor Reference Manual for Software Development and Optimization.* Intel Corp., 2003.

66. Kostiakov, S. NexGen Processors: A Real Alternative to Pentium. *CompUnity,* 1995, #1(2), pp. 69–74.

67. Climan, D. Three Rivals of Pentium. *PC Magazine,* Rus. Ed., 1995, Issue 5, pp.117–34.

68. *AMD-K5 Processors.* Advanced Micro Devices, Inc., 1996.

69. *AMD-K6 MMX Processor. Product Overview.* Advanced Micro Devices, Inc., 1997.

70. Oberman, S., Favor, G., Weber, F. AMD 3DNow! Technology: Architecture and Implementations, IEEE Micro, 1999, March–April.

71. Huynh, J. The AMD Athlon™ MP Processor Technology and Performance Leadership for x86 Microprocessors. White Paper. Advanced Micro Devices, Inc., August 27, 2002.

72. AMD Eighth-Generation Processor Architecture. White Paper. Advanced Micro Devices, Inc., October 16, 2001.

73. *Cyrix 5x86. Architectural Overview.* Cyrix Corporation, 1995.

74. *6x86 Processor.* Data Book, (Order Number: 94175-01), Cyrix Corporation, March 1996.

75. *6x86-P200+ Processor.* Data Book Addendum, Cyrix Corporation, May 1996.

76. *Cyrix Launches 6x86MX Processor.* Cyrix Corporation, May 1997.

77. *Cyrix 6x86MX Processor.* Cyrix Corporation, July 1997.

78. *6x86MX Processor Performance Benchmarks.* Product Info, Cyrix Corporation, 1997.

79. *IDT WinChip C6 Processor.* Data Book. Centaur Technology, Inc., March 1998.

80. *WinChip2. IDT WinChip 2.* Processor data sheet. Integrated Device Technology, Inc., September 1998.

81. Poluvialov, A. VIA Cyrix III (Samuel 2) 600 and 667 MHz, *iXBT Hardware,* 5 January 2000.

82. *Crusoe Processor Model TM3200 Feature.* Transmeta Corporation, 2000.

83. *Crusoe Processor Model TM5400 Feature.* Transmeta Corporation, 2000.

84. *Digital Semiconductor Alpha 21164PC Microprocessor,* Product Brief, Digital Equipment Corporation, June 1997.

85. *Digital Semiconductor Alpha 21264 Microprocessor,* Product Brief, Digital Equipment Corporation, May 1997.

86. Gwennap, L. Digital 21264 Sets New Standard Clock Speed, Complexity, Performance Surpass Records, But Still a Year Away, *Microdesign Resources.* 1996, October 28.

87. *Alpha Architecture Handbook,* (Order Number EC—QD2KB—TE), Digital Equipment Corporation, October 1996.

88. *Digital Semiconductor 21164 Alpha Microprocessor.* Data Sheet, (Order Number: EC—QP98B—TE), Digital Equipment Corporation, February 1997.

89. Bannon, P. Alpha 21364: A Scalable Single-chip SMP. Compaq Computer Corporation, Microprocessor Forum. 13 October, 1998.

90. Borzenko, A. Next Generation I/O vs. Future I/O/, *PC WEEK,* 9(183), 1999.

91. *SPARC Architecture. 8th Edition.* User Manual. SUN Microsystems, 1990.

92. The UltraSPARC Architecture. The UltraSPARC Processor, Technology White Paper, Sun Microsystems, Inc., 1995.

93. Horel, T., Lauterbach, G. UltraSPARC-III: Designing Third-Generation 64-Bit Performance. *IEEE Micro,* 1999, May–June.

94. *Introduction to the MAJCTM Architecture.* Sun Microsystems Inc., 1999.

95. Tremblay, M., *Microprocessor Architecture for Java Computing.* Sun Microsystems Inc., 1999.

96. Lesartre, G., Hunt, D. *PA-8500: The Continuing Evolution of the PA-8000 Family.* Hewlett-Packard Company, 1997.

97. Gwennap, L. PA-8000 Combines Complexity and Speed, *Microprocessor Report.* 1994, Vol. 8, No. 15.

98. Gwennap, L. HP Pumps Up PA-8x00 Family, *Microprocessor Report 1996*, Vol. 10, No. 14.

99. PA-RISC 8x00 Family of Microprocessors with Focus on PA-8700. Technical White Paper. April 2000, *www.jpn.hp.com/products/servers/parisc/technology/pa_risc/pdfs/wpaper.pdf*.

100. Paap, G., Silha, E. PowerPC: A Performance Architecture, Proceedings of COMPCON 1993, pp. 104–08.

101. *PowerPC 620 Microprocessor*. Product Overview. Motorola, Inc., 1994.

102. *AltiVec Technology*. Fact Sheet. Motorola, Inc., 1998.

103. *PowerPC 6xx or 7xx RISC Microprocessor Technical Summary*, Motorola 1998.

104. Motorola Launches PowerPC G4, *www.theregister.co.uk*.

105. Slegel, T., Averil, R. III, Check, M., Giamei, B., et al. IBM's S/390 G5 Microprocessor Design. *IEEE Micro*, 1999, March–April.

106. Kuzminskiy, M. IBM Goes Mainstream. *ComputerWorld Russia*, 17 November 1998.

107. Tendler, J., Dodson, S., Fields, S., Le, H., Sinharoy, B. POWER4 System Microarchitecture. IBM Server Group. October 2001.

108. Yeager, K. The Mips R10000 Superscalar Microprocessor. *IEEE Micro*, 1996, Vol. 16, pp. 28–40.

109. *R10000 Microprocessor*. Product Overview. MIPS Technologies, Inc., October 1994.

110. Babayan, B., Kim, A., Sakhin, Yu. Russia's Domestic Universal Microprocessors of MSTC-R Series. Electronics: *Science, Technology, Business*. 2003, No. 3, pp. 46–51.

111. Kane, G., Heinrich, J.*MIPS RISC Architecture*. Prentice Hall, 1992.

112. *www.niisi.ru/otd12.htm*.

3 Signal, Communications, and Media Microprocessors

3.1 GENERAL INFORMATION ABOUT DIGITAL SIGNAL PROCESSING

Digital signal processing (DSP) is defined as a real-time arithmetic processing of a sequence of signal amplitude values sampled at uniform periods of time. Some examples of DSP follow:

- Signal filtering
- Convolution (mixing) of two signals
- Calculating values of a two-signal correlation function
- Signal amplification, limiting, or transformation
- Direct/inverse Fourier signal transforms

In most cases, analog signal processing, which has traditionally been used in many radio devices, is a cheaper way of obtaining the necessary result; however, when the circumstances require either high processing precision, a miniature device, or the stability of device characteristics over various temperature conditions, DSP proves to be the only acceptable solution.

An example of using an analog circuit to filter a signal is shown in Figure 3.1. The operational amplifier used in the filter allows the dynamic range of the processed signal to be expanded. The shape of the filter's amplitude-frequency response curve is determined by the values of R_f and C_f. It is difficult to obtain a high figure of merit value for an analog filter because its characteristics greatly depend on the temperature conditions. Filter components introduce additional noise into the resulting signal. Analog filters are difficult to adjust over a wide frequency range.

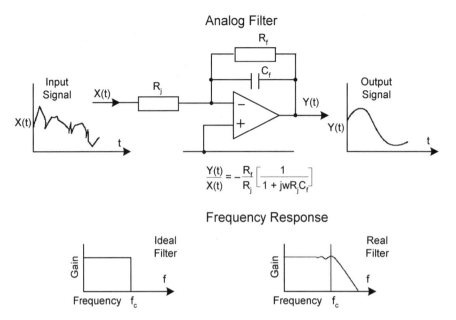

FIGURE 3.1 Analog signal processing.

Analogous signal processing results can be obtained by using a digital circuit, such as the one shown in Figure 3.2. The circuit is made of the following components:

- Low-pass filters (LPF) that remove extra harmonics from the frequency spectrum before the AD and after the DA conversions
- AD and DA signal converters
- The finite impulse response (FIR) digital filter proper

The filter's frequency response characteristic is determined by the values of the $C(k)$ filter coefficients. Varying the number of the coefficients (that is, the filter's length) and their values, a filter of any required frequency response can be obtained. The introduced noise (quantization noise) depends on the frequency and width of the AD and DA converters, as well as the precision of the computations.

The mathematically defined schematics of the conversion performed on the sequence of a signal samples can also be represented graphically as digital filter block diagram.

By the type of their impulse, response filters are classified into filters with FIR and those with infinite impulse response (IIR). Block diagrams of FIR and IIR filters are shown in Figure 3.3 and Figure 3.4, respectively.

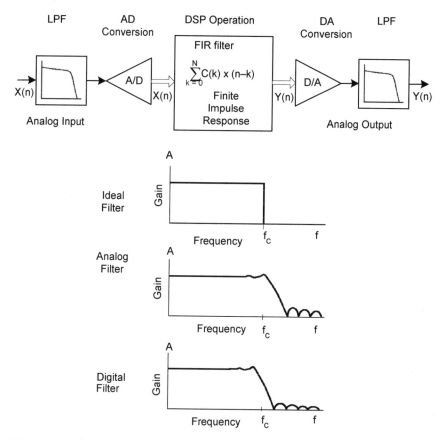

FIGURE 3.2 Digital signal processing.

The conventional notations represent the following:

T—One cycle delay unit

×—Multiply unit

+—Add unit

To implement digital filtration algorithms efficiently, basic DSP operations must be hardware supported: multiplication with accumulation (MAC), modular address arithmetic, and normalization of arithmetic operation results.

Another frequent signal conversion is the Discrete Fourier Transform (DFT)(direct and inverse) [115].

Any signal can be represented in both a time domain (a collection of charts in time/amplitude coordinates) and a frequency domain (a sequence of charts in frequency/amplitude coordinates). Depending of the processing algorithms used, either frequency or temporal signal representation can be chosen. The FT allows moving between different signal representations.

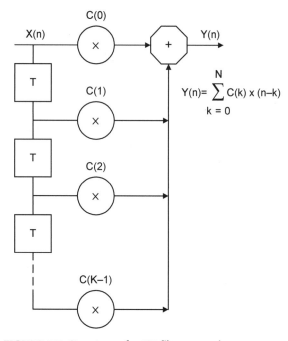

C(0)
X(n)
Y(n)
T
C(1)
Y(n)= ∑ C(k) x (n–k)
N
k = 0
T
C(2)
T
C(K–1)

FIGURE 3.3 Structure of a FIR filter cascade.

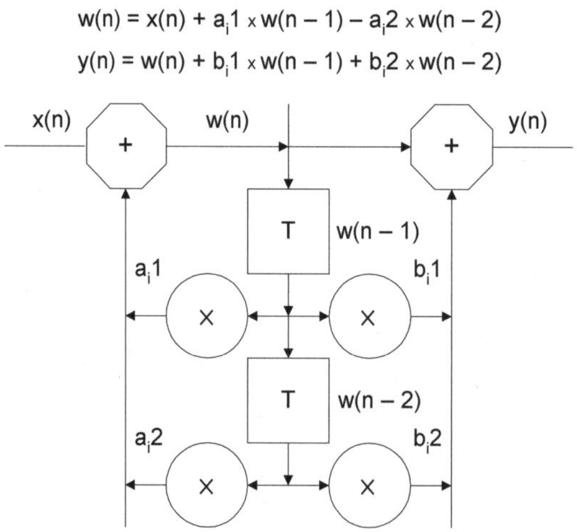

w(n) = x(n) + a_i1 x w(n – 1) – a_i2 x w(n – 2)
y(n) = w(n) + b_i1 x w(n – 1) + b_i2 x w(n – 2)
x(n)
w(n)
y(n)
T
w(n – 1)
a_i1
b_i1
T
w(n – 2)
a_i2
b_i2

FIGURE 3.4 Structure of an IIR filter cascade.

The DFT is analytically expressed by the following equation:

$$X(f) \approx \tilde{X}(f) = T \sum_{n=-\infty}^{+\infty} \chi(nT)e^{-j2\pi fnT},$$

where $\chi(nT)$ is a signal sample sequence.

There are various ways to implement the DFT. Several algorithms use techniques that allow the calculation volume to be reduced. These algorithms are given the common name of the Fast Fourier Transform (FFT).

In practice, the summation interval is limited by a certain number of temporal samplings, N, that depends on the conversion precision required. In this case the equation looks like this:

$$X(f) \approx \tilde{X}(f) = T \sum_{n=0}^{N-1} \chi(nT)e^{-j2\pi fnT},$$

where N is the number of conversion samples.

To reduce the number of multiplication operations when performing the DFT, a method called time subsampling is used. The essence of this method consists in that the FT over an N sample sequence can be expressed by transforms performed on subsequences of this sequence, each of which is $N/2$ samples long. Because the number of multiplications is proportional to the number of samples, performing double transform over $N/2$ samples with the subsequent combining of the results takes less time. Applying the subsampling procedure to a sample sequence recursively, the computation scheme shown in Figure 3.5 is obtained.

In Figure 3.5, k/N means multiplying by the $e^{-jk\frac{2\pi}{N}}$ coefficient.

In implementing this conversion scheme, bit operations are employed, along with multiplication and addition operations. The resulting samplings have inverse bit order (see Figure 3.5), that is, the order in which element's position is determined by an inverse of the element's index binary representation. Samplings are reordered by either reordering the array elements or by performing a bit inversing operation on the index when referencing an array element. The latter method takes less time to perform but requires that data addresses can be manipulated.

In most real-life applications, the considered basic DSP algorithms must be performed in real-time mode, which demands increased performance from the processor on which they are executed. Hardware support of the basic DSP algorithms is a typical feature of the signal processors.

The main digital signal microprocessor families present on the world market will be considered.

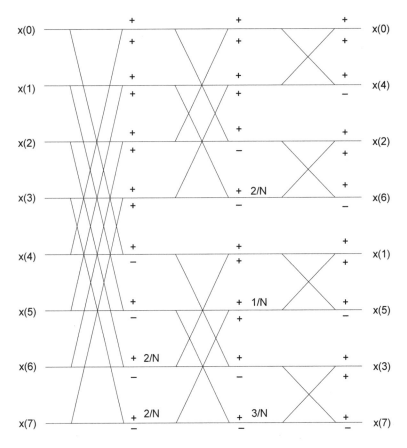

FIGURE 3.5 An eight-point FFT scheme.

3.2 SIGNAL PROCESSING MICROPROCESSORS

Specialized microprocessors—digital signal processors—are used to build digital signal processing systems. Universal microprocessors cannot be used or are inefficient for DSP purposes because of their low performance in the given area and their excessive capabilities for the task at hand.

The distinctive features of signal processors are their processing of short floating-point numbers (40 bits and less), prevalent use of fixed-point numbers 32 bits and less wide, as well as the orientation toward simple processing of large data arrays.

The distinctive feature of DSP tasks is the real-time streaming processing of large volumes of data. This demands high-performance processors and intensive data exchange with external devices. These requirements are currently met by the specific architecture of the signal processors and their problem-oriented instruction sets.

Signal processors are highly specialized. Techniques for shortening the instruction cycle, also typical for universal RISC processors, are widely employed in signal processors. Among these are such as pipelining on the level of individual microoperations and instructions, placing operands for most instructions in registers, using shadow registers for saving the calculations state when switching processes, and separate instruction and data memories (Harvard architecture). At the same time, a typical signal processor feature is a hardware multiplier that can multiply two numbers in one cycle. Universal microprocessors usually take several cycles to do a multiplication operation in a sequence of shifts and additions. Other signal microprocessor features are inclusion into the instruction set of such operations as multiply with accumulate (MAC; $C := A \times B + C$), address bit inversion, and bit operations. Signal microprocessors provide hardware support of program loops, ring buffers, interrupt processing, and so on.

The one-cycle multiplication as well as instructions' using contents of memory cells as operands is the reason for the low operating frequencies of these processors.

Signal processors from various manufacturers fall into two classes with substantially different prices. To the first class belong low-price, fixed-point microprocessors; to the other belong more expensive microprocessors with hardware support of floating-point operations.

The need to use floating-point data in signal processing stems from several factors. For many integral and differential conversion tasks, computation precision is of special importance. The exponential data representation provides this precision. Algorithms of compression, decompression, and adaptive filtration in the DSP involve determining logarithmic dependencies and are sensitive to the data-representation precision over a wide dynamic range of values. Using floating-point data substantially simplifies processing, because it does not require performing rounding and data normalization operations or tracking the loss of accuracy or overflow situations.

The price for this convenience is high complexity of floating-point functional units, the need to use more sophisticated microcircuit manufacturing technologies, and, as a result, high prices of signal microprocessors.

Another approach to obtaining high performance has also become popular nowadays. A large number of on-chip transistors can be used to build a symmetrical multiprocessor system with simple processors operating on integer operands. Examples of such processors, called mediaprocessors, are the Mediaprocessor from MicroUnity, the TriMedia from Philips®, the Mpact Media Engine from Chromatic Research®, the NV1® from Nvidia®, and the MediaGx from Cyrix.

These signal processors were designed based on the needs for real-time video and audio data processing in multimedia PCs, game stations, and domestic radio electronics. Because of their less-complex circuitry, as compared to the traditional signal processors, media processors' prices are quite low (about $100) whereas their performance for the dollar ratio is two to three orders of magnitude higher. Peak performance of media processors is several billion integer operations per second.

Some of the largest signal microprocessor manufacturers are Motorola, Texas Instruments®, Analog Devices®, and Lucent Technologies®. Each of these companies produces a wide range of devices oriented toward solving a wide range of tasks. When choosing a DSP microprocessor for a specific project, many parameters must be taken into consideration. Using products from one or another company is to a great extent determined by the preferences of the designers; however, specific advantages offered by individual microprocessor families should also be considered.

3.3 SIGNAL MICROPROCESSOR FROM TEXAS INSTRUMENTS

There are two classes of Texas Instruments [116] signal processors: processors for fixed-point number processing and those for floating-point number processing (see Figure 3.6). The first class is represented by three processor families, whose base models are the TMS320C10, the TMS320C20, and the TMS320C50, respectively.

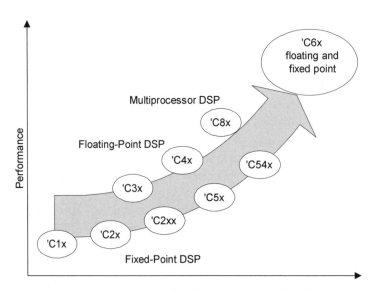

FIGURE 3.6 Texas Instruments signal microprocessor families.

To the second class belong the TMS320C30 and TMS320C4 processors. The TMS320C80 also supports floating-point operations and is a single-crystal multiprocessor system. The TMS320C6x family comprises both fixed-point and floating-point processors.

Later generation processors of a family inherit the main architectural features of the preceding generations and are upward compatible with respect to the instruction sets (which cannot be said about different family processors).

3.3.1 TMS320C1x Family Microprocessors

The first processor of the family, TMS320C10, was introduced in 1982, and owing to several successful engineering approaches received a widespread use [117]. The structure of a typical member of the family, the TMS32C15 microprocessor, is shown in Figure 3.7.

The processor is based on the modified Harvard architecture. This architecture's difference from the traditional Harvard architecture is the capability to exchange data between data and instruction memories, which makes for greater device flexibility.

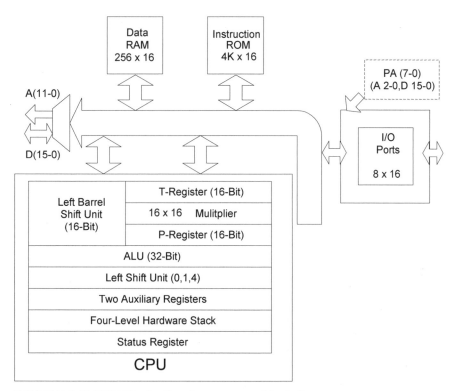

FIGURE 3.7 TMS320C1x family microprocessor block diagram.

The TMS320C10 is a 16-bit processor. Its address space is 4 Kwords instruction memory and 144 words data memory. The words are 16 bits wide. The processor instruction cycle is 160–200 ns long.

Arithmetic functions in the processor are implemented by hardware. It has a hardware multiplier, a shifter, and hardware-supported autoincrement/decrement of data's address registers (AR0, AR1).

The processor has eight 16-bit I/O ports for interaction with external devices. It has the capability to process external interrupts.

The other members of the family (C14–C17) have analogous architectures and differ in the instruction cycle length, memory configuration, and presence (or absence) of additional peripheral devices. (For example, C17 implements a μ-/A-law data codec and a converter of logarithmic pulse-code modulation (PCM) to linear PCM.

3.3.2 TMS320C2x Family Microprocessors

The microprocessors of the TMS320C2x family have the same architecture as the TMS320C1x, but they have higher performance and greater functional capabilities [118]. All microprocessors of the family can use 64 Kwords of data and instruction memory and have 16 16-bit I/O ports and a serial port.

The structure of the TMS320C2x microprocessor is shown in Figure 3.8.

The TMS320C2x family microprocessors use an external DMA controller. In addition to the multiplication operations, the microprocessors' multiplier can perform a squaring operation in one cycle. Processors are equipped with hardware support of multiple instruction execution. They also support inverted indirect binary addressing mode, which is used for efficient execution of the FFT.

The main technical characteristics of the second-generation processors are listed in Table 3.1.

TABLE 3.1 Main Technical Characteristics of the TMS320C2x Family Signal Processors

| Processor | Techno-logy | Cycle (ns) | Internal Memory | | | External Memory | | I/O | | |
			RAM	ROM	PROM	Data	Code	Serial	Parallel	DMA
TMS32020	NMOS	200	544			64K	64K	1	16 × 16	Yes
TMS320C25	CMOS	100	544	4K		64K	64K	1	16 × 16	Yes

TABLE 3.1 Main Technical Characteristics of the TMS320C2x Family Signal Processors *(Continued)*

Processor	Techno-logy	Cycle (ns)	Internal Memory			External Memory		I/O		
			RAM	ROM	PROM	Data	Code	Serial	Parallel	DMA
TMS320C25-50	CMOS	80	544	4K		64K	64K	1	16 × 16	Yes
TMS320E25	CMOS	100	544		4K	64K	64K	1	16 × 16	Yes
TMS320C26	CMOS	100	1568	256		64K	64K	1	16 × 16	Yes

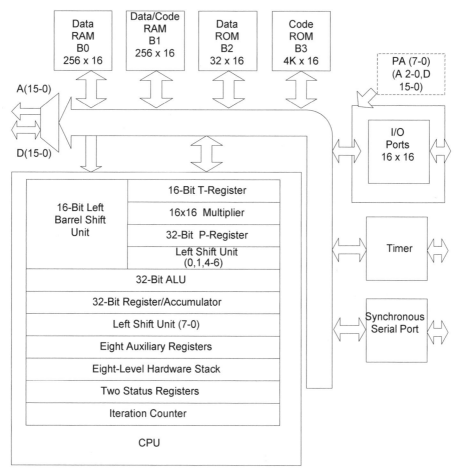

FIGURE 3.8 TMS320C2x signal microprocessor block diagram.

The main features that make the TMS320C2x computer architecture different from the TMS320C1x architecture follow:

■ Multiplication and saving the results operations are performed in a single instruction cycle

■ The instruction set supports floating-point calculations

■ An internal masked 4 Kwords ROM (for the TMS320C25) or an ultraviolet 4 Kwords EPROM (for the TMS320E25)

■ Programs are executed from the on-chip program memory; the memory size is 544 words, 256 of which can be used as data memory

■ Extended external memory is 128 Kwords large (64 Kwords for programs and 64 Kwords for data)

■ Block data transfer instructions

■ An external interface for organizing interprocessor connections and synchronization means for accessing the shared memory

■ A capability to introduce wait cycles when accessing slow external memory or peripheral devices

■ An on-chip timer and a serial port

■ Five (in TMS320C20) or eight (in TMS320C25) auxiliary registers and a special arithmetic unit for them

■ A four-word (for TMS320C20) or eight-word (for TMS320C25) hardware stack and a capability to expand the stack in the data memory by software

■ Bit data-processing instructions

■ Three user-masked interrupts

■ DMA mode (only for the TMS320C25)

3.3.3 TMS320C5x Family Microprocessors

Processors of the generation following the TMS320C2x inherit the general architecture features of the previous generations and are binary compatible with them. However, they differ from the previous generations by their greater functional capabilities, increased clock frequency, and lower power consumption [119].

A block diagram of the TMS320C5x processor structure is shown in Figure 3.9.

The processor has hardware-implemented circular buffers; two independent circular buffers can also be created in the data memory by software. The processor supports multiple-program-block execution. It has 11 shadow registers that are used for rapid saving and restoring of the main registers when processing software or hardware interrupts. The processor's parallel logic unit can perform bit and logic operations on operands stored in the memory or various registers.

The processor can address 244 Kwords of memory: 64 Kwords program memory, 64 Kwords data memory, 64 Kwords 16-bit I/O ports, and 32 Kwords

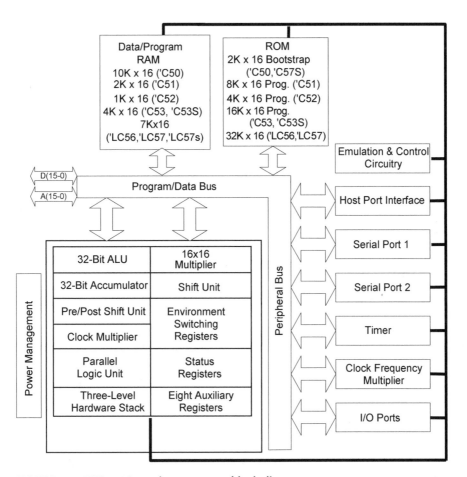

FIGURE 3.9 TMS320C5x microprocessor block diagram.

global memory. The programmable wait cycle generator provides slow memory operation capability. Memory ready request and acknowledge signals allow the processor to be used in shared memory multiprocessor systems. The processors of the TMS320C5x family differ from each other mainly by their on-chip memory configurations.

In addition to the 16-bit I/O ports, this family's processors are also equipped with two serial ports (the TMS320C52 with only one), a timer, and a JTAG interface for testing and debugging.

During development of the processors of this family, substantial attention was given to providing energy-saving operating modes. Processors are produced for 3 V and 5 V power supply voltages. They support active, peripheral, and sleep modes. Current consumption in the active mode is 1.5 mA/MIPS at 3 V and 2.5 mA/MIPS at 5 V. In the peripheral mode, the CPU halts and only the periphery continues to operate, reducing current consumption to 0.25 mA/MIPS and 0.4 mA/MIPS for 3 V and 5 V, respectively. In the sleep mode, the processor halts until it receives an interrupt signal. Current consumption in this mode is 5 mcA.

3.3.4 TMS320C2xx Family Microprocessors

The TMS320C2xx architecture [120, 121] is based on the TMS320C5x family architecture. The main features of the TMS320c2xx microprocessors follow:

■ Binary compatibility with the 'C1x and 'C2x families
■ Extended instruction set to speed up DSP algorithms and support high-level language constructions
■ High performance (up to 40 MIPS)
■ Low power consumption provided by the energy-saving mode

The modified Harvard architecture, which provides for separate instruction and data buses, allows instructions and operands to be fetched simultaneously; the capability to conduct exchange between the data and program memories makes the processor more flexible, thus, coefficients that are located in the program memory can be sent to the data memory, thereby saving memory allocated to coefficients

The processor has a larger, as compared to the previous families, on-chip memory and reprogrammable nonvolatile flash memory.

TMS320C2xx can, on average, execute one instruction per cycle, thanks to its four-stage pipeline. The 'C2xx processors have facilities to control interrupts, to re-execute operations, and to call subprograms and functions.

A typical structure of 'C2xx microprocessors is shown in Figure 3.10.

All microprocessors of the family have the same processor core, but differ in memory configurations and on-chip periphery. With the exception of the TMS320C209, all devices have one synchronous and one asynchronous serial port.

The synchronous port is intended for exchanges with other processors, codecs, and external peripheral devices. The port has two four-word FIFO memory buffers with interrupt-generation mechanism. The synchronous port's maximum exchange speed is half of the processor's operating clock frequency (at 40 MHz, the exchange speed is 20 Mb/s).

The asynchronous serial port is intended for data exchange with other devices. Exchange data format is eight data bits, one start bit, and one or two stop bits. The exchange speed can reach up to 250,000 10-bit characters per second.

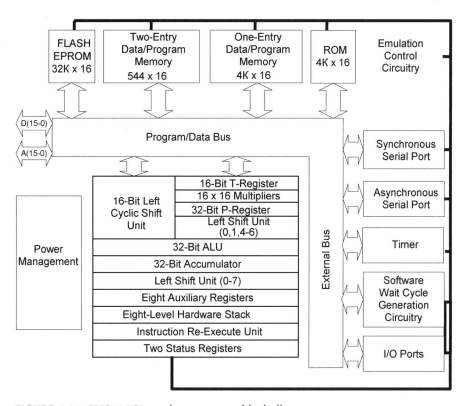

FIGURE 3.10 TMS320C2xx microprocessor block diagram.

Owing to their better value-for-the-dollar ratio among all microprocessor families ($0.12 per MIPS), the TMS320C2xx microprocessors have received widespread use and are employed in communications systems, multimedia devices, industrial automation equipment, military equipment, and so on.

3.3.5 TMS320C54x Family Microprocessors

The distinctive feature of TMS320C54x processors is the combination of the modified Harvard architecture with one internal instruction and three data buses [122, 123]. This internal processor organization makes for a great degree of instruction-execution parallelism. Typical features of this family are a specialized instruction set, on-chip auxiliary peripheral devices, and increased internal memory. All of this makes it possible to achieve increased flexibility and performance.

The three data buses are used to read operands, to write operation results, and to fetch instructions simultaneously in a single processor cycle.

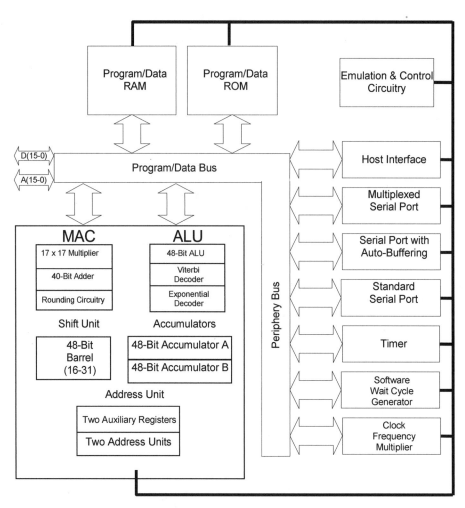

FIGURE 3.11 TMS320C54x microprocessor block diagram.

The overall size of memory that the processor can address is 192 16-bit words. The memory is divided into three specialized segments: instructions, data, and I/O, each of which can hold up to 64 Kwords. Some processor models can have an internal ROM up to 48 words in size and up to 10 Kwords of dual-ported RAM. The processor has a data-protection feature preventing scanning of the internal memory. When this feature is enabled, the contents of the on-chip memory cannot be accessed by any instruction.

To speed typical signal-processing operations, in addition to the standard for the DSP processors barrel shift and address arithmetic units, the processor has several additional functional modules that enhance its flexibility and performance.

The MAC unit performs operations of the $S := S + a \times b$ type on 17-bit operands in one processor cycle. Operations of this type are typical for filtration, compression, and correlation function calculation algorithms.

For fast calculations of the $y = \exp(x)$ values, the microprocessor has an exponent function calculation unit. This unit uses the accumulator value as the value of the function's x argument and calculates the corresponding $y = \exp(x)$ value in one cycle. The CMPS Operation unit is used for efficient implementation of the Viterbi operation; it performs an Add/Compare Selection operation in one cycle.

The microprocessor's ALU can perform arithmetic or Boolean operations on complex numbers (using two registers/accumulators ACCA and ACCB). It can also function as two 16-bit ALUs capable of performing two 16-bit operations simultaneously. The ALU and MAC unit can perform operations simultaneously in a single cycle.

The barrel rotator shifts data 0–31 bits to the left or 0–16 bits to the right in a single cycle. Jointly with the exponent function calculation unit, it can also normalize the accumulator contents in a single cycle. The extended shift capabilities allow the processor to scale data, extract bits from numbers, and guard against overflow and loss of accuracy conditions.

All microprocessors of the TMS320C54x family have the same structure (Figure 3.11), differing only in the suites of on-chip periphery devices. The on-chip periphery comprises the following:

- Software-controlled wait cycle generator
- Software-controlled memory bank switch
- Parallel I/O ports
- Hardware timer and clock generator

The wait cycle generator allows increasing the number of the external bus cycles for operations with slow external memory or peripheral devices.

The memory bank switch allows one cycle to be added automatically when crossing memory bank boundaries inside the program address space or when switching from the instruction address area to the data address area. This additional cycle allows the memory unit to relinquish the bus before the other device receives control over it, thereby avoiding memory access conflicts.

The microprocessors of the family have 64K of I/O ports. These ports are intended for communication with external devices using a minimum of additional external decoding circuitry. The host port interface (HPI) is an 8-bit parallel port connecting the signal processor to the system's host processor. The term *host processor* is used because it seems that using *main processor* or *control processor* is not quite adequate. Data exchange between the host and signal processors is carried over the on-chip HPI 2-Kword 16-bit memory, which can be used as either instruction or data memory. The exchange speed over the HPI reaches up to 160 MBps.

The microprocessors of the family have high-speed duplex ports, which allow connections with other microprocessors, codecs, and other devices. The following varieties of serial ports are implemented in the processors:

- Universal
- Time-multiplexed
- Auto-buffered

The universal serial port uses two memory-mapped registers: a data-send register and a data-receive register. Transmission and receiving of data are accompanied by a masked interrupt, which can be processed by software. The time-multiplexed port can service up to seven devices. The buffered serial port allows direct exchange between devices and the memory without using the processor resources in the process. The maximum exchange speed over the serial port can reach up to 40 MBps.

Like in the TMS320C5x and TMS320C2xx families, the processor implements an effective three-level power management system.

High performance (up to 66 MIPS) and expanded functional capabilities combined with the low price assure the microprocessor a wide application area: cellular and mobile phones, personal radio paging systems, personal digital assistants (PDAs), and wireless data transmission devices (for radio networks).

3.3.6 TMS320C3x Microprocessors

The first representative of the floating-point processor class was the TMS320C30 [124]. At the time it was introduced at the end of the 1980s, TMS320C30 had substantially higher performance than processors from other signal processor manufacturers. The processor has a flexible instruction set, excellent hardware support of floating-point operations, powerful addressing system, expanded addressing space, and supports execution of C language constructions on hardware level.

The processor was produced on the 0.7-micron CMOS technology with three metallization layers. All operations in the processor are executed in a single cycle. With a cycle being 60 ns, the TMS320C30's operating speed is about 33 MFLOPS. The processor's high performance on DSP algorithms is provided by the hardware implementation of several specific functions that in other processors are executed by either software or microcode. The processor has a pipelined register-oriented architecture and can simultaneously perform multiply- and arithmetic-logic operations on fixed- and floating-point numbers in a single cycle. The processor block diagram is shown in Figure 3.12.

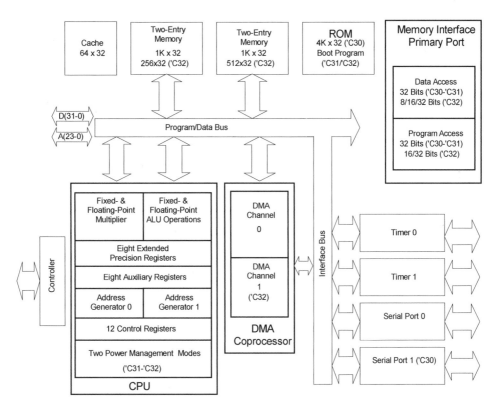

FIGURE 3.12 TMS320C30 microprocessor block diagram.

The processor has a 32-bit instruction bus and a 24-bit address bus. It has two RAM units of one 32-bit Kword each, a 32-bit floating-point multiply unit, a 64 32-bit word code cache, eight extended-precision registers, two address generators, and a register file. The processor implements various addressing modes. The processor's 40-bit ALU supports both fixed- and floating-point operations. The integrated DMA controller allows simultaneous calculation and memory exchange operations to be performed. The multiprocessor interface, two external interface ports, two serial ports, and the extended interrupt system simplify construction of systems based on TMS320C30. Owing to its high performance and ease of use in computer systems, the processor can be used as either a host processor or as a specialized processor.

The 'C3x processors differ, in general, by the number of serial ports (one in the 'C31 and 'C32, two in the 'C30) and the DMA channels (two in the 'C30 and 'C32, one in the 'C31).

Processors of the family enjoy a great popularity among system developers. Taking into account a considerable amount of the developed software for the TMS320C3x microprocessors, TI consequently reissued this family; however, the new version was produced on the 0.18-micron technology, has the internal memory size increased to 34 Kwords, has higher operating clock frequency, and consumes less power. The performance of the renovated microprocessor is 150 MFLOPS.

Main application areas of the 'C3x family microprocessors follow: digital audio, 3D graphics, videoconference communications, industrial robots, copiers and duplicators, and telecommunication systems.

3.3.7 TMS320C4x Family Microprocessors

The TMS320C4x [125, 126] family microprocessors were the next representatives of the TI floating-point signal processors.

Because of their unique architecture, the TMS320C4x microprocessors received wide use in multiprocessor systems and practically pushed out the previously dominating this area Inmos transputer families, which are considered in Chapter 4.

The TMS320C4x processors are compatible with the TMS320C3x in terms of the instruction sets but have higher performance levels and better communications capabilities.

The TMS320C40, TMS320C44, and TMS320LC40 processors comprise the TMS320C4x family.

The performance of TMS320C40 is 30 MIPS/60 MFLOPS; its I/O subsystem throughput is 384 MBps. The 'C40 has six on-chip high-speed (20 MBps) communications ports and six DMA channels, 2 Kwords of memory, a 128-word program cache, and a bootstrap ROM. Two external buses provide access to 4 Gwords of combined address space.

The TMS320C44 processor is a cheaper version of the 'C40. It has four communications ports and 32-Mword addressing space. However, its performance and processor throughput are the same as those of the 'C40.

The TMS320LC40 is an architectural analog of the TMS32C40. However, it has lower power consumption, increased performance level (40 MIPS/80 MFLOPS), and a greater throughput (488 MBps).

The structure of the TMS320C40 microprocessor is shown in Figure 3.13.

The CPU of the TMS320C4x has a pipelined register-oriented architecture. It has the following components:

- Multiplier
- ALU
- 32-bit barrel rotator

- Internal buses
- Additional register arithmetic modules
- Register file

The multiplier operates on 32-bit fixed-point data and 40-bit floating-point data. Multiplication takes one cycle (25 ns) to perform on data of any type; moreover, it is done in parallel with data processing in other functional units of the microprocessor (the ALU, for example).

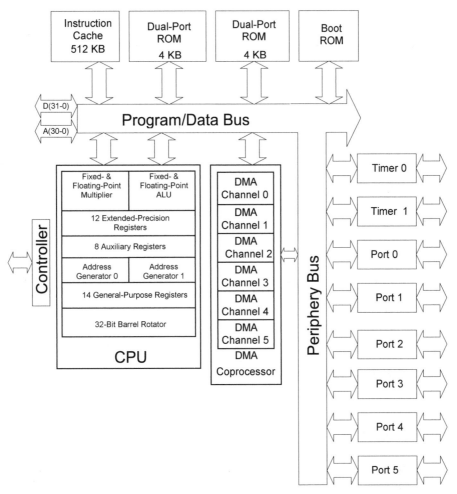

FIGURE 3.13 TMS320C40 microprocessor block diagram.

The ALU performs operations on 32-bit integer and logic data and on 40-bit floating-point numbers, including the data representation format-conversion operations. Operations take one cycle to execute. The microprocessor implements division and square root functions in hardware. The barrel rotator scan shift data from 1 to 32 positions to the left or right in a single cycle.

The two additional register arithmetic modules (address generators) function in parallel with the multiplier and the ALU and can generate two addresses per cycle. The processor supports based indexed, cyclic, and bit-reversed addressing.

The primary register file of the CPU is a multiaccess type and has 32 registers. All its registers can be used by the multiplier and the ALU as general-purpose registers. The registers can perform some special functions. For example, the 12 extended-precision registers can be used for storing floating-point operation results; the eight auxiliary registers can be used for some indirect addressing methods as well as in function of general-purpose integer and logic registers. The rest of the registers provide such system functions as addressing, stack control, interrupts, processor status indication, and reexecution of instruction blocks.

The extended-precision registers are intended for storing and processing 32-bit integers and 40-bit floating-point numbers. The auxiliary registers are available to both the ALU and the two address arithmetic units. The main function of these registers is to generate 32-bit addresses. They can also be used as loop counters or as general-purpose registers.

The microprocessor can address four 32-bit Gwords. The crystal contains two dual-access RAM0 and RAM1 units of 4 KB each as well as a dual-access bootstrap ROM.

The processor's 128 32-bit word instruction cache stores the most-often-used code blocks, which makes it possible to lower the average instruction fetch time. The high performance of the TMS320C4x is obtained owing to the internal process parallelism and the multibus organization of the processor. The separate buses make it possible to simultaneously perform instruction and data fetch and DMA operations.

The six (four in the 'C44) high-speed (160 Mb/s) communications ports provide efficient data exchange among processors. Transmitted and received data are buffered during the exchange process. The synchronization of all exchange operations between the channels, the CPU, and the DMA coprocessor is also automatically controlled. The six channels of the DMA coprocessor have their own address generators, counters, and input and output registers; they provide conflictless memory accesses and data exchange with slow memory modules and peripheral devices without negatively affecting the processor performance. The distinctive feature of the DMA coprocessor is its capability to automatically initialize the channels after performing an exchange.

The Ly link, $y = 0$, … …, 5, is made of an 8-bit bidirectional data bus Dy (7–0) and bidirectional one-bit control lines for the following signals:

- **REQy**—Token request by a communication port
- **ACKy**—Acknowledgement of granting the link for data transmission
- **STRy**—Communication port strobe accompanying placing data on the data lines
- **CRDYy**—Communication port ready signal; issued by the receiving 'C4x upon completing receiving the previous portion of data

Data and control lines are bidirectional. This necessitates coordinating the port states of the links connecting two microprocessors: Either one port is in output mode and the other in the input mode or both ports are in the high-impedance state, which excludes electrical signal exchange between them. This requirement must be complied with at the initial setup; afterwards, it is supported by the link operation protocol. The output port has the token, which is passed to the other port when the link's exchange direction changes. It takes four cycles to pass the token.

PORT QUEUES

Each port has an input and output queues of the FIFO type. The processor or the DMA channel sends data to the end of the port's output queue for transmitting over the link. The received data are taken out from the beginning of the input queue. Both queues have eight elements, each of which stores one 32-bit word. When two 'C4x are connected, a queue of 16 elements is created: eight elements at each end of the link.

LINK INTERFACE

Data exchange over the links is controlled by software by writing corresponding codes to the status registers and controlling the links. Each 'C4x link is allocated a part of the address space 16 words in size. The functions of the words follows:

- The first word holds the port's control register
- The second word is element 0 of the input FIFO queue
- The third word is element 7 of the output FIFO queue
- The remaining words are reserved

Fields and individual bits of the control register determine the following:

- Port's exchange direction: input or output
- Terminating port's input status and switching it into output

■ Terminating port's output status and switching it into input
■ Output queue filled indicator
■ Input queue filled indicator

LINK FUNCTIONING

When a port terminates functioning as the input port, it does not issue a ready-to-receive signal after receiving the first byte. The data transmission ceases until the port is switched into the input port mode or a reset signal is received. No bytes are lost in the subsequent resumed transmission.

A communication port does not acknowledge a request for the token in the following cases:

■ It no longer functions as the input port.
■ Its input FIFO queue is filled.

In this case, the port, keeping the token, can continue functioning as the output port.

If a communication port ceases functioning as the input port at the moment it receives a request for the token, the token request is acknowledged before it stops.

A port's ceasing functioning as the output port results in the following:

■ If the output port does not have the token and it no longer is functioning as the output port, the token request is not sent.
■ If the output port has the token and is conducting a transmission, after the currently transmitted word is sent, the next word is not transmitted.
■ If the output port has the token but is still functioning as the input port while having ceased functioning as the output port, it must pass the token when it is requested.
■ When a port is functioning as the output, it resumes transmission if it has the token; otherwise, transmission must be requested the regular way.

The main synchronization mechanism is based on the *ready/not ready* signals. If a DMA channel or the CPU attempts to read from an empty input queue or write to the full output queue, a *not ready* signal is issued and the DMA channel or the CPU resume reading or writing after receiving a *ready* signal.

A ready signal for the output channel is the OCRDY (Output Channel Ready), which also is an interrupt signal. A ready signal for the input channel is the ICRDY (Input Channel Ready) signal, which also is an interrupt signal.

Each port can generate four different interrupt signals:

- Input queue full
- Input channel ready
- Output channel ready
- Output queue full

The CPU can process all four interrupt signals, whereas DMA channels can process only the ready signals.

Two 32-bit timers can operate from either internal or external synchronization, outputting signals about the tracked time slots and internal events to the processor or to the external environment.

3.3.8 TMS320C8x Family Microprocessors

The TMS320C80 microprocessor, introduced at the end of 1994 [127], is also called Multimedia Video Processor (MVP) because of its high performance in image-processing tasks, in virtual-reality systems, video and audio data compression and decompression, and communications-information processing.

The TMS320C80 processor is a new approach to raising performance and functionality of digital signal processors. It comprises four enhanced advanced digital signal processors (ADSP), each of which can perform several RISC operations per cycle and a fifth master processor (MP): a 32-bit processor with a high-performance FPU.

In addition to the processor core, the crystal has the following units integrated:

- Transfer Controller (TC), which is an intelligent DMA controller supporting DRMA and SRAM interface
- Video Controller (VC)
- JTAG testing and debugging port
- 50 KB of SRAM

A simplified version of the microprocessor, TMS320C82, is also produced. It has a smaller memory, only two ADSPs, no video controller, and, correspondingly, costs less.

The structure of the microprocessor is shown in Figure 3.14.

The overall performance of TMS320C80 in register operations reaches 2 billion RISC-like instructions per second. Because of such high performance, in some applications TMS320C80 can replace more than 10 high-performance signal microprocessors or universal microprocessors available on the market before its appearance.

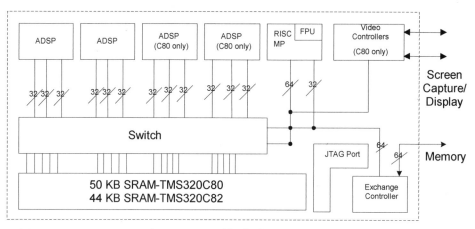

FIGURE 3.14 TMS320C8x microprocessor block diagram.

The technical characteristics of the TMS320C80 microprocessor follow:

- 40 MHz or 50 MHz operating frequency
- More than 2 billion operations per second performance
- SRAM and DRAM DMA
- 2.4 GBps bus throughput for data and 1.8 GBps for instructions
- 4 GB addressing space
- Support of up to four external interrupts
- 3.3 V power-supply voltage
- About 4 million transistors
- CMOS 0.5-micron process technology

TMS320C80 MICROPROCESSOR STRUCTURE

The architecture of the TMS320C80 processor is of the MIMD type. The processors comprising TMS320C80 are independently programmable and can execute different as well as a joint tasks. The processors exchange data via the shared on-chip memory. Access to the memory is provided by the Crossbar switch, which also functions as a monitor when several processors simultaneously access the same memory segment.

The architecture of the TMS320C80 processors needs to be considered in more detail.

ARCHITECTURE OF THE MASTER PROCESSOR

The master processor has a RISC architecture and an on-chip FPU. Like other RISC-architecture processors, the MP uses load/store instructions to access memory data. It also executes most integer, bit, and logic instructions on register operands within one cycle.

The FPU is pipelined and operates on single- and double-precision data. Multiplication, multiplication/accumulation, and load/store operations are combined in the pipeline. The FPU uses the same register file as the ALU. Its performance is about 100 MFLOPS.

The special Scoreboard mechanism tracks registers' availability and provides for their conflictless use.

The structure of the MP is shown in Figure 3.15.

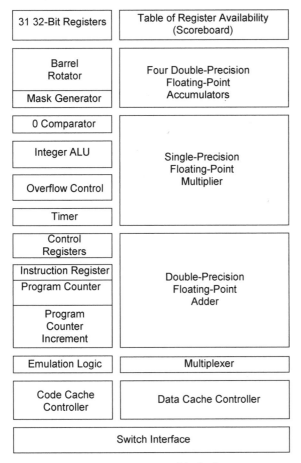

FIGURE 3.15 Master processor block diagram.

The main components of the MP follow:

- 31-register register file (of 32-bit registers)
- Barrel rotator
- Mask generator
- Timer
- Integer ALU
- Control register
- Four double-precision floating-point accumulators
- Floating-point multiplier
- Floating-point adder
- Cache controller

Each of the on-chip MP's code and data caches is 4 KB. The caches are controlled by the controller that is a part of the MP.

The MP can service up to four external interrupts. Three of them are requested by a signal transition whereas the interrupt handler for the fourth one is called by a level change.

For different processing parallelization schemes within the microprocessor framework, the MP can be used either as a controlling or as a universal arithmetic-logic and graphics processor.

The MP's instruction set consists of the following instruction types:

- Arithmetic
- Logic
- Comparison
- Floating-point
- Arithmetic conversion
- Vector arithmetic
- Vector multiply/accumulate
- Vector conversion
- Vector double-precision multiply/accumulate
- Branch and context switching
- Control
- RAM read/write
- Shift

ADSP PROCESSOR ARCHITECTURE

The architecture of the TMS320C80 ADSP processors is optimized for 2D and 3D graphics-, video-, and audio-processing applications. ADSP can perform a multiplication operation, an arithmetic-logic operation (a shift add, for example), and two memory accesses, all in a single cycle. The internal ADSP parallelism allows speeds of more than 500 MOPS to be achieved for some algorithms.

ADSP manipulates 32-bit words; its instructions are 64 bits wide. The processor uses flat, immediate, and 12 types of indirect addressing.

The following features are characteristic of the ADSP architecture:

- Three-stage pipeline
- 44 user registers (10 address, six index, eight data, and 20 other registers)
- 32-bit, three-way ALU
- Bit replicator
- Two address units
- 32-bit barrel rotator
- Mask generator
- Conditional operation unit (to lower the branch times)

The structure of the ADSP processor is shown in Figure 3.16. In the drawing, the A/S notation designates the align/sign expand unit; the Rep notation designates the replicator.

TRANSFER CONTROLLER

The transfer controller handles processor and memory transfer operations both inside the crystal (via the switch) and outside the crystal. For these tasks it uses its interface circuitry, which supports all common memory standards (DRAM, VRAM, SRAM) and can change the bus width dynamically over the range from eight bits to 64 bits. Using the priority system of servicing DMA memory accesses, the transfer controller can exchange data at the speed of up to 400 MBps without interrupting calculations.

The transfer controller supports linear and coordinate memory addressing for efficient exchanges when working with 2D and 3D graphics.

VIDEO CONTROLLERS

The two TMS320C80 microprocessor's on-chip video controllers have video capture and display capabilities in both vertical and horizontal scanning modes. The capture/scan modes can be set for each controller independently.

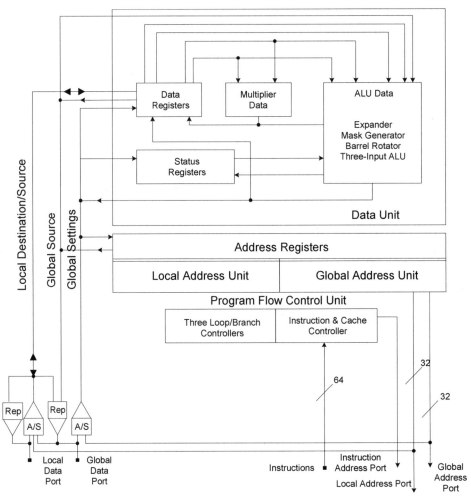

Figure 3.16 ADSP processor block diagram.

TMS320C80 MICROPROCESSOR APPLICATION AREAS

Application areas of the microprocessor are much wider that its name would imply. The processor is used in multimedia information processing, video conferencing, 2D and 3D graphics processing, virtual-reality emulation, and data-transmission systems.

3.3.9 TMS320C6x Family Microprocessors

The new digital signal processor family from TI, the TMS320C6x, includes both fixed- and floating-point processors [128–130]. The first representative of this family, the TMS320C6201 processor, operates only on fixed-point data.

Operating at 200 MHz frequency, processor's performance is up to 1.6 billion operations per second. The processor is used in the following areas:

■ Wireless data transmission systems
■ Remote medical diagnostics means
■ Base mobile communications stations
■ Modem pools and remote servers
■ xDSL and cable modems
■ Multichannel telephone platforms, office ATS, voice message systems
■ Multimedia systems

In addition to the processor core, the TMS320C6201 crystal contains the following units:

■ 1 Mb memory (512 Kb for programs and 512 for data)
■ 32-bit external memory interface supporting SDRAM, SBSRAM, and SRAM standards
■ Two expanded buffered serial ports
■ 16-bit CPU port
■ Two data memory access channels with bootstrap capability
■ Time-slot generator

Built on the VelociTI architecture, developed by TI, the 'C62xx processor is the first from VLIW signal processors to use instruction-level parallelism to raise performance.

The structure of the TMS320C6201 microprocessor is shown in Figure 3.17.

The TMS320C6201 processor comprises three main parts: the CPU (processor core), peripheral devices, and memory.

A VelociTI VLIW processor is the core of the TMS320C6201. It has eight functional units, including two multipliers and six ALUs. The units interact via two register files of 16 32-bit registers. The CPU can execute up to eight instructions per cycle.

Program parallelism is detected during the compilation stage; hardware does not analyze data dependencies during execution. Code is executed in independent functional units in the order specified by the program.

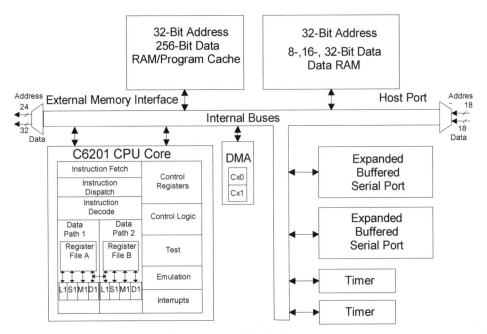

FIGURE 3.17 TMS320C6201 microprocessor block diagram.

The processor packs instructions, reducing the code size and instruction fetch time. The 256-bit data memory bus allows eight 32-bit instructions to be selected in a single cycle. All instructions contain conditions for their execution, which allows the processor performance expenditures on branch executions to be lowered and the degree of the processing parallelism increased.

The processor can operate on 8-, 16-, and 32-bit data. For applications requiring high-precision computations, there is a capability to operate on 40-bit data. Results of all basis arithmetic operations are rounded and normalized. The processor can perform bit field operations, such as extract, set, clear, and bit counting.

The CPU has two data-processing modules, each of which has functional units L, S, M, and D and a 16-register 32-bit register file. The functional units perform shift, multiply, logic, and address operations. All operations are executed on registers. Two data-addressing units (D1 and D2) exclusively handle data transfers between the register file and the memory. The control register file determines various aspects of the processor's functioning.

The processing of a VLIW starts with fetching a 256-bit packet from the program memory. The instructions are bundled for joint execution into an execute packet up to eight instruction long. The lower instruction bit of each instruction determines whether the following instruction belongs to the same execute packet.

The instruction fetch/decode/dispatch unit can send up to eight 32-bit instructions per cycle to the functional units over each data-processing path (A and B).

The 'C62xx implements flat and cyclic (for the A4-A7 and B4-B7 registers) addressing mode. The addressing mode is determined by the address mode register (AMR).

The 'C62xx family processors have 14 interrupts; these correspond to the Reset, NMI, and number 4–15 interrupts.

'C62xx has on-chip memory that can be used as program memory or as cache. The processor's external memory interface combines the internal and external memories into a unified memory space.

The on-chip memory is split into data and program memory. The 'C62xx family processors have two 32-bit data memory ports and one 256-bit program memory port. The TMS320C6201's on-chip data and program memories are 64 KB each. The data memory is of the interleaved type (four 16-bit banks) to increase the fetch speed by accessing different memory banks simultaneously.

The 'C62xx family processors can also have on-chip an external memory interface, a DMA controller, host port interface, power-management devices, expanded buffered serial ports, and 32-bit timers.

The next representative of the TMs320C6x family is the TMS320C6701 microprocessor. It supports floating-point operations, has a wider, 128-bit, external memory interface and lower, 167 MHz, operating frequency.

The following functional units were added to the TMS320C6701 microprocessor:

- M unit for multiplying 24 32-bit integer numbers or 32 64-bit floating-point numbers
- L unit for arithmetic operations on 32 40-bit integers or 32 64-bit floating-point numbers
- S unit for shift, branch, and item operations with 32 64-bit floating-point numbers
- D unit for addressing operations

The processor's peak performance is 688 MFLOPS.

3.4 SIGNAL MICROPROCESSORS FROM ANALOG DEVICES

There are two families of Analog Devices microprocessors: the ADSP-21xx and the ADSP-210xx [131, 132].

The ADSP-21xx family comprises one-crystal 16-bit microprocessors with common base architecture optimized for executing DSP algorithms and other applications requiring high-performance fixed-point computations. To date, there are 15 microprocessors in this family, which mainly differ by on-chip peripheral devices, such as cache, timers, and ports.

The ADSP-210xx family comprises 32-bit microprocessors oriented at floating-point signal algorithms. The family consists of the ADSP-21010, ADSP-21020, ADSP-21060, ADSP-21062, ADSP-21160, and ADSP-TS001 microprocessors.

Microprocessors' instruction sets are upward compatible within each family. The latter representatives of the family have higher performance levels and contain additional on-chip functional units.

In the further presentation discussions of the Analog Devices signal microprocessors, the general architectural and constructive features of each family, and the most significant different features of individual family representatives will be pointed out.

3.4.1 ADSP-21xx Family Microprocessors

The ADSP-21xx family microprocessors [131] successfully compete with the analogous products from Motorola and TI, owing to their high-performance-at-a-low-price combination, as well as because of existence of advanced hardware and software application system development tools. The processors' high performance in executing signal algorithms is achieved by flexible multifunction instruction set, hardware implementation of most of the typical signal processing operations, high level of process parallelism, and shortening of the instruction cycle. The ADSP-21xx microprocessors have a modified Harvard architecture, which allows individual access to the split program and data memories. The architecture has become the de facto DSP standard and its analogs are used in many other processors, including the TMS320xxx (see Section 3.3).

The generalized structure of the ADSP-21xx microprocessors is shown in Figure 3.18.

Each microprocessor of the family has three independent functional units: an ALU, a MAC unit, and a barrel rotator. Each unit directly operates on 16-bit data and provides hardware support to variable-precision computations.

Microprocessors have an instruction address generator and two data address generators that provide addressing of both the internal and external data and instructions. The address generators function in parallel, which shortens the instruction execution cycle by allowing an instruction and two operands to be fetched per cycle.

The serial ports provide an interface with most of the standard serial devices as well as with hardware data compression/expansion devices operating under the A- or μ-companding laws.

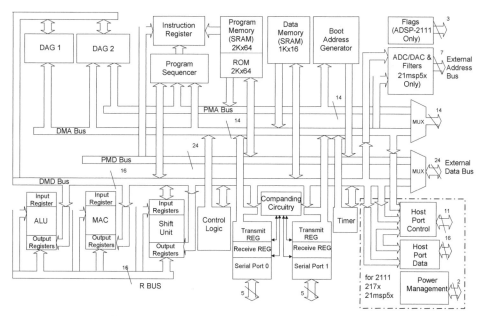

FIGURE 3.18 Generalized structure of ADSP21xx microprocessor.

The host processor interface port does not need additional interface circuitry to interact with the system's host processor, which can be either a given family processor or some other processor.

A distinctive feature of the ADSP-21msp5x microprocessors is the analog interface for the processed signal. The interface units are comprised of an ADC, a DAC, digital and analog filters, and a parallel interface.

The ADSP-2181 microprocessor has an internal DMA port and a byte DMA port, which provide rapid exchange with the internal memory. The internal DMA port supports asynchronous exchange with the program memory while the byte port is capable of reading and writing both instructions and data.

The distinctive feature of the microprocessors from Analog Devices is their high degree of internal operation parallelism. The processors can do the following per cycle:

- Generate the address of the next instruction
- Fetch the next instruction from the memory
- Perform one or two data transfers
- Refresh one or two data pointers
- Execute an operation

In the same cycle, a microprocessor equipped with the proper unit can do the following:

■ Receive and/or send data via the serial ports
■ Receive from and/or send data to the host processor
■ Receive and/or send data over the analog interface

The main characteristics of the ADSP-21xx family microprocessors are listed in Table 3.2.

TABLE 3.2 Main Characteristics of ADSP-21xx Family Processors

Capabilities	2101	2103	2105	2115	2111	2171	2173	2181	2183	21msp58
ALU	+	+	+	+	+	+	+	+	+	+
MAC Unit	+	+	+	+	+	+	+	+	+	+
Shift	+	+	+	+	+	+	+	+	+	+
Data Address Generator	+	+	+	+	+	+	+	+	+	+
Instruction Address Generator	+	+	+	+	+	+	+	+	+	+
Data RAM (Kwords)	1	1	0.5	0.5	1	2	2	16	16	2
Instruction RAM (Kwords)	2	2	1	1	2	2	2	16	16	2
Timer	+	+	+	+	+	+	+	+	+	+
Multichannel Serial Port	+	+	–	+	+	+	+	+	+	+
Serial Port 1	+	+	+	+	+	+	+	+	+	+
Host Interface Port	–	–	–	–	+	+	+	–	–	+
DMA Port	–	–	–	–	–	–	–	+	+	–

TABLE 3.2 Main Characteristics of ADSP-21xx Family Processors *(Continued)*

Capabilities	2101	2103	2105	2115	2111	2171	2173	2181	2183	21msp58
Analog Interface	–	–	–	–	–	–	–	–	–	+
Power Supply Voltage (V)	5	3.3	5	5	5	5	3.3	5	3.3	5
Performance (MIPS)	20	10	13.8	20	20	33	20	33	33	26

The microprocessor core shared by all ADSP-21xx family microprocessors is shown in Figure 3.19. The microprocessor's ALU performs a standard set of arithmetic and logic functions, including division. The MAC unit performs a multiple/accumulate operation in a single cycle. The shift unit performs arithmetic and logic shifts, normalization, and exponentiation. The microprocessor's functional units can exchange operation execution results over the internal result bus.

The internal functional units are interconnected by the following five buses: data memory address (DMA), program memory address (PMA), data memory data (DMD), program memory data (PMD), and the internal result (R) buses. The first four buses have multiplexed external interface in the form of the address and data buses (see Figure 3.19).

Instruction sets of all microprocessors are upward compatible. Some microprocessors of the family (ADSP-2171, ADSP-2181, and ADSP-21msp5x) have additional and expanded instructions. Each instruction takes one cycle to execute. Microprocessor's multifunctional instructions combine several data transfers with arithmetic-logic processing.

All microprocessor's devices are 16 bits wide and operate on fixed-point data. Numbers are either unsigned or represented in complement code. Logic operations can be performed on bit strings.

Enhancements to the microprocessors of this family are being made in the direction of increasing the operating frequency, lowering power consumption, and expanding the processor's communications capabilities.

The new microprocessor of the ADSP-219xx family [133] has a modified core that Analog Devices considers the key element in the technology aimed at developing promising 16-bit general-purpose signal processors and processors for embedded applications.

The core is intended to be used as a base upon which special circuitry and software will be custom built for specific customer requirements and parameters.

For applications requiring high-performance, processors with several cores on one crystal will be built. Four-core processors with the performance of 1.2 MAC operations per second per square inch of the crystal are planned for production in the future.

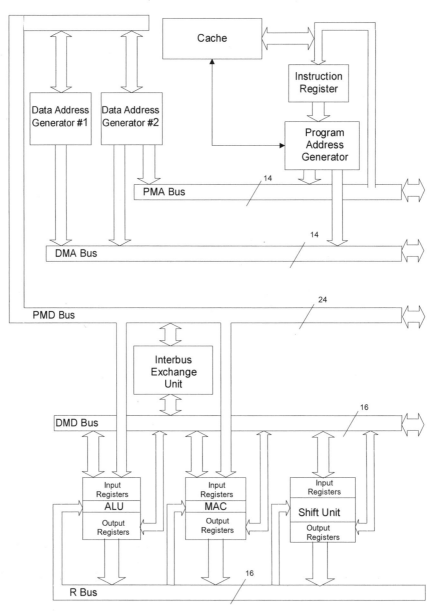

FIGURE 3.19 ADSP-21xx microprocessor core block diagram.

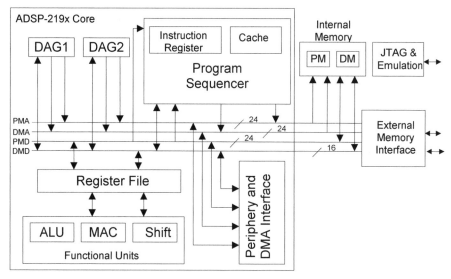

FIGURE 3.20 ADSP-219x microprocessor block diagram.

The structure of the ADSP-219x microprocessor is shown in Figure 3.20.

As compared with the 14-bit address bus of the ADSP-218x family microprocessors, the ADSP-219x family uses a 24-bit bus, which allows flat addressing of 64 Kwords or page addressing of 16 Mwords. The address generators of the 219x family processors support all old and five new addressing modes.

The ADSP-219x is binary compatible with the previous microprocessors of the ADSP-21xx family. It has a higher performance level (up to 300 MIPS) and lower power consumption (0.4 mW/MIPS).

3.4.2 ADSP-21xxx Family Microprocessors

Another microprocessor family from Analog Devices is the ADSP-21xxx. Its microprocessors are targeted at floating-point data processing [132]. The general core structure of these microprocessors is similar to the previously considered ADSP-21xx core structure; however, the buses and functional devices are wider and hardware support for floating-point data processing was added.

The capability to perform both fixed- and floating-point computations led to increased complexity of the functional devices and, as a result, made the microprocessor's manufacture more complicated and increased its cost.

The following microprocessors belong to this Analog Devices microprocessor family: ADSP-21020, -21010, -21060, -21062, and -21160.

The distinctive features of the ADSP-21xxx microprocessor architecture follow:

Fast and flexible arithmetic units—All instructions are executed in a single cycle. In addition to the traditional arithmetic operations, the microprocessor's instruction set includes such as 1/x, 1/R(x), shift, cyclic shift, and combined add/subtract-multiply operations.

Independent data streams input into and output from computation units—During one processor cycle, the ADSP-21xxx can simultaneously read/write two operands to/from the register file, load two operands to the ALU, and place two operands into the multiply unit. During the same cycle, the ALU and the multiply unit can produce two results (or three if the ALU performs the operation concurrently with an add/subtract operation. The processor's 48-bit instruction word allows specifying concurrent arithmetic instruction execution and data exchange.

Extended precision and expanded dynamic range of performed operations—All representatives of the family operate on 32-bit floating-point data, 32-bit integers (signed and in supplementary code), and 40-bit extended-precision floating-point data. The extended precision of calculations is achieved by reducing the rounding error of the computation units. The accumulator for 32-bit floating-point data is 80 bits wide.

Two address generators—They pre- and post-generate flat and indirect data addresses and perform modular and bit-inverse operations on addresses.

Efficient means for generating instruction sequences and organizing program loops—Loop initialization, loopback, and exit are performed within one processor cycle in up to six nesting levels. The processor has hardware support for executing branch and delayed branch instructions.

The 33.3 MHz ADSP-21020 was the first microprocessor of the ADSP-21xxx family. Its performance was 66 MFLOPS. The processor's ALU operates on 32-bit data that are expanded to 40 bits for floating-point operations. Fixed-point data is processed in 32-bit format using 80-bit accumulator.

The next representative of the family was the less-expensive ADSP-21010 microprocessor. It operated at 21.5 MHz and on only 32-bit data.

The structure of the ADSP-21020 microprocessor is shown in Figure 3.21.

The processor's universal ALU, barrel rotator, and the universal multiply unit function independently, thereby providing a high level of internal operation parallelism. The general-purpose register file is used in data exchange between the computation units and the internal bus as well as for storing intermediate results. It has 32 registers (16 primary and 16 secondary) and 10 ports and, in tandem with the processor's Harvard architecture, allows efficient data exchange between the computation units and the memory. The processor's extended Harvard architecture allows fetching an instruction and up to two operands within one cycle. The instruction cache operates selectively: only those instructions are cached, fetching those that conflict with fetching data from the program memory.

The address generators (DAG1 and DAG2) provide hardware implementation of circular buffers. Circular buffers are used by digital filtering algorithms and the Discrete Fourier Transforms, which require cyclic change of the addresses of the processed data. Physically, a circular buffer can be located starting from any memory address; register pointers are used to reference its contents. The two DAGs have 16 primary and 16 secondary registers, which make it possible to simultaneously work with 32 circular buffers.

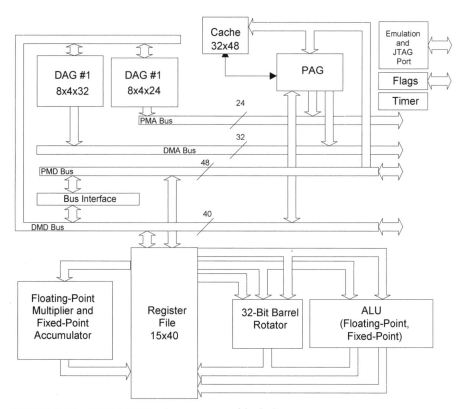

FIGURE 3.21 ADSP-21020 microprocessor block diagram.

3.4.3 SHARC Architecture ADSP-2106x Family Microprocessors

The next generation of the ADSP-21xxx family is represented by the ADSP-21060 and ADSP-21062 microprocessors. These microprocessors have core structures similar to the previously considered ADSP-210xx microprocessors, and their instruction sets are compatible. However, these microprocessors have significant architectural differences, because of which they were placed into a separate architecture family: Super Harvard Architecture Computer (SHARC). The SHARC architecture (Figure 3.22) continues developing the transputer direction in the microprocessor technology and sets new standards for integrating signal processors into multiprocessor systems. This architecture is an example of a harmonious union of the construction principles of distributed and linked systems: it combines the simplicity and effectiveness of distributed systems scalability with the programming convenience of shared memory systems.

A SHARC microprocessor comprises a high-performance floating-point processor core, a host processor interface, a DMA controller, serial ports, communication links, and a shared bus.

At 40 MHz processors' performance is 80 MIPS and 120 MFLOPS [134].

The bus switch connects the processor core with the independent I/O processor, dual-ported memory, and the multiprocessor system bus port.

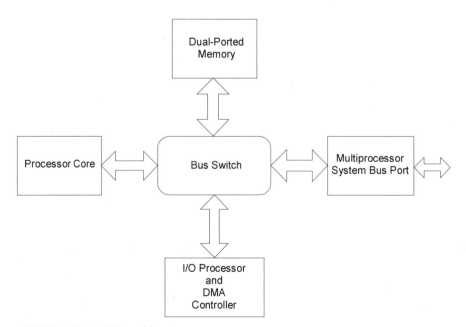

FIGURE 3.22 SHARC architecture.

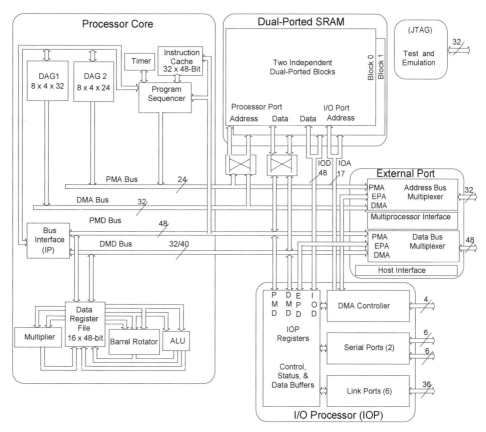

FIGURE 3.23 ADSP-2106x microprocessor block diagram.

Computation modules based on the ADSP-2106x microprocessor are produced as microprocessor clusters: cards with ISA, PCI, or VME interface have from three to eight nodes. Mezzanine boards SHARCPAC and TRANSPAC that are installed into special motherboard connectors are also produced.

Computation modules based on SHARC microprocessors are used in radio- and hydrolocation systems, speech recognition and image-processing systems, telecommunications, and medical diagnostic systems.

The structure of the ADSP-2106x microprocessors is shown in Figure 3.23.

ADSP-21060 microprocessors have on-chip 4 Mb static dual-ported RAM that can be configured to work with 16-bit words (256 Kwords), 32-bit words (128 Kwords), and/or 48-bit words (80 Kwords). All memory can be split into program and data memory of any size.

The other member of the family, the ADSP-21062 microprocessor, differs from the ADSP-21060 by its smaller size on-chip memory (2 Mb).

The microprocessor's overall addressing space is 4 Gwords. The on-chip external memory controller allows setting a variable number of the generated wait cycles and supports page exchange with dynamic memory.

The host processor interface provides easy connection to standard 16- and 32-bit microprocessor buses. Data is transferred asynchronously over the interface; the transfer speed is limited only by the microprocessor's operating clock frequency. The host interface is accessible via an external port and is mapped into the microprocessor's address space. The four channels of the controller provide data and instruction exchange via the host interface with the minimal participation of the processor core.

The on-chip 10-channel DMA controller provides data exchange between the internal and external memories, peripheral devices, host processor, serial ports, and microprocessor links.

The ADSP-21060 has two synchronous serial ports for communications with various peripheral devices. Its maximum data transfer speed is 40 Mb/s. In the DMA mode, exchange can be conducted simultaneously in both directions. During the exchange process, supplementary data conversion can be performed, such as, for example, μ- or A-companding.

The ADSP-21060 provides developers with a wide range of possibilities in creating multiprocessor signal systems. The common addressing space can be shared by several processors. An automatic semaphore support is provided for read/modify/write memory operation sequences. The on-chip distributed bus logic allows creating systems of up to six ADSP-21060 processors and the host processor. Interprocessor control is performed by using interrupt vector mechanism.

The six 4-bit links provide the processor with additional I/O capabilities. The links transfer data at the rising and falling clock edges, thereby providing a transfer of eight bits per clock. The links are used in multiprocessor systems to create point-to-point connections. Data can be transferred over the links in 32- or 48-bit words directly to the processor core or, by using DMA channels, to the internal memory. Each link has its own input and output data buffer registers. The minimal interprocessor exchange speed over the links or an external port is 240 MBps. Processors can be booted either from the ROM, host processor, or via one of the links.

MICROPROCESSOR LINKS

Each Lx link ($x \in \{0, 1, ..., 5\}$) is made of four bidirectional data lines LxDAT ($x \in \{0, 1, 2, 3\}$) as well as two bidirectional control lines, LxCLK and LxACK, that provide asynchronous data transfer in a handshaking mode.

One of the link's ports is configured for input and the other for output or both ports are placed into the high-impedance state.

For data exchange, each link's ports can use one of the six buffers. Data is read out of or written to the buffers under control of either the DMA controller or the CPU. Buffers of links 5, 4, 3, 2, 1, and 0 are supported by DMA channels 7, 6, 5, 4, 3, and 1, respectively.

The DMA controller is programmed for buffer work by specifying buffer size, initial memory address, address increment, and the transfer direction. When the DMA controller completes the operation, an interrupt is generated individually for each of the 10 DMA channels.

The buffers can be read or replenished by the processor by read/write operations in the external device memory area. If an attempt to read from an empty buffer is made, the processor must switch into the wait state until external data are received. Naturally, a write operation to a filled-up buffer also needs to be postponed until free room becomes available in the buffer.

In applications for which the delay introduced by the DMA controller is unacceptably long, the processor can work directly with the buffers. In this case the DMA controller must be placed into an inactive state.

The following interrupts are generated during links' operation:

- If a DMA channel is enabled, upon completion of the message transmission the DMA controller generates a masked interrupt.
- A masked interrupt is generated if the DMA controller is disabled while the input buffer has data in it or the output buffer is empty. Upon an interrupt, a buffer `read` or `write` operation needs to be performed.
- A masked interrupt other than those generated by the buffers is generated upon an external access to the disabled link's port.

ASYNCHRONOUS HANDSHAKING

The transmitter asserts the LxCLK line high when placing next nibble on the LxDAT lines. The receiver uses the falling edge of this signal to latch the nibble. The receiver asserts the LxACK high when next word is received into the buffer.

The transmitter examines the LxACK at the beginning of each new word transmission. The word is not transmitted if there is no required LxACK level at this time.

If there is no LxACK, the transmitter continues to assert the LxCLK high. When the LxACK goes high, the LxCLK goes low and the transmission of a new word begins.

The input buffer may be filled if a high-priority DMA or chain-loading operation is being performed. The LxACK is not asserted in this case; however, it is asserted immediately after the operation completes. The data is latched in the input buffer by the falling LxCLK edge. A data nibble is sent every ADSP-2106x cycle; if the special LCKx2 bit is sent, two nibbles per cycle are sent.

BUFFERS

Each buffer is made of an internal and an external register. During transmission the internal register is used to receive data from the internal memory under the control of the DMA controller of the CPU. The external register is used for un-packing nibbles for the link port (the high nibble comes first). These two registers place data into a FIFO queue. A buffer can hold two words before a filled signal is generated. If a register is empty, the LxCLK signal goes low.

During reception the external register packs the received nibbles into words and, under the control of the DMA controller of the CPU, sends them to the memory over the internal register. If the DMA controller or the CPU has not managed to extract data from the internal register and both buffer registers are busy, the LxACK signal is not asserted and the reception terminates.

The register width is configured by software and can be 32 or 48 bits. The register width must be 48 bits to work with 40-bit data or instructions.

NEGOTIATING LINK USE

Before two processors can start interacting over the connecting link, it must be determined which of them will be the transmitter and which the receiver. This is done by exchanging the token: a software modified flag. Upon the initialization, the token is set in one of the processors, designating it as the link master and permitting it to transmit.

If the input port wants to become the link master to transmit data, it must assert the LxACK (data request) to the current link master. The master uses software protocol to determine when LxACK is asserted to acknowledge data and when to request the token.

If the current master decides to pass the token, it sends back the identifier designated by the user as the token and clears its own token. At the same time, the slave processor examines the received data, and if they contain the necessary word, it sets its own token, thereby switching into the master state.

If the received data does not contain the necessary identifier, the slave processor must recognize that the master processor is beginning a new data transfer.

The master processor can also request to receive data by software protocol. It does so by passing the token to the slave without receiving a request for it.

The DMA controller and link ports exchange data using the same handshaking protocol that is used by all I/O ports.

HOST COMPUTER INTERFACE

The interface allows connecting to standard 16- and 32-bit microprocessor buses using minimum additional equipment. The interface serves as an external port that is mapped into the address space and provides asynchronous data transfer.

The host processor generates a bus request. The ADSP-2106x grants the bus upon completion of the current cycle, issuing a bus grant signal and a ready signal. Thereafter, the host processor can read from and write to the ADSP-2106x memory.

TRANSMISSION ERROR DETECTION

A special control register stores information on the nibble counter status of each link port. If the counter is not zeroed out upon transmission completion, a transmission error signal is generated. Special protocols at the transmitting and receiving link ends are used to control this signal.

At the end of the transmission of a word block, the transmitter must configure its port for reception. The exchange protocol provides for transmitting one extra blank word. This allows the receiver to send a corresponding message to the transmitter. After receiving a data block, the reception protocol generates an interrupt to read the control register and to send a corresponding message to the transmitter.

3.4.4 SHARC Architecture ADSP-2116x Family Microprocessors

The next product from Analog Devices was the 32-bit SHARC architecture ADSP-2116x family microprocessors. They were introduced in 1998 [135].

The main difference between the ADSP-2116x family and the ADSP-2106x family is that the ADSP-2116x family microprocessor core has two processor elements. The processor elements can execute an instruction simultaneously, each on its own data (the SIMD mode). SIMD data processing allows the processor performance to be substantially increased with negligible changes in the software model. Another difference from the previous series microprocessors is wider internal buses.

Each processor element has the following set of functional devices: a fixed- and floating-point ALU, a floating-point multiplier with an 80-bit register/ accumulator, a barrel rotator, and a register file of 32 40-bit registers. The ALU and multiplier operate on 32-bit fixed- and floating-point data and on 40-bit extended-precision floating-point data.

The structure of the ADSP-2116x microprocessors is shown in Figure 3.24.

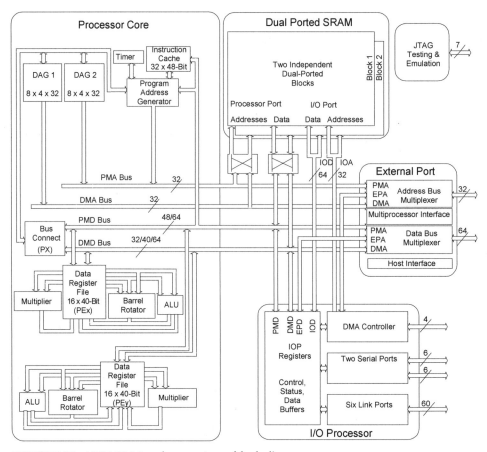

FIGURE 3.24 ADSP-2116x microprocessor block diagram.

The new microprocessor family is binary compatible with the ADSP-2106x family. The ADSP-2116x programming system, ADI Visual DSP, comprises translators, debuggers, optimized libraries, as well as hardware development support tools.

The ADSP-21160 microprocessor is the first representative of the new family. It has a 4 Mb internal memory, 14 DMA channels, and six links. The microprocessor's main features follow:

- Executing instructions, in both computation blocks, in a single cycle in SIMD mode
- A dual address generator
- Efficient loop-sequencing instructions
- 100 MHz operating frequency
- Peak performance of 600 MFLOPS
- 2.5 V power supply voltage

At relatively low operating clock frequency, the high performance of the ADSP-2116x microprocessors is achieved by use of an efficient instruction set. For example, a FIR filter program for the TMS320C67x microprocessor has 100 instructions, whereas the same program for the ADSP-2116x needs only 25 instructions.

Table 3.3 lists benchmark results of testing SHARC microprocessors on common DSP algorithms.

TABLE 3.3 *SHARC Microprocessors Performance*

Function	ADSP-21065L SHARC	ADSP-21160 SISD	ADSP-21160 SIMD
Clock Frequency	66 MHz	100 MHz	100 MHz
Clock Duration	15 ns	10 ns	10 ns
MFLOPS Average	132	200	400
MFLOPS Peak	198	300	600
1,024 Point FFT	274 mcs	180 mcs	90 mcs
FIR Filter Link	15 ns	10 ns	5 ns
IIR Filter Link	60 ns	40 ns	20 ns
Matrix Multiplication			
[3x3] * [3x1]	135 ns	90 ns	45 ns
[4x4] * [4x1]	240 ns	160 ns	80 ns
Division	90 ns	60 ns	30 ns
Square Root	135 ns	90 ns	45 ns

3.4.5 TigerSHARC ADSP-TS001 Microprocessor

Analog Devices sees static detection of instruction-level parallelism as a way of further increasing performance.

The ADSP-TS001 is the first TigerSHARC microprocessor with the new static superscalar architecture [136]. The TigerSHARC processors combine DSP capabilities with the RISC and VLIW features.

The ADSP-TS001 has a high level of support of such typical DSP processors features as short processor cycle of a determinate length, rapid reaction to interrupts, and efficient peripheral device interface. One way used to achieve this is using such VLIW approaches to scheduling functional units loading as detecting instruction-level parallelism during the compilation stage and specifying independent functional unit loading order in programs. Another way is the RISC approaches to instruction execution, such as fixed instruction structure, pipelined execution of up to four 32-bit operations on data in the register file per cycle, branch prediction, and so on.

Running at 150 MHz, the ADSP-TS001 has the highest performance level among the SHARC family processors, operating on both fixed-point and floating-point data. The ADSP-TS001 can process 8-, 16-, and 32-bit fixed- and floating-point data; moreover, the execution speed rises as the data width decreases.

The main characteristics of the ADSP-TS001 microprocessor follow:

- Performance
  - 1,200 MMAC/s (Million MACs per second) at 150 MHz with 16-bit fixed-point data
  - 300 MMAC/s at 150 MHz with 32-bit floating-point data
  - 900 MFLOPS with 32-bit floating-point data
- Memory
  - 6 Mb on-chip unified SRAM (not the traditional Harvard architecture)
  - Up to 4 GB addressing space
  - On-chip SDRAM controller
- Communications capabilities
  - 600 MBps external bus bandwidth
  - 600 MBps total data bandwidth over four communications ports
  - Capability to join up to eight processors into a multiprocessor configuration without using additional interface circuitry
  - Four general-purpose I/O ports

The internal structure of the TigerSHARC processor is shown in Figure 3.25.

The following units comprise the microprocessor: two computation units (X and Y), memory, two address arithmetic ALUs, program sequencer, and peripheral components.

Each of the two computation units has a 32-register register file, a multiplier, an ALU, and a 64-bit shift unit.

Efficient loading of the microprocessor's functional units is scheduled in accordance with the VLIW approach at the program compilation stage. This allows the processor to execute in a single cycle eight 40-bit multiply operations with accumulation on 16-bit data, two 40-bit multiply operations with accumulation on 16-bit complex numbers, or two 80-bit multiply operations with accumulation on 32-bit data.

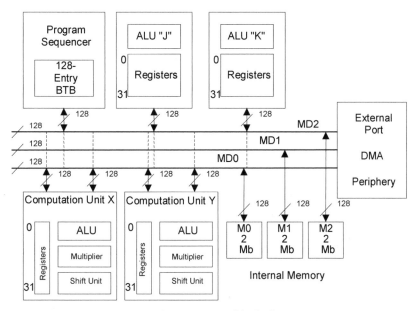

FIGURE 3.25 ADSP-TS001 microprocessor block diagram.

The two ALUs, J and K, are intended for generating addresses but can be used to perform operations on data. Each ALU has its own 32-item register file and supports circular buffers and bit-inverse addressing.

The program sequencer provides such instruction execution order where each next instruction is executed based on the results of the previously specified condition. Moreover, one instruction can be executed by two computation blocks simultaneously using different data (SIMD processing).

To reduce performance losses caused by the need to reload execution pipelines during branches, the microprocessor employs a branch target buffer and static branch prediction.

The processor's internal and external memories share the same addressing space. The internal memory is divided into three 128-bit blocks of 2 Mb each. This makes it possible to read from the memory into the register file quad, long, and regular words as well as to fetch up to four 32-bit instructions per cycle. Simultaneously, 256 data bits can be loaded into the register file or written to the memory. Eight-, 16-, and 32-bit long words can be packed and written to the memory sequentially. At 150 MHz, the internal memory bandwidth is 7.2 Gb/s.

The microprocessor's bus structure provides for simultaneous transfer of two result operands. Three 128-bit buses form a high-speed data-transfer channel between the internal functional units and external peripheral devices. The high-speed

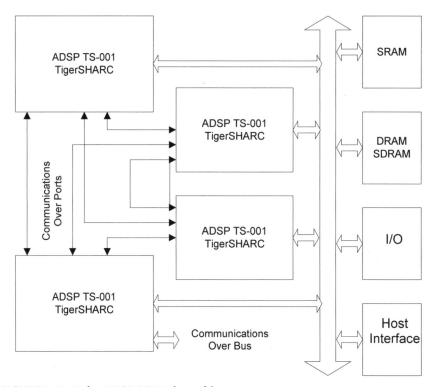

FIGURE 3.26 Using ADSP-TS001 in multiprocessor systems.

64-bit external bus interface allows up to eight-processor multiprocessor systems built based on the ADSP-TS001. Figure 3.26 shows a possible multiprocessor system version.

Along with the external bus, communications ports can be used for processor interaction. Data is transferred over the communications ports by a separate I/O processor and does not require interference from the CPU. This method has no restrictions on the number of interacting processors and is more flexible; however, it provides lower throughput.

New microprocessors of the TigerSHARC family are planned to be produced using lower figures process technology, which will allow the operating frequency to be raised. Moreover, different members of the family will have different sizes of the on-chip memory and different suites of integrated peripheral devices.

Main application areas of the TigerSHARC microprocessors are telecommunications requiring high-performance data processing: third-generation cellular network base stations, Voice over IP (VoIP) applications, servers, and network concentrators.

3.4.6 ADSP-21535 Blackfin Microprocessor

The ADSP-21535 microprocessor is the first representative of the Blackfin family microprocessors. It has microsignal architecture that was developed jointly by Analog Devices and Intel [137]. The processor has been produced since the beginning of 2002 in two versions: the 300 MHz ADSP-21535PKB-300 and the 200 MHz ADSP-21535PKB-200.

A distinctive feature of this architecture is combining in one processor signal processing, SIMD multimedia data processing, and RISC-like instruction set capabilities.

What makes the ADSP-21535 stand out is its extremely low power consumption combined with quite high performance level and a wide range of integrated peripheral devices.

Processor's performance in the MAC operation is 600 MMACS at 300 MHz and 400 MMACS at 200 MHz.

In addition to the microprocessor core, the crystal contains 256 KB of SRAM, a DMA controller, an interrupt controller, a universal bus interface unit, boot loader memory, an external memory controller, a PCI bus controller, USB interface, a universal asynchronous interface (UART), and a JTAG emulation and control unit.

The ADSP-21535 core has three components: the address calculation unit, the control unit, and the data-processing unit (Figure 3.28).

The address calculation unit has two data address generation units (DAG0 and DAG1) and a register file they share. The register file has four sets of registers: index, offset, length, and base. Eight additional 32-bit registers can be used jointly with the main index registers as stack and memory pointers.

The control unit has a program sequencer, instruction fetch and decode units, and a loop instruction buffer (for local instruction storage to reduce the number of the program memory accesses).

The data-processing unit has nine functional units: two MAC operation units, two 40-bit ALUs, four video ALUs, and a barrel rotator. The functional units operate on 8-, 16-, or 32-bit data in the register file.

Each MAC instruction multiplies 16-bit data in a single processor cycle and forms a 40-bit result.

Each ALU can execute a standard collection of arithmetic and logic instructions, most of them in a single processor cycle. The ALU can interpret contents of eight 32-bit registers as 32-bit data items or as pairs of 16-bit data items. The two ALUs can produce up to four operation results on 16-bit data per cycle.

The barrel rotator can perform simple and circular shifts, 40-bit operand normalization, and data extraction and storing.

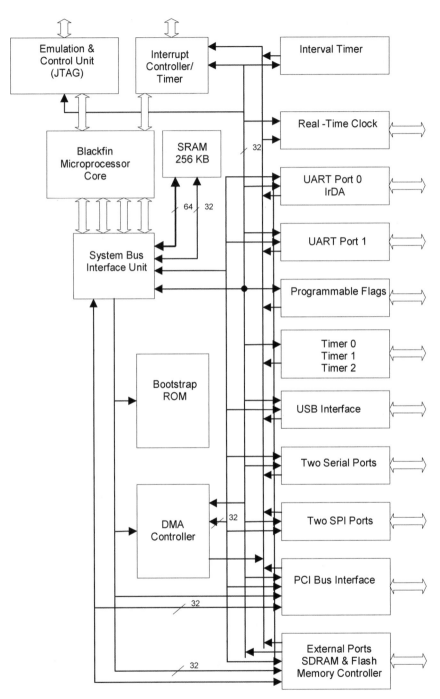

FIGURE 3.27 Blackfin microprocessor block diagram.

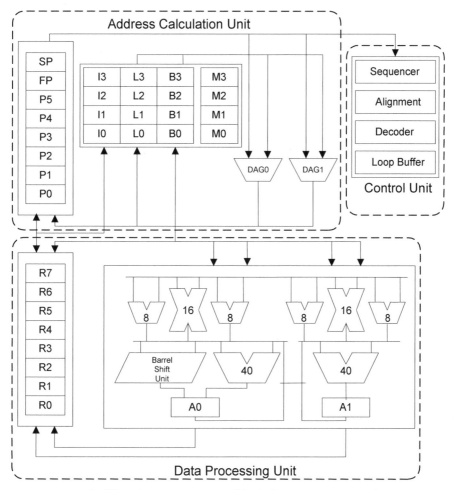

FIGURE 3.28 Blackfin microprocessor core block diagram.

The microprocessor utilizes modified Harvard architecture with hierarchic memory. All microprocessor memory is considered as a unified addressing space up to 4 GB in size.

The first level of the memory hierarchy operates at the processor core frequency and has the lowest access time. The program memory of this level (16 KB on-chip) stores only instructions. The first-level data memory (two on-chip banks 16 KB each), in addition to data, stores the stack and local variables. The internal 4 KB scratchpad memory also belongs to this hierarchy level. With the exception of the scratchpad memory, the first-level memory can be configured as fast access memory or as cache.

The second level in the memory hierarchy is the internal (256 KB SRAM) and external memory, which takes several processor cycles to access. This memory is shared by instructions and data.

When developing the Blackfin family, its designers gave a great deal of attention to reducing power consumption. The processors have a flexible dynamic power-management system: the power supply voltage and the processor frequency change depending on the intensity of the calculations.

Further development of the family is going in the direction of raising performance, reducing power consumption, and using problem-oriented peripheral device suites.

Other representatives of the Blackfin family follow:

ADSP-21532—300 MHz operating frequency, equipped with audio (I2S) and video (CCIR-656) interfaces, has 640 Kb SRAM and an on-chip voltage regulator.

ADSP-BF533—The highest performance microprocessor in the family, at 600 MHz performs 1,200 MMAC/s. Has on-chip 1.2 Mb static high-speed memory, voltage regulator, and a suite of peripheral devices oriented at a wide range of multimedia data-processing tasks. Power consumption is 280 mW.

The new Blackfin processors supports embedded operating systems, such as Linux Embedded, ThreadX, and Nucleus.

The application area of the Blackfin family microprocessors is mobile devices, multimedia devices, game consoles, and Internet applications (video telephones, WWW terminals, NetTV, and so on).

3.5 SIGNAL MICROPROCESSORS FROM MOTOROLA

Signal processors from Motorola are divided into the 16-bit fixed-point, 24-bit fixed-point, and 32-bit floating-point family groups. To the first group belong the DSP561xx, DSP566xx, and DSP568xx families; the second group comprises the DSP560xx and DSP563xx families; the third group has one family member: the DSP960xx.

3.5.1 Fixed-Point 24-Bit Microprocessors

The Motorola's 24-bit microprocessor line consists of two families: DSP560xx and DSP563xx [138, 139]. The main architectural principles that Motorola signal microprocessors are based on were developed for and embodied in the DSP560xx family. The further development of enhancing the signal processors has been conducted in three directions:

- Increasing performance of the 24-bit processors by pipelining functional units and raising the clock frequency
- Building inexpensive 16-bit microprocessors with expanded means for interacting with peripheral devices
- Developing high-performance processors with a floating-point unit

These three directions will be considered using the most popular family representatives as examples. The most important processor differences within families will also be pointed out.

3.5.2 DSP560xx Family Microprocessors

The DSP56000/DSP56001 microprocessors are the first representatives of the Motorola signal processor line [138]. Their architecture is oriented at maximizing the throughput in DSP applications with intensive data exchange. This is achieved by using expandable architecture with complex integrated periphery and a universal I/O subsystem. These properties, as well as the low power consumption, minimize the complexity, cost, and the lead times of application systems based on the DSP56000/DSP56001 microprocessors.

The microprocessors operate at frequencies up to 33 MHz and have about 16 MIPS performance. This makes it possible to perform a 1,024-sample FFT in 3.23 ms.

The differences between the family's microprocessors consist in their internal memory types. For embedded application use, the DSP56000 microprocessor can work under control of a program stored in the 3.75 Kwords PROM. There also is a version of the DSP56000 processor with the internal program memory protected from unauthorized reading. The DSP56001 has an on-chip 512-word RAM and a 32-word bootstrap ROM. It also has two memory modules of which one is preprogrammed as Mu-Law/A-Law expansion tables and the other as sine wave conversion tables.

The structures of the DSP56000 and DSP56001 microprocessors are shown in Figure 3.29 and Figure 3.30, respectively.

The functions of the main components of the microprocessor structure will be described when considering the general structure of the DSP560xx microprocessor family.

The subsequent development of the DSP560xx microprocessor family proceeded within the framework of the concept of a processor core common to all family members. The family comprises 24-bit fixed-point processors DSP56002,

4, 7, 9, and 11 [139] differing by the configuration of the internal memory and peripheral devices.

The processors of this family are characterized by high throughput and expanded data width, which makes for high computation precision and a wide dynamic range of the processed data. They also have energy-saving features.

A typical structure of a DSP560xx family representative microprocessor is shown in Figure 3.31. The microprocessor's main components follow:

- Data buses
- Address buses
- Data ALU
- Address generation unit (AGU)
- Program control unit (PCU)
- Memory expansion (port A)
- On-chip emulation circuitry (OnCE^{TM})
- Frequency multiplier circuitry

The processor has three independent functional units: the program control unit, the address generator unit, and the data ALU. Data between registers of the functional units is sent over bidirectional 24-bit buses: the data bus X (XDB), data bus Y (YDB), program data bus (PDB), and global data bus (GDB). Some instructions use X and Y data buses as a unified 48-bit bus. To raise the operating speed, operands for instructions are loaded into the ALU from the X and Y memory modules over the independent XD and YD buses, whereas the instructions themselves are loaded over the PDB. Data exchange with peripheral devices is carried over the GDB.

The bus structure supports main data transfers of the *register/register, register/memory*, and *memory/register* types. Two 24-bit and one 56-bit word can be transferred in a single cycle. Exchange between buses is accomplished via the internal switch, which allows any two buses to be connected without introducing wait cycles. Addresses for the internal X and Y data memories are transferred over the bidirectional 16-bit XA and YA buses; program memory addresses are transferred over the bidirectional program address bus. External memory is addressed using the unidirectional bus that is the output of the three-input switch of the XA, YA, and PA buses.

The bit manipulation unit is physically located in the bus switch unit, which provides access to any memory area and allows performing bit operation on data in the memory, registers, and contents of the address and control registers.

The data ALU performs all arithmetic and logic operations; it has four 24-bit source registers, two 48-bit accumulator registers, two eight-bit accumulator expansion registers, a shift unit, two data shifter/limiter units, and a parallel (non-pipelined) one-cycle MAC unit.

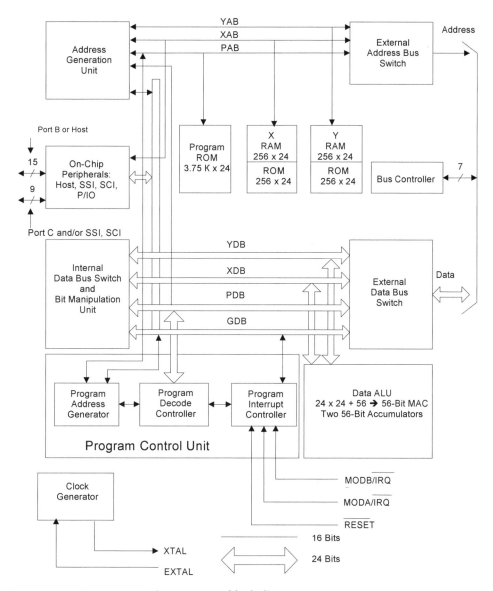

FIGURE 3.29 DSP56000 microprocessor block diagram.

The A and B accumulators are used as buffer registers for the XD and YD buses; their 8-bit extension registers are used by the shifter/limiter circuitry to detect and process overflow situations arising from arithmetic or shift operations.

The data ALU can perform double-precision multiplication. This mode is specified by setting the corresponding bit in the processor's status registers.

Multiplication of two 48-bit operands produces a 96-bit result, which is stored in four 24-bit registers.

The address generation unit works in parallel with other processor's components. Using two identical 16-bit arithmetic units, each of which can perform linear, modular, and cyclic arithmetic operations, the AGU takes one cycle to calculate the needed data addresses.

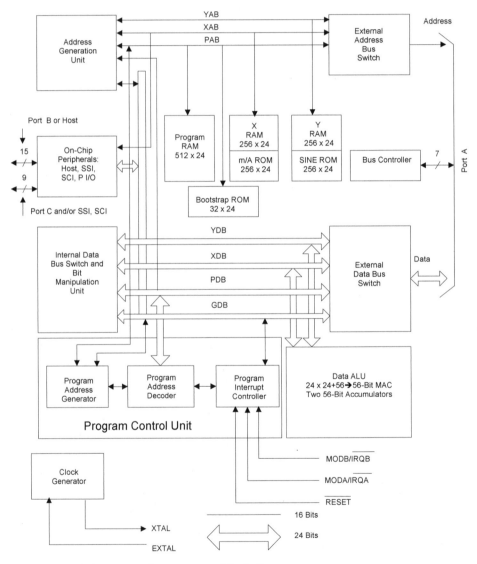

FIGURE 3.30 DSP56001 microprocessor block diagram.

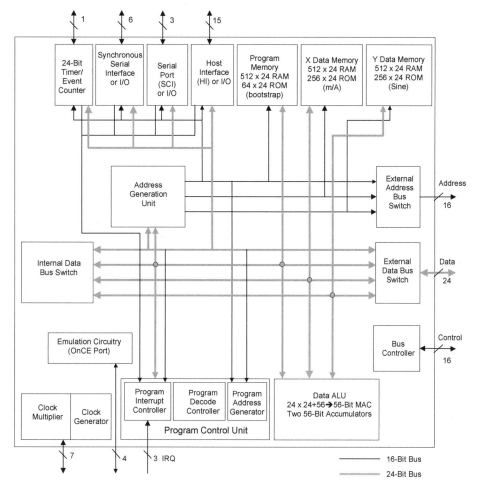

FIGURE 3.31 DSP560xx family microprocessor block diagram.

The A and B accumulators are used as buffer registers for the XD and YD buses; their 8-bit extension registers are used by the shifter/limiter circuitry to detect and process overflow situations arising from arithmetic or shift operations.

The data ALU can perform double-precision multiplication. This mode is specified by setting the corresponding bit in the processor's status registers. Multiplication of two 48-bit operands produces a 96-bit result, which is stored in four 24-bit registers.

The address generation unit works in parallel with other processor's components. Using two identical 16-bit arithmetic units, each of which can perform linear, modular, and cyclic arithmetic operations, the AGU takes one cycle to calculate the needed data addresses.

Each address ALU had three four-register sets associated with it: the address (R0–R3 and R4–R7), offset (N0–N3 and N4–N7), and modifier (M0–M3 and M4–M7) registers. The registers are used in triplets: R0:N0:M0, R1:N1:M1,... ..., R7:N7:M7. An address is formed from the contents of an address register and an offset register, taking into account the type of arithmetic determined by the contents of the modifier register.

The program control unit generates program addresses, decodes instructions, and performs hardware loop control and internal and external interrupt processing. It has a 15-level 32-bit system stack (SS) and six directly addressed registers: program counter (PC), loop counter (LC), loop address (LA), status (SR), operating mode (OMR), and stack pointer (SP) registers. The 16-bit PC can address up to 65,536 instructions. The SS holds the PC and the SR during procedure calls, interrupt handling, and software loop executions.

Processor instructions are executed in the three-stage pipeline (prefetch, decode, execute), followed by checking for five possible processor states: *normal, exception, reset, wait,* and *stop.*

The PCU comprises three hardware blocks: the program decode controller, the program address generator, and the program interrupt controller. The PDC decodes instructions loaded into the instruction buffer and generates all the control signals necessary for instruction execution. The contents of the instruction buffer are duplicated to execute REP and JMP instructions more efficiently.

The main purpose of the PGA block is hardware generation of loop addresses. When a loop is initialized, its starting address is pushed onto the stack, the loop-control variable is in the LC register, and the ending address is in the LA register. When next iteration is completed, the jump address is popped off the stack and is not software-generated, which speeds up the processing substantially.

The PIC receives all interrupt requests, arbitrates among them, and generates the address of the interrupt vector. Priority level 0 (the lowest priority level), 1, and 2 interrupts are maskable; priority level 3 (the highest level) interrupt is non-maskable.

The external memory port A provides asynchronous data exchange with various memory types and external devices over the 24-bit data bus. The port can work with both high- and low-speed memories as well as with other universal or signal processors in the master/slave mode.

The on-chip emulation circuitry allows the status of registers, memory, and peripheral devices to be interactively analyzed and the program debugging process controlled. Debugging can be allowed for system developers but other users can be forbidden access to the internal processor resources.

The clock frequency multiplier allows the processor to operate at increased internal clock frequency, providing synchronization of the internal and external clocking pulses as well as lowering the frequency in the power-saving mode.

Opcode	Operands	XDB	YDB
MAC	X0, Y0, A	X: (R0)+, X0	Y:(R4)+,Y0

FIGURE 3.32 DSP560xx microprocessor instruction structure.

The microprocessor's programming model is represented as three units functioning in parallel: the ALU, the AGU, and the PCU. The microprocessor's instruction set is oriented toward efficient support of C language and is organized in such a way so as to provide these units with workload for the duration of each cycle, thereby achieving maximum program execution speed.

Microprocessor instructions can be one or two 24-bit words long. A typical DSP560xx instruction consists of four fields. The operation code field specifies the corresponding action by the ALU, AGU, or PCU. The operand field, obviously, holds the operand. The remaining two fields specify transfers over the XD and YD buses, which the microprocessor carries out concurrently with the main operation. An example of a MAC instruction is shown in Figure 3.32.

Owing to its high performance and low cost, the DSP560xx microprocessor family is widely used in the most diverse areas—communication systems, digital audio systems, robotics, military electronic, and so on.

3.5.3 DSP563xx Family Microprocessors

The microprocessors of this family [140] are built around a new type of processor core called New DSP Engine (NDE). The processing in the functional units of the new core is pipelined, which allows an instruction to be executed in a single cycle. This doubles the performance level as compared to the DSP560xx core; both cores are binary compatible.

The DSP563xx core has a data ALU, a memory address generation unit, a program control unit, an instruction cache controller, a DMA controller, a frequency multiplier, and memory and periphery expansion buses. To reduce the cost of the systems built on its basis, the DSP563xx is provided with universal external memory interface that supports various memory types, such as DRAM, SRAM, and SDRAM. The parallel six-channel DMA controller provides the high data-transfer speed needed for DSP applications. Special attention in the microprocessor's design was given to lowering power consumption in both active and inactive operating modes.

The main technical characteristics of the DSP563xx microprocessor family follow:

- 66/80/100 MIPS performance at 66/80/100 MHz, respectively
- Fully pipelined parallel MAC unit
- 56-bit parallel barrel rotator
- 16-bit arithmetic support

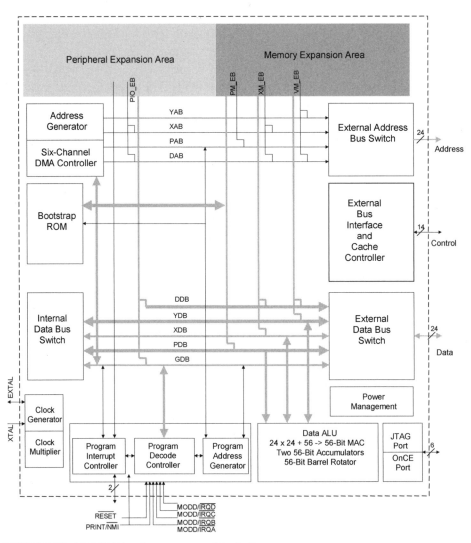

FIGURE 3.33 DSP56300 microprocessor block diagram.

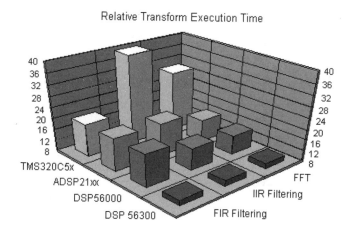

Relative Transform Execution Time

FIGURE 3.34 Comparing efficiencies of signal microprocessor architectures.

- Highly parallel instruction set
- Memory expandable hardware stack
- Hardware support of nested loops
- On-chip user-controlled instruction cache
- JTAG test access port

The structure of the DSP56300 microprocessor is shown in Figure 3.33.

The diagram in Figure 3.34 compares architectures of signal microprocessors by the number of cycles needed to perform typical DSP algorithms: FIR and IIR filtering and a 1,024 sample FFT. Examination of the values shows that the architecture of Motorola DSP563xx processors is more efficient and has better task-execution time, with the microprocessors' clock frequencies being the same.

3.5.4 16-Bit Fixed-Point Microprocessors

The DSP561xx, DSP566xx, and DSP568xx are Motorola's 16-bit microprocessors.

3.5.5 DSP561xx Family Microprocessors

The DSP561xx family consists of the DSP56156 and DSP56166 [141] 16-bit fixed-point microprocessors. They are built on the same processor core and differ by the internal memory configurations and peripheral devices.

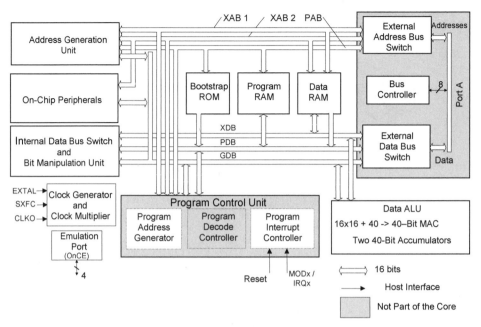

FIGURE 3.35 DSP561xx microprocessor core block diagram.

The DSP56156 was developed specifically for use in cellular phones. Its instruction set is oriented primarily toward processing telecommunications protocols and speech signals. Some instructions and addressing methods had to be sacrificed for the sake of reducing the instruction-code length.

The microprocessor has an on-chip sigma-delta decoder with direct input of the analog signal and 2 Kwords of on-chip memory.

The structure of the microprocessor core is shown in Figure 3.35.

The main technical characteristics of the family microprocessors follow:

- 30 MIPS performance at 60 MHz
- One-cycle 16x16 MAC unit
- Two 40-bit accumulators with an extension byte
- Variable precision floating- and fixed-point arithmetic
- Highly parallel instruction set with a flexible addressing system
- Hardware nested-loop implementation, including infinite loops
- Three 16-bit internal data buses and three 16-bit internal address buses
- Programmable access time to the external bus
- Interface mapped into the peripheral device memory
- On-chip emulation and test unit (OnCE)
- Low power consumption and power management facilities.

3.5.6 DSP566xx Family Microprocessors

Like the DSP561xx family, the DSP566xx microprocessor family was developed for use in cellular phones. In this application area it is important not only to be able to process signals over a wide dynamic intensity range but also to have a highly efficient energy-saving system.

The family comprises the DSP56603 and DSP56603 microprocessors, which differ by their memory configurations and the peripheral device suites [142].

The microprocessors of this family are highly integrated devices. Their crystals both contain a 16-bit core, a pipelined MAC unit, problem-oriented PROM, and separate memory modules for storing application program data and the bootstrap program.

The instruction set of the DSP566xx microprocessors is compatible with the DSP563xx.

The structure of the DSP566xx family microprocessors is shown in Figure 3.36.

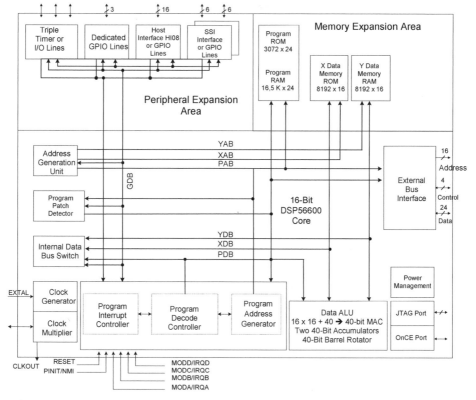

FIGURE 3.36 DSP566xx microprocessor block diagram.

The main technical characteristics of the microprocessors of this family follow:

- Two 40-bit accumulators have an extension bit to control result overflow
- 40-bit barrel rotator
- Hardware nested-loop implementation and a fast return from interrupt handlers
- The microprocessors' peripheral circuitry includes three general-purpose ports with the total of 31 lines that can be used as an 8-bit parallel interface to the host microprocessor or a synchronous serial interface
- About 60 MIPS performance
- Operating power supply voltage from 1.8 V to 3.3. V; consumed current from 0.55 mA to 0.85 mA

3.5.7 DSP568xx Family Microprocessors

Microprocessors of this family combine functions of high-efficiency signal processors and intellectual controllers. Currently the family comprises the DSP56L811 and DSP56L812 microprocessors [143]. The application area of these microprocessors is quite wide: wireless communications equipment, radio modems, digital answering machines, and so on. Owing to the high number of functional units integrated on the chip, systems built around these microprocessors have not only high performance levels but also low costs.

The main technical characteristics of the family's microprocessors follow:

- Up to 25 MIPS performance at 40 MHz
- One-cycle 16 × 16 MAC unit
- Two 36-bit accumulators with extension bits
- 16-bit barrel rotator
- Hardware nested-loop support and instruction reexecution
- Two external interrupt inputs
- Instruction-set supporting DSP and controller functions
- Unlimited length software stack for subprograms and interrupts
- Extended on-chip emulation and testing facilities
- Efficient energy-saving system

The structure of the DSP568xx microprocessors is shown in Figure 3.37.

The microprocessors are 16-bit all around: The processor's instruction length also is 16 bits; therefore, it is not binary compatible with the families of 24-bit microprocessors considered earlier. Data memory is one 2-Kword block. The program memory has a bootstrap ROM and memory for storing programs loaded from an outside source.

TABLE 3.4 Lists Characteristics of Other Representatives of the DSP568xx Microprocessor Family

Processor	Program ROM	Program RAM	Data RAM	Data ROM	Bootstrap ROM	Timers	I/O	Serial Interface	A/D	Power Supply Voltage (V)	Bus Frequency (MHz)
DSP56824	128×16	32K×16	2K×16	2K×16	—	Three 16-bit	16	1 SPI	—	2.7	70
DSP56F801	1K×16	8K×16 flash	1K×16	2K×16 flash	2K×16 flash	One quad timer unit	11	SPI SCI	Two 4 × 12-bit ADC	3.3	80
DSP56F803	512×16	32K×16 flash	2K×16	4K×16 flash	2K×16 flash	Two quad timer units	16	SPI SCI CAN	Two 4 × 12-bit ADC	3.3	80
DSP56F805	512×16	32K×16 flash	2K×16	4K×16 flash	2K×16 flash	Two quad timer units	32	SPI2 SCI CAN	Two 4 × 12-bit ADC	3.3	80
DSP56F807	2K×16	60K×16 flash	4K×16	8K×16 flash	2K×16 flash	Four quad timer units	32	SPI2 SCI CAN	Four 4 × 12-bit ADC	3.3	80

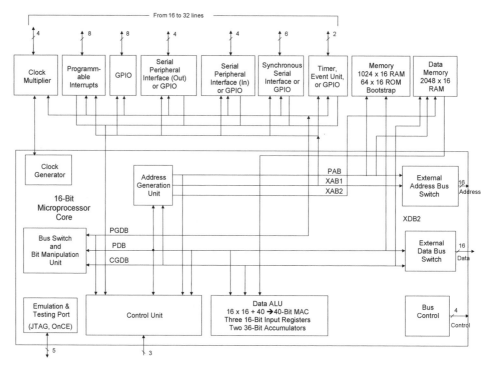

FIGURE 3.37 DSP568xx microprocessor block diagram.

The processor's programmable universal I/O subsystem can be configured to meet the requirements of a specific task: Subsystem's outputs can be individually configured for input or output, multiplexed between microprocessor's peripheral blocks, or used as outputs of the interrupt-processing subsystem. This type of universal use of the processor output pins allows their number to be reduced.

3.5.8 DSP9600x Family Floating-Point Microprocessors

Microprocessors of the given family are intended for floating-point data processing.

This family comprises the 32-bit DSP96002 microprocessor [144]. This is a single-crystal universal microprocessor with an FPU. The DSP96002 has 1,024 words of memory split evenly between data memories X and Y, 1,024 words of program memory, two data PROMs, a two-channel DMA controller, a bootstrap subsystem, and on-chip emulation and test facilities.

The structure of the DSP96002 microprocessor is shown in Figure 3.38.

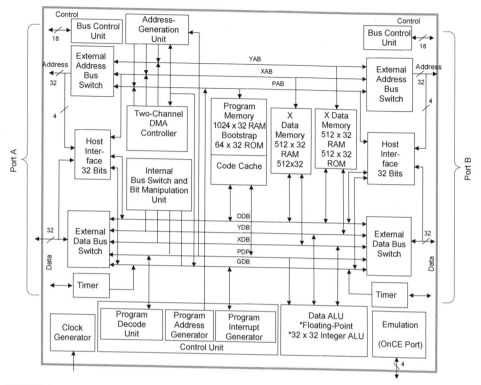

FIGURE 3.38 DSP96002 microprocessor block diagram.

The CPU has three 32-bit execution devices functioning in parallel: the data ALU, the address generation unit, and the program control unit. The processor has two identical memory expansion ports providing an interface with various types of memory (SRAM, DRAM, VRAM). Each port can be converted into a host interface that provides the capability to easy integrate processors into a multiprocessor system.

The microprocessor's main characteristics follow:

- One-cycle 32 × 32 MAC unit
- Highly parallel specialized instruction set
- Hardware nested-loop and fast return from interrupts support
- Expanded to 1 Kword instruction cache
- Five 32-bit address buses: the internal unidirectional address buses X and Y, the program address bus, and two external address buses
- Seven 32-bit data buses: the internal bidirectional data buses X and Y, the internal bidirectional global data bus, the internal bidirectional DMA data bus, the internal bidirectional program data bus, and two external data buses

- On-chip memory including: 1,024-word program RAM, two independent data RAMs of 512 words each, two independent ROMs of 1,024 words each, and the 64-word bootstrap ROM
- Ability to address up to 4 GB of 32-bit words of external data and program memory each
- Performance at 40 MHz is 200 MIPS

3.6 COMMUNICATIONS PROCESSORS

Another category of specialized processors quite close in architecture to the signal processors, but with some considerable differences, comprises the communications processors. What makes them related to the signal microprocessors is the application area they both share: communications systems. The differences lie in the specific areas that communications processors are used in within these systems and the tasks placed upon them.

If the traditional signal microprocessors are oriented toward efficient implementation of physical and channel-level protocols, the communications microprocessors are mainly designed to process network and transport-level protocols. Both the signal and communications microprocessors have the same nature of the processing: real-time stream data processing.

Another microprocessor class will be considered in Section 3.7: microprocessors for mobile communications devices. These microprocessors combine features of the signal, communications, and multimedia microprocessors, because they must perform functions typical for all these types of devices.

3.6.1 MPC8260 Microprocessor

Motorola's MPC8260 PowerQUICC II™ [145] microprocessor is a universal communications processor. It integrates in a single crystal a high-performance PowerPC architecture RISC computation core, a flexible system-integration unit, and a variety of peripheral controllers.

Devices for various communications applications can be built based on this microprocessor, including the following:

- Remote servers
- LAN-WAN bridges
- Cellular base stations
- Local computation network routers

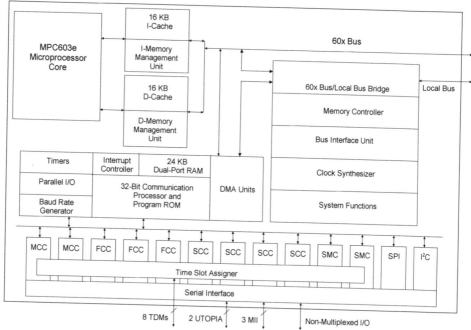

Figure 3.39 MPC8260 microprocessor block diagram.

The microprocessor's structure is shown in Figure 3.39.
The microprocessor consists of three main functional blocks:

- Microprocessor core with the memory and data/instruction cache controllers
- System interface unit
- Communications processor unit

The MPC8260 microprocessor has two buses for servicing the communication needs of the high-performance processor core and of the high-speed communications channels: the 60x microprocessor bus (the 64-bit data bus and the 32-bit address bus) and the local bus (the 32-bit data bus and the 18-bit address bus). The core and the communications processor module are connected to the 60x bus.

Both the microprocessor core and the communications processor module have their own frequency multipliers, which allow the components to be independently adjusted to the optimal operating frequency.

The processor core is an on-chip version of the MPC603e 32-bit integer RISC PowerPC microprocessor operating at up to 200 MHz and with four-way set associative split instruction and data caches of 16 KB each. The microprocessor does not support floating-point operations.

The microprocessor core can be disabled. In this case, the microprocessor functions as a peripheral device to the external core.

The microprocessor's system interface unit comprises two memory controllers supporting most existing memory types (SDRAM, DRAM, EPROM, flash, and so forth) and other peripheral devices, which turn the microprocessor into a full-featured single-crystal system.

The system interface unit has the following components:

- 64-bit system bus compatible with the 60x bus
- Local bus (32-bit data bus, 32-bit internal and 18-bit external address buses) operating at the same frequency as the 60x system bus; this bus is for the communication controller operations
- Memory controller making up to 12 memory banks available; these can be connected both to the system and to the local microprocessor buses
- JTAG test and debugging port
- Real-time timers and a time slot timer
- L2 cache interface

The communications processor module is a collection of circuits that implements interaction with the peripherals and provide various hardware-level interface functions, including support of communications protocols (such as Fast Ethernet and ATM), multiplexed full-duplex channels, and so on.

The communications processor module contains:

- Communication processor (CP), which is an on-chip 32-bit RISC processor; the communication processor is connected to a separate local bus and, consequently, does not affect the microprocessor core, which is connected to the 60x bus; the communication processor performs low-level tasks and initiates processing of DMA requests, thereby freeing the microprocessor core for performing high-level tasks; the instruction set of the communication processor is optimized for communications tasks, and, at the same time the communication processor can also execute general-purpose applications, thereby unloading the main processor.
- Two independently functioning serial DMA controllers optimized for block transfers over the local and the 60x buses.
- Three full-duplex high-speed serial controllers. These support the ATM protocol providing 155 Mb/s transfer speeds (UTOPIA interface) and the IEEE 802.3, Fast Ethernet, and HDLC protocols providing up to 45 MBps transfer speeds. Transparent transfers of bit streams also are possible.
- Two multichannel controllers that can jointly process up to 256 HDLC streams received over eight interface channels with speeds of up to 64 Kb/s;

a multichannel controllers can also process super channels with speeds over 64 Kb/s and subchannels of 64-Kb channels.

■ Four full-duplex serial communications controllers supporting IEEE 802.3/Ethernet, HDLC, Local Talk, UART, BISYNC, protocols and transparent data transfers.

■ Two full-duplex serial managing controllers supporting GCI and UART protocols and transparent data transfers.

■ Serial Peripheral Interface and I^{2}C bus controllers.

■ Time-slot assigner; this device switches data streams from these sources— four full-duplex communication controllers, three high-speed serial controllers, and two serial managing controllers.

The microprocessor's internal circuitry is powered by 2 V, the I/O subsystems are fed 3.3 V; the total power consumption is about 2.5 W. Processor's performance at 200 MHz is 280 MIPS; SPECint95 benchmark index is 5.1.

3.6.2 Network Microprocessors from Intel

The first representative of Intel's network processors is the IXP1200 [146]. This is a highly integrated hybrid microprocessor oriented toward use in communication applications, for which it is important to have efficient access to the fast memory subsystem and I/O devices and, at the same time, possess high performance in processing variable-length data (bits, bytes, words, and doublewords).

The structure of the IXP1200 microprocessor is shown in Figure 3.40.

The processor's main components are the following: the StrongARM microprocessor core, the six data processing microengines (DME), the IX bus interface unit, the PCI bus interface unit, and the SDRAM and SRAM external memory control units.

A DME is a 32-bit multithread RISC processor operating at 162 MHz. DMEs process and exchange data independently of the processor's computation core.

Each DME has four program counters, an ALU, a shift unit, 128 32-bit general-purpose registers, 128 32-bit data-exchange registers, and a local 4 KB memory. The highly specialized instruction set of the DME's microprocessors, hardware support of context switching and process synchronization, operation execution by the ALU and by the shift unit combined in one cycle, together all of these provide the performance level of over 10^9 OPS.

The IX bus interface unit links the microprocessor with the network peripherals connected to this bus. The IX bus can be configured as one bidirectional 64-bit bus or as two unidirectional 32-bit buses; its maximum bandwidth is 4.2 GBps at 66 MHz and 5 GBps at 80 MHz.

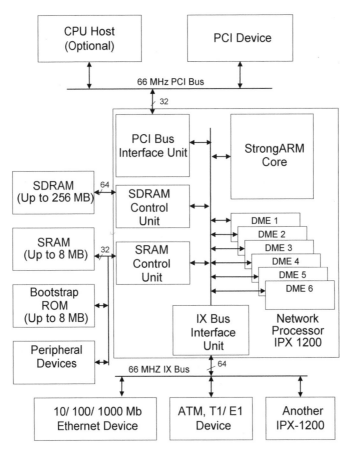

FIGURE 3.40 IXP1200 microprocessor block diagram.

The IXP1200 microprocessor supports two types of memory: static (SRAM) and dynamic (SDRAM). Each memory type is controlled by the corresponding on-chip controllers.

The static memory is used as the fast memory to store search tables; the dynamic memory is intended for storing current information and exchange queues. The IXP1200 microprocessor can address up to 256 MB of dynamic memory.

The SDRAM control unit services data read/write requests from the PCI unit (including DMA support), from the microprocessor core, and from the DMEs. At 81 MHz, the SDRAM interface peak bandwidth is 684 MBps.

The SRAM control unit handles memory requests from the microprocessor core and the DMEs and controls the SRAM itself (up to 8 MB), the bootstrap ROM (up to 8 MB), and the peripheral device ports.

At 81 MHz and 32-bit SRAM access, the peak throughput is 334 MBps.

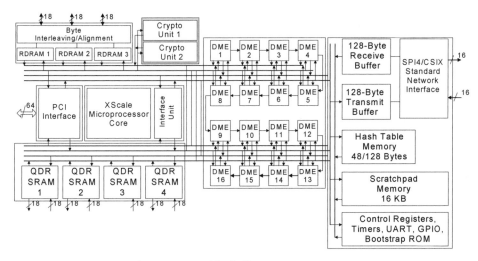

FIGURE 3.41 IPX2850 microprocessor block diagram.

The PCI bus interface unit of the IXP1200 processor controls the 32-bit PCI bus that connects the host processor (if one is used) and other peripheral devices with PCI interface. The PCI 2.1 specification (32 bits, 66 MHz) is supported; the peak throughput is 264 MBps.

Representatives of the next generations of the network processors are the IXP2xxx family microprocessors: IXP2400, IXP2800, and IXP2850 [147].

The IXP2850 microprocessor (Figure 3.41) contains the Intel® XScale™ core operating at 700 MHz and 16 independent 32-bit multithread DMEs operating at 1.4 GHz and providing overall performance of over 23.1×10^9 OPS. The DMEs provide the computing power necessary to solve types of tasks for which expensive high-speed custom-made microchips were needed earlier.

The crystal also incorporates the four-channel QDR SRAM controller, the three-channel RDRAM controller, the PCI interface, the network interface, the peripheral device interface, the hash table memory, and the scratchpad memory.

The IXP2850 has two on-chip blocks that implement common cryptographic algorithms (such as 3DES/DES, AES, and SHA-1), which makes the processor well suited for use in devices operating in networks with IPsec and TCP/SSL protocols. The processor can perform cryptographic data processing at up to 10 Gb/s speed.

The representatives of the next family, IXP42x, are the IXP420, IXP421, IXP422, and IXP425 microprocessors [148]. They are intended for use in batch communications system equipment. The microprocessors integrate on-chip the high-performance Intel XScale microprocessor core, network devices, and a wide range of peripheral device interfaces. Microprocessor functional units are connected by internal advanced high-performance bus (AHB) and advanced peripheral bus (APB).

The structure of the IXP422 microprocessor is shown in Figure 3.42.

The main technical characteristics of the IXP42x family microprocessors follow:

- 266 MHz and up core clock frequency
- Integrated hardware acceleration of common cryptographic algorithms, such as SHA-1, MD5, DES, 3DES, and AES
- Two integrated 10/100 Base-T Ethernet controllers
- SDRAM controller supporting from 8 to 256 MB of memory
- Low power consumption, from 1.0 to 1.5 W
- USB 1.1 controller
- Two high-speed UARTs supporting 921 Kbaud each
- 16 output pins for general-purpose ports
- 33/66 MHz PCI 2.2 interface for connection of up to four processors
- 16-bit configurable expansion bus

The individual members of the family differ by their clock speeds and the suites on the on-chip components.

The microprocessors are used in wireless access points, VPN network equipment, network gateways, routers and switches, network printers, and so on.

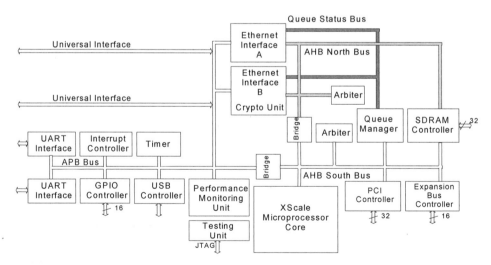

FIGURE 3.42 IXP422 microprocessor block diagram.

3.7 INTEL PCA ARCHITECTURE MICROPROCESSORS

A relatively new direction for Intel was developing microprocessors for mobile communications devices: cellular phones, pocket computers, communicators, and Internet terminals. The architecture Intel developed specifically for this type of devices is called Intel Personal Client Architecture (Intel PCA). Microprocessors from the PXA25x and PXA26x [149] families belong to this type of microprocessor.

The main features of PCA microprocessors are the wide range of on-chip communication devices, core-level signal-processing support, low power consumption, and small dimensions combined with quite high performance.

The internal structure of the PXA255 microprocessor is shown in Figure 3.43.

Both microprocessor families are built around the XScale microprocessor core, which is a development of the StrongARM microprocessor core. The XScale microprocessor core is not used as an independent product but as a computation block in problem-oriented microprocessors and microcontrollers. The main structure blocks of the XScale microprocessor core are shown in Figure 3.44.

The MAC unit has a 40-bit accumulator and a 16-bit SIMD processing unit.

The memory control unit handles the program and the data memory. It supports split memory access and mapping of the virtual addressing space into physical. Individual memory areas can be assigned different attributes defining how they are to be accessed: cache, buffered memory, line-allocation method, write method, and so on. TLBs are used to speed up converting virtual addresses into physical instruction and data.

The program cache is 32 KB in size. It is 32-way, set associative, with the cacheline 32 bytes long. The processor has the capability to keep critical code parts cached. In addition to the main program cache, there is an additional 2 KB cache for the processor's debugging operating mode.

The processor has a 128-entry BTB.

Data cache consists of the main 32-way set associative 32 KB cache and the 2-way 2 KB cache for debugging purposes. The processor allows a part of the data cache to be reconfigured for use as RAM for storing special tables or frequently used variables.

The power-management unit allows the operating clock frequency and the microprocessor power supply voltage to be changed according to the intensity of the computations.

The on-chip Performance Monitoring, Debug, and JTAG units are for use by software developers.

The XScale microprocessor is downward binary compatible with the Advanced RISC Machine (ARM) family microprocessors. At the same time, in addition to the standard ARM instruction set, the XScale microprocessor supports the Thumb instruction set. These are 16-bit instructions that perform the same functions

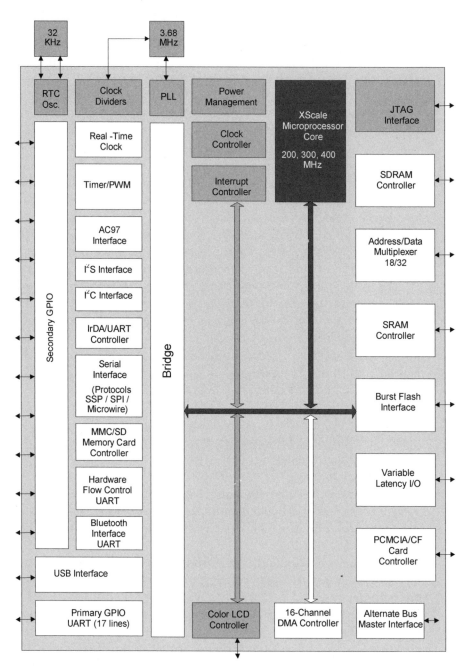

FIGURE 3.43 PXA255 microprocessor block diagram.

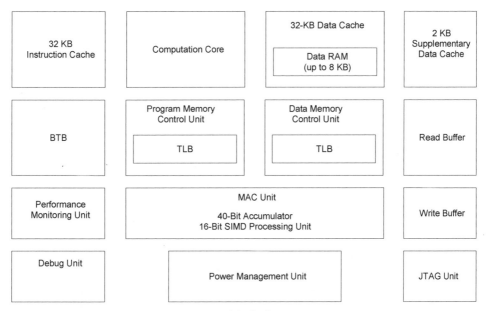

FIGURE 3.44 XScale microprocessor core block diagram.

as the 32-bit ARM instruction set but allow more compact program code to be obtained. The XScale instruction set is also supplemented with floating-point instructions and specialized DSP instructions that are executed in the coprocessor. Instructions operate on data contained in 16 32-bit registers.

One of the main differences between the XScale and the StrongARM is the execution pipeline. The XScale is a superscalar processor: it has a seven-stage main pipeline, an eight-stage operation pipeline, and a six-stage MAC pipeline. The first four stages of these pipelines are shared by all of them. The longer pipeline depth as compared with the StrongARM allows the processor to operate at higher frequencies. Having several pipelines allows out-of-order instruction execution (provided there are no data dependencies between instructions).

The PXA25x microprocessors are produced on 0.18-micron technology for 200 MHz, 300 MHz, and 400 MHz processor core frequencies. The microprocessors have on-chip wireless data-transfer interfaces, USB interface, memory card controllers, and MMC/SD and PCMCIA/CF controllers.

Microprocessors of the PXA26x family are built on the system-in-a-package (SIP) technology, which is based on the multi-level-cell (MLC) flash memory technology.

The processors of this family differ from the processors of the PXA25x family by their smaller dimensions, additional communication ports, and the Intel StrataFlash® Wireless Memory located in the same package as the processor.

The PXA261 and PXA262 microprocessors are integrated products for mobile devices. These microprocessors combine the processor and the StrataFlash flash memory in the same packaging. The PXA261 package comprises the 200 MHz processor core and 128 MB of memory. The core of the PXA262 processor can operate at 200 MHz or 300 MHz clock frequency and has two 128 MB memory crystals.

3.8 MEDIA PROCESSORS

The growing popularity of multimedia technologies has engendered the corresponding increase of applications for them and companies developing those applications. This has induced microprocessor developers to devote more attention to supporting signal-processing algorithms on the microprocessor instruction set level [150]. Currently, a tendency is observed to shift the attention from the purely numerical operations to operations on new data types that are typical of processing video and audio information (the x86 architecture instruction-set extensions such as MMX, SSE, and 3DNow!).

To date, microprocessors that provide multimedia support on the hardware level can be broken into two classes. These are universal microprocessors with multimedia instruction set extensions and multimedia processors. Above all, this classification reflects processor orientation toward different application areas.

Use of universal microprocessors with multimedia instruction set extensions is practical in applications with large proportions of digital signal processing. Use of multimedia microprocessors is more effective in applications in which multimedia operations dominate the traditional numerical operations.

The microprocessors adapted to the multimedia processing requirements, which were considered in the previous sections, belong to the first microprocessor class. Some of these are Pentium MMX, Pentium III and 4, VIA Cyrix MIII, AMD Athlon, SUN UltraSPARC, DEC Alpha, and others.

The second class comprises microprocessors with hybrid architectures characteristic to both the traditional signal and universal microprocessors. These microprocessors are called media processors and are intended for processing audio signals, graphics, and video images, as well as for solving several communications area problems. In this they are close to the communications processors, which were considered earlier.

To date, the following microprocessors can be said to comprise the media processor class:

- Mediaprocessor from MicroUnity
- TriMedia from Philips

- Mpact Media Engine from Chromatic Research
- NV1 from NVidia
- MediaGx from Cyrix

Despite the expanded multimedia capabilities of the universal microprocessors, the application area of media processors remains quite broad. To date, they are used in entry-level PCs, pocket computers, communicators, game stations, Internet terminals, and so on.

3.8.1 Mediaprocessor from MicroUnity

The MicroUnity Mediaprocessor is oriented toward use in multimedia and wideband communications systems [151]. In various designs, Mediaprocessor is used together with the ADC MediaCodec and the external cache interface MediaBridge supplementary devices, also developed by MicroUnity.

The ADC MediaCodec allows the organization of a wide-band communication channel interface, whereas the MediaBridge can play the role of interface with the PCI bus or the main memory in multimedia computers.

The Mediaprocessor was produced on four-layer CMOS 0.5-micron technology for operating clock frequencies from 300 MHz to 1 GHz.

The processor's multithread architecture allows up to five different tasks be performed, using for this five sets of 64 64-bit registers. For each task, a virtual 200 MHz processor is assigned (the microprocessor's clock frequency being 1 GHz).

The processor has a 64 KB unified on-chip data/instruction cache. The dedicated 8-bit I/O bus has up to 1 GBps bandwidth and is used for connecting the microprocessor with the MediaCodec and MediaBridge units.

The Mediaprocessor has a compact instruction set, which includes special signal-processing operations and extended mathematical operations. For example, one instruction in a single cycle multiplies four 32-bit operands by other four 32-bit operands, sums the obtained values with the wider operands, and returns the resulting four values in the 32-bit format.

3.8.2 TriMedia Microprocessor from Philips

The TriMedia multimedia microprocessor from Philips can be used two ways. The first use is as a DSP coprocessor, unloading the main processor of a multimedia system. It can also be used as an embedded universal processor in various multimedia devices: game stations, DVD and video CD players, and so forth. [151, 152].

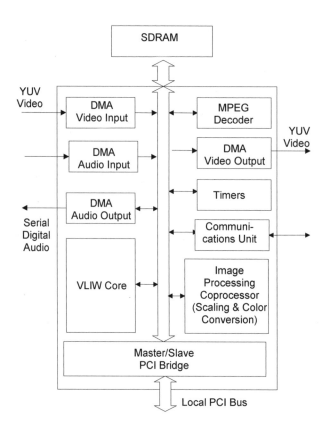

FIGURE 3.45 The TriMedia microprocessor block diagram.

The structure of the TriMedia microprocessor is shown in Figure 3.45. It is centered around the 400-Mb/s-bandwidth high-speed bus that interconnects the main microprocessor units: the processor core, the video input and output, the audio input and output, the MPEG VLD, the video-processing coprocessor, and the communications unit.

The TriMedia microprocessor core is built on the VLIW architecture principles and is capable of processing five RISC instructions in a single cycle (Figure 3.46). The processor's deep pipelining of microoperations is achieved by the 27 functional modules (including an integer ALU, several floating-point ALUs, and several DSP units). The microprocessor has a 48 KB split cache (32 KB for instructions and 16 KB for data).

The other microprocessor's modules interact with the VLIW core and perform processing specific for the given type of data. Using the DMA mode allows data be prepared for processing in several processor modules concurrently.

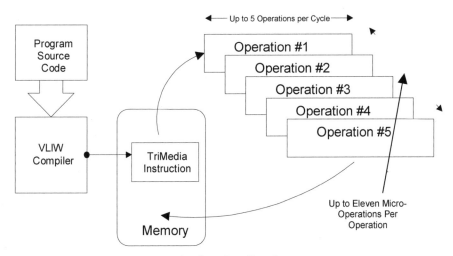

FIGURE 3.46 Instruction execution by TriMedia microprocessor.

The multiprocessor instruction set is optimized for efficient multimedia application execution and allows performing simultaneous audio and video data MPEG decoding only using 22 percent of the processor's computational resources and 12 percent of the memory resources.

3.8.3 Mpact Media Engine Microprocessor from Chromatic Research

The main application area of the Mpact Media Engine microprocessor is multimedia PCs built on the x86 family microprocessors [151–153]. In these systems, the Mpact Media is entrusted with the functions of graphics accelerator, 3D graphics coprocessor, MPEG codec, audio card, fax modem, or telephone expansion card.

The Mpact Media is a VLIW processor with the core optimized for real-time multimedia application execution. For example, output of 2D or 3D images, music synthesis, and modem data transfer can all be performed concurrently.

The Mpact Media has 1.4 million transistors. The processor's CPU has five ALUs, one of which is used to track changes in images, an important activity in video data coding.

The ALUs are interconnected by the 792-bit trunk bus capable of transferring up to eight million integers per second. The 4 KB eight-port on-chip memory is connected to the external RDRAM by the bus with a 500 Mb/s bandwidth. This architecture makes for high-intensity data stream input to the ALU and the performance of one to two billion operations per second in most multimedia applications.

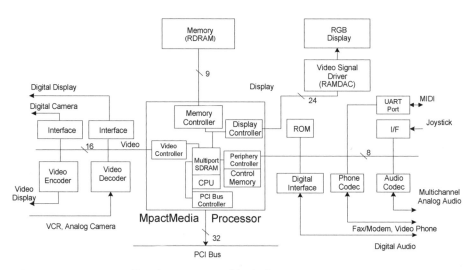

FIGURE 3.47 MpactMedia microprocessor block diagram.

The structure of the microprocessor is shown in Figure 3.47.

The processor's instruction word is 72 bits long and contains two instructions from three to five bytes long. The instruction set contains SIMD instructions for processing data arrays.

3.8.4 NV1 Microprocessor from Nvidia

The NV1 processor is primarily oriented at use in game stations [154]. This crystal was designed specifically for executing game programs, which require fast execution of audio and video data-conversion algorithms. The processor's architecture is shown in Figure 3.48.

One of the processor's components is a 3D graphics accelerator with a Sound Blaster emulation unit. Most 3D graphics accelerators only approximate curves by a number of straight lines substituting curve sections and the resulting curves are jagged, because generating smooth curves requires performing intensive calculations to create a large number of curve sections. The NV1 employs the non-uniform rational B-spline (NURBS) algorithm to round out rectangular sides. This makes it possible to improve the quality of the created image and to reduce the volume of calculations due to the reduced number of control points needed to be calculated for the curve being created.

The processor crystal also integrates the I/O processor (for joystick or other game manipulators). All processing blocks of the microprocessor are interconnected with a unidirectional circular bus. The bus controller accepts transactions and forwards them to the necessary component.

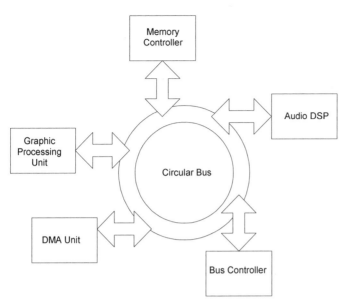

FIGURE 3.48 NV1 microprocessor block diagram.

The NV1 has hardware support of most multimedia algorithms. This feature, on one hand, makes them impossible to modify but, on the other hand, reduces the processor's dimensions and cost. The processor multimedia instruction performance is about 350 MIPS. Its shortcoming is the absence of an MPEG decoder.

3.8.5 MediaGX Microprocessor from Cyrix

On February 20, 1997, Cyrix presented its MediaGX media processor [155, 156]. The processor was produced on CMOS 0.5-micron technology with three metallization layers for 133 MHz, 150 MHz, 166 MHz, and 180 MHz frequencies.

The microprocessor is aimed at use in inexpensive multimedia computers. It performs both system and multimedia functions. The processor has Windows®-compatible software that implements virtual system architecture. This allows hardware implementation of multimedia functions to be replaced with software. Consequently, MediaGX-based systems do not require audio or video expansion cards. In systems, MediaGX is used jointly with the peripheral controller Cx5510 microchip.

An example of a computer system based on the MediaGX is shown in Figure 3.49.

The processor has a 64-bit data bus and is binary compatible with the x86 family; moreover, it can be used as a high-performance virtual video card. The processor's structure is shown in Figure 3.50.

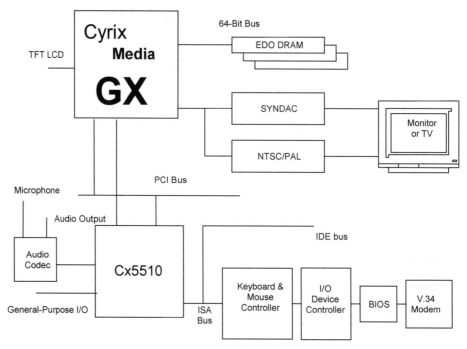

FIGURE 3.49 Block diagram of MediaGX-based computer system.

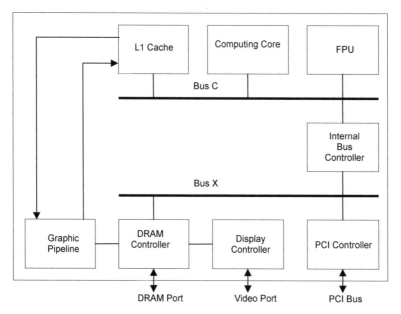

FIGURE 3.50 MediaGX microprocessor block diagram.

The microprocessor comprises the following main components:

- Processor core
- FPU
- L1 cache
- Internal bus controller
- PCI bus controller
- Graphic pipeline
- Display controller
- Dynamic memory controller

Data and instructions are fetched from the memory and sent to the L1 cache over the internal buses X and C. Instructions are issued for execution in the processor core or the FPU from the external memory or cache. Video data is processed in the graphics pipeline and the video controller. From the video controller, digital data is either sent directly to an LCD or converted into analog signal by the RAMDAC and sent to a CRT display. The video controller operates at the processor core frequency, supports all VGA and VESA modes, and supports maximum resolution of 1280×1024 for eight-bit color or 1024×768 for 16-bit color.

The PCI controller handles the internal bus that connects the MediaGX to the Cx5510. This is not a system bus; ISA bus is used for peripheral interface. The Cx5510 microchip provides the I/O.

The processor core fetches, decodes, and executes x86 integer instructions. These steps are all executed in a single cycle. The decoupled load/store unit can process several memory requests in a single cycle.

The instruction execution cycle in the processor pipelined core consists of the following six stages:

- Instruction fetch (IF)
- Instruction decode (ID)
- Address calculate 1 (AC1)
- Address calculate 2 (AC2)
- Execution (EX)
- Writeback (WB)

The execution result is written to the cache and not directly to the main memory.

The processor employs branch prediction mechanism with an 80 percent success rate. The FPU executes floating-point instructions concurrently with integer instructions. The FPU has a 64-bit data interface and four-entry input and output queues.

The processor is equipped with 16 KB writeback four-way set associative unified code and data cache. The internal bus controller bridges the C and X buses.

The graphics controller provides full-scale GUI support, including rendering, bit-map operations, and high-speed block transfers. Systems based on MediaGX do not have separate video memory and use the main memory for the graphic buffer.

In computer systems, MediaGX is used jointly with a satellite Xc5510 chip. The Xc5510 chip performs the following functions:

- Audio controller with FM sound synthesis, MPU-401 MIDI interface, and a capability to connect a supplementary wave table synthesis device
- PCI-ISA bridge
- DMA controller
- Interrupt controller
- Keyboard and mouse controller
- E-IDE controller
- Timer
- Power management

The structure of the Cx5510 satellite chip is shown in Figure 3.51.

Computers based on the MediaGX processor and motherboards for them are produced by several well-known manufacturers, including Compaq, LG Electronics®, Daewoo Telecom®, SCI System®, and Tatung®. In terms of performance, these systems are compatible with the same operating frequency as Pentium-based systems but are substantially less expensive.

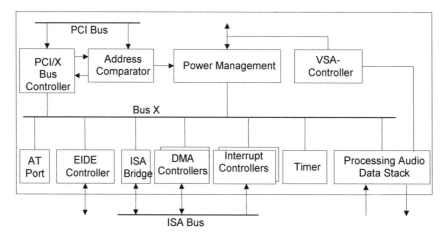

FIGURE 3.51 Block diagram of Cx5510 integrated controller.

REVIEW QUESTIONS TO CHAPTER 3

1. Give a definition of digital signal processing.
2. Point out the main advantages and disadvantages of processing analog signals digitally.
3. Give an example of a DSP task and describe the devices that perform it.
4. State the main characteristics of the DSP task.
5. What are the typical functional, architectural, and structural features of digital signal microprocessors?
6. Name the firms that manufacture signal microprocessors.
7. Cite the classification of signal microprocessors from Texas Instruments.
8. Name the main architectural and structural features of a fixed-point signal microprocessor family from Texas Instruments.
9. Name the internal buses and main functional units of a fixed-point signal microprocessors from Texas Instruments. Describe their functions.
10. What is the difference between the modified and traditional Harvard architectures?
11. Cite the manufacturer-recommended application areas of the TI fixed-point signal microprocessors.
12. Name the main representatives of the TI floating-point signal microprocessor family; describe their main architectural and structural features.
13. Name the internal buses and main functional units of the floating-point signal microprocessors from Texas Instruments (TMS320C3x, 'C4x). Describe their functions.
14. Name the functional unit of the TMS320C80 microprocessor. Describe its functions.
15. What is the difference between the traditional VLIW and VelociTI architectures?
16. Which structural features of the TI floating-point microprocessors stem from the specifics of the parallel data processing?
17. Cite the manufacturer-recommended application areas for TI floating-point signal microprocessors.
18. Cite the classification of the signal microprocessors from Analog Devices.
19. Name the members of the 16-bit microprocessor family from Analog Devices. Describe their main architectural and structural features and state the application area.
20. What architectural approaches are used to shorten the instruction cycle in the microprocessors from Analog Devices?

21. Name the internal buses and main functional units of the microprocessors from Analog Devices. Describe their functions.

22. Explain the purpose of the HIP port of the microprocessors from Analog Devices.

23. Name the members of the 32-bit microprocessor family from Analog Devices; describe their main architectural and structural features and state the application area.

24. What is the distinctive feature of the extended Harvard architecture implemented in microprocessors from Analog Devices?

25. Name the main characteristics and architectural features of the ADSP-21060/62 microprocessors.

26. Point out the main features of the SHARC architecture implemented in the ADSP-21060/62 microprocessors.

27. What elements of the SHARC microprocessor structure allow it to be used to build multiprocessor systems?

28. Name the main features of the TigerSHARC family microprocessors.

29. What is the distinctive feature of Analog Devices' BlackFin architecture?

30. Cite the classification of the Motorola signal microprocessors.

31. Name the members of the Motorola 24-bit signal microprocessor family, their characteristics, and application areas.

32. What is the difference between the microprocessor cores of the DSP563xx and the DSP560xx families?

33. Name the members of the Motorola 16-bit microprocessor family, their main architectural and structural features, and the application area.

34. Name the internal buses and main functional units of Motorola signal microprocessors. Describe their functions.

35. Describe the purpose and the implementation specifics of the general-purpose I/O subsystem of the Motorola signal microprocessors.

36. State the main structural and architectural features and the technical characteristics of the DSP96002 microprocessor.

37. Point out the main common features of the signal, communications, and media microprocessors and their differences.

38. To what range of applications are the PCA architecture microprocessors oriented?

39. List the classes of microprocessors equipped with hardware support of multimedia data processing. Name the members of these families. State the areas of their efficient use.

40. Name the main structural and architectural features of media microprocessors.

41. Cite the main technical characteristics of the following media microprocessors: Mediaprocessor (MicroUnity), TriMedia (Philips), Mpact Media (Chromatic Research), NV1 (Nvidia), MediaGX (Cyrix).

42. What system functions are placed on media microprocessors? Give an example of a computer system structure built based on a media processor.

ENDNOTES

113. *MIPS Architecture Single Chip Microprocessor 1B812.* Technical Manual. *www.niisi.ru/o/1b578omp_short.doc.*

114. Introduction to Digital Filters. Trans. Bogner, R., Konstantinidis, A., editors. Moscow: *Mir,* 1976.

115. Guy, R., Sohie, L., Chen, W. *Fast Fourier Transforms on Motorola's Implementation on Digital Signal Processors,* Motorola, Inc., 1993.

116. *TMS320 Digital Signal Processor Solutions.* Texas Instruments Inc., 1997.

117. *TMS320C1x Digital Signal Processors.* Production Data, Texas Instruments, Inc., 1993.

118. *TMS320C2x Digital Signal Processors.* Production Data, Texas Instruments, Inc., 1994.

119. *TMS320C5x Digital Signal Processors.* Production Data, Texas Instruments, Inc., 1995.

120. *TMS320C2xx User's Guide.* Texas Instruments, Inc., January 1997.

121. *TMS320C203, TMS320C209, TMS320VC203 Digital Signal Processors.* Advance Information, Texas Instruments, Inc., 1995.

122. *TMS320C54x DSPs.* Product Bulletin, Texas Instruments, Inc., 1996.

123. *TMS320C54x, TMS320LC54x, TMS320VC54x Fixed-Point Digital Signal Processor.* Data Book, Texas Instruments, Inc., February 1996.

124. *TMS320C30, TMS320C30 Digital Signal Processors.* Production Data, Texas Instruments, Inc., April 1996.

125. *TMS320C44 Digital Signal Processor,* Data Book, Texas Instruments Inc., 1995.

126. *TMS320C4x, User's Guide,* Texas Instruments Inc., March 1996.

127. *TMS320C80. Multimedia Video Processor (MVP).* Technical Brief, Texas Instruments, Inc., 1994.

128. *TMS320C62xx.* Technical Brief. Texas Instruments, Inc., January 1997.

129. *TMS320C62xx CPU and Instruction Set.* Reference Guide, Texas Instruments, Inc., January 1997.

130. *TMS320C6201 Digital Signal Processor.* Product Preview. Texas Instruments, Inc., January 1997.

131. *ADSP 21xx Family Manual.* Analog Devices, Inc., 1995.

132. *ADSP-21000 Family Applications Handbook.* Analog Devices, Inc., 1995.

133. *ADSP-219x Family Manual.* Analog Devices, Inc., 2002.

134. *ADSP-2106x SHARC DSP Microcomputer Family.* Analog Devices, Inc., 1996.

135. *ADSP-21160 SHARC Technical Specification.* Analog Devices, Inc., 1998.

136. King, K, Girling, G, Vorin, K, Levin, N, Morris, D., Kester, W. *Digital Signal Processor Hardware.* Analog Devices, Inc., 2002.

137. *ADSP-21535.* Preliminary Technical Data. Analog Devices, Inc., June 2002.

138. *DSP56000/DSP56001 User's Manual.* Motorola, Inc., 1995.

139. *DSP56000 24-bit Digital Signal Processor.* Family Manual. Motorola, Inc., 1995.

140. *DSP56300 Family Manual.* Motorola, Inc., 1995.

141. *DSP56100 Family Manual.* Motorola, Inc., 1995.

142. *DSP56600 Family Manual.* Motorola, Inc., 1996.

143. *DSP56800 Family Manual.* Motorola, Inc., 1996.

144. *DSP96002 Product Documentation.* Motorola, Inc., 1996.

145. *MPC8260 PowerQUICC II Technical Summary.* Motorola, 9/98, p. 20.

146. *IXP1200 Network Processor.* Product Brief, Order Number: PB-0010-0899-3K, *www.level1.com.*

147. *Intel® IXP2850 Network Processor.* Product Brief. Intel Corporation. 2002.

148. *Intel® IXP422 Network Processor.* Product Brief. Intel Corporation. 2003.

149. *Intel PXA26x Processor Family,* Developer's Manual, Intel Corporation, 2003.

150. Slater, M. The Microprocessor Today, *ComputerWeek-Moscow,* 1997, No. 17–18, pp. 34–39, 46, 47.

151. Halfhill, T., Montgomery, J. Multimedia Chips Will Dominate the Technical Talk at this Microprocessor Forum, *Byte, Chip Fashion.* 1996.

152. Hars, A. Hot Chips, *Cool Media, Byte.* 1995, Vol. 20, pp. 45–50.

153. *Mpact Media Processor.* Data Sheet, Chromatitc Research, Inc., 1997.

154. Andrews. Hot Chips, Tough Choices. *Byte,* 1995, Vol. 20, p. 25.

155. *Cyrix Announces the MediaGX Processor.* Cyrix Corporation, February 1997.

156. *Cyrix MediaGX Processor.* Data Book, Cyrix Corporation, 1997.

4 ▌Transputers: Basic Building Blocks of Multiprocessor Systems

4.1 MAIN FEATURES OF TRANSPUTERS

The parallelism concept has the potential of raising performance and reliability of computer systems. As such it has been attracting attention of computer specialists for a long time. Since the 1960s, theoretical, experimental, and industrial research and design in this direction, which did not escape notice by the American high-technology export-control specialists, have been conducted in Russia [157, 158]. Theoretical foundations for constructing massively parallel systems from elementary computer machines based on LSI circuits was presented in the treatise *High Performance Homogeneous Universal Computer Systems* [44] by E. Yevreinov and Y. Kosarev. The outlook for further performance increases is associated exactly with this type of system. Historically, the first industrial design oriented at massively parallel systems was a transputer [21, 25].

A *transputer* is a microcomputer with its own internal memory and channels for connecting with other transputers. The interconnecting channels are often called *links*. The term *transputer* is compounded from *trans* in transistor and *puter* in computer. As such, it reflects the main application area of its name bearer: use as the basic computing element in constructing massively parallel computer systems.

Some specialists consider the term transputer as the name of the specific product from Inmos Company; others interpret it as a generic name for microprocessors with built-in interprocessor exchange channels. The term *transputer-like microprocessor* also is used when referring to non-Inmos products. On one hand, it underlines this fact, while on the other, points out that the microprocessor has built-in links for constructing parallel systems. It is quite possible that the rash developments in microelectronics will not allow the term transputer to take, and it will be absorbed by the more general term *microprocessor*, as all microprocessors will have the transputer's trademark—built-in interprocessor links—in one form or another.

Currently, specifications InfiniBand, Rapid I/O [161, 162], and several others have been developed that introduce point-to-point connections and switches. Construction of computer systems on the basis of point-to-point connections is founded on the ideas that have been approved after having been tested in transputer systems.

The first transputer, T414, was introduced by Inmos, Inc. (Bristol, Great Britain) in 1983. Its main characteristics follow:

- Digit width of 32 bits
- 2 KB internal memory
- Four links
- Link exchange speeds of 5, 10, or 20 Mbps
- 5 MHz external clock frequency and 15 MHz internal
- Performance of 10 MIPS

Transputers have become commonly available and widely known since 1985. With modifications—larger, 4-KB internal memory and higher clock frequency—the following models were produced:

T-4 family—Members T424 and T425; 20-MHz, 25-MHz, and 30-MHz clock frequencies

T-2 family—Members T212 and T222 are 16-bit versions

T-8 family—Members T800, T801, and T805; have an on-chip FPU, and performance reaches up to 30 MIPS and 4.3 MFLOPS

Along with the processors, several peripheral devices for the transputer families were produced. Some of these were the M211 ST506 HDD controller chip, the G412 graphic RGB monitor controller chip, the C004 32-channel programmable link switch chip, and other devices, such as signal processors.

The high degree of functional autonomy of the transputer, ease of integration, and presence of peripheral devices allow systems on their base to be built in short time. Transputer links can perform data exchange simultaneously with calculations, without lowering the processor's performance. Due to this property of transputers, systems built on their base are very scalable and have a high performance-to-cost ratio. Transputers were used to build the first massively parallel systems with the number of processors over several hundreds and up to several thousands.

Currently, following the Inmos transputers, transputer-like universal and signal processors Power4, Alpha 21364, TMS 320 C4X, and ADSP 2106X have been produced (see Chapter 2 and Chapter 3).

4.2 ARCHITECTURE AND STRUCTURE OF INMOS TRANSPUTERS

4.2.1 Architecture of the T-2, T-4, and T-8 Families

Transputers pertain to the RISC class processors. Transputer instruction set is oriented toward supporting OCCAM programming language. This high-level language is named after the Middle Ages philosopher William of Occam, who espoused the principle of using only the minimum number of entities, cutting off any superfluities. This principle is known as Occam's Razor. OCCAM programming language is highly parallel and allows parallel calculations to be specified according to the Communicating Sequential Process language model [164]. A program written in OCCAM is a collection of asynchronous, concurrent, interacting processes. A process is the execution flow of a code sequence of a program or a part of a program. The processes interact by exchanging data under the rendezvous principle: exchange takes place only when each of the interacting processes reaches in the course of its execution respective instructions to transmit or receive data. Regardless of whether it is transmitting or receiving data, the process that reaches the exchange instruction first is placed into the wait state until the other process reaches the corresponding receive or transmit instruction. Use of only these two instructions to provide synchronization of and communication between processes without resorting to any other coordinating means for this purpose is the crux of following the Occam principle.

In transputers, this model of parallel computing is supported by a hardware scheduler that uses time slicing to run parallel processes. There is no limit on the number of threads that can be executed concurrently. This method of organizing computations can be considered nowadays as one of the ways to implement multithread architecture.

A transputer-based multiprocessor system is a collection of transputers connected by their links via communication lines (directly or using a switch).

Programming transputer-based computer systems is simple, because the same parallel computation model is supported both inside individual transputers as well as over the transputer system as a whole. Because of this property, a multiprocessor system program can be designed and debugged on a single processor and then scaled to a network of transputers without significant changes. The only exception is the limited number of communication channels a process on one transputer can have to processes being executed on other transputers. However, this limitation was overcome in the T-9000 transputer.

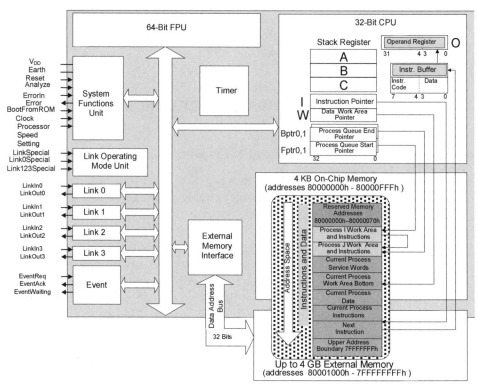

FIGURE 4.1 T-800 transputer block diagram.

The generalized structure of the T800 transputer, the processor's architectural registers, the process control structures and allocation of their memory work areas are shown in Figure 4.1.

Depending on what family it belongs to, a transputer comprises a 32- or 16-bit central processor, external memory interface, two or four bidirectional channels/links, a programmable event channel, a timer, 2 or 4 KB internal RAM, a link control unit, and a system functions unit. Some models can have a 64-bit FPU and/or peripheral device controllers, for such devices as hard drives, monitors, and network adapters. The latter are usually put on a chip in place of two links.

4.2.2 Central Processor

The transputer's 32-bit CPU operates at clock frequencies up to 30 MHz, which are obtained from the external 5 MHz base clock by the internal frequency multiplier circuit. The internal structure of the CPU is shown in Figure 4.2.

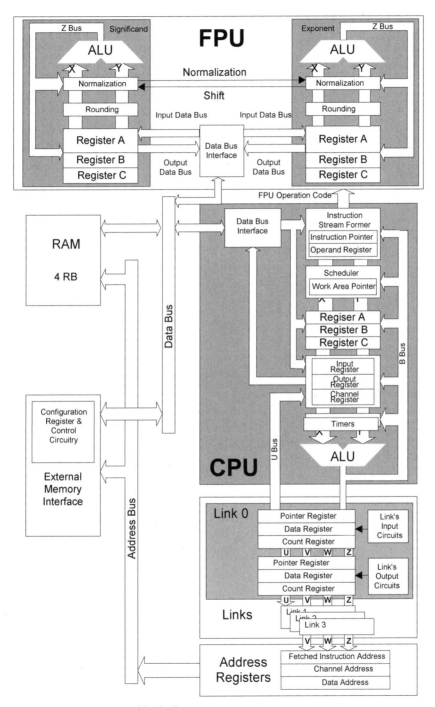

FIGURE 4.2 Transputer block diagram.

The CPU's hardware scheduler organizes the execution of collections of processes in time-splice mode. At the moment of its creating, each process is allocated a work memory area (aligned at the word boundary) in the transputer's addressing space and assigned a priority. There are two priority levels: 0 is the high priority level, 1 is the low priority level. Work area address and the priority are combined to form the process descriptor or context.

The CPU has two register timers for low- and high-priority-level processes. The high-priority timer is incremented every 1 mcs; the low-priority timer is incremented every 64 mcs.

The transputer's W register points to the working area of the process; the instruction pointer register I contains the address of the instruction to be executed next.

The general-purpose registers A, B, and C make up the FIFO register stack. The ALU performs logic and arithmetic operations on operands contained in the register stack. Operands are placed into the stack through its apex, register A. When A is loaded with data, its previous contents are moved to B, and the contents of B are moved to C, whose contents are lost. Operation results are also formed in the A register, with the contents of the C register pushed into the B register and becoming indeterminate.

4.2.3 Transputer Instruction Set

All transputer instructions are one-byte long and are executed in a single processor cycle. Instructions have two fields (see Figure 4.1). The first four bits are the instruction code; the lower four bits are used to form either the instruction's operand or the instruction code in the operand register O.

Operations implemented by the transputer instructions are divided into primary and secondary. The set of primary operations includes 13 most common operations, such as addition to a 0–15-range constant, unconditional branches, word load/store, and so on. Three instructions, `pfix`, `nfix`, and `opr`, are used to perform secondary operations.

4.2.4 Instruction Execution

Instructions are fetched from the transputer's memory and placed into the CPU's buffer. In one cycle, two instructions are fetched in the T-414 and four instructions in the T-800.

Most instructions are executed in the following three stages:

1. Instruction's operand is placed into the lower four bits of the operand register O.

2. Operation specified by the instruction code is performed; contents of the O register are interpreted as operand.
3. The operand register O is cleared.

The `pfix`, `nfix`, and `opr` instructions are an exception from this sequence.

In stage 2, the `pfix` instruction shifts the contents of the O register four bits to the left and does not clear the O register in this stage. The `nfix` instruction is executed the same as the `pfix`, but the contents of the O register are two's complemented prior to being shifted.

Because the contents of the O register are not cleared after the `pfix` and `nfix` instructions, a sequence of these instructions allows operands up to 32 bits wide to be formed in it. The sequence of the primary instructions necessary to form the secondary operation is generated by the compiler. As a rule, the length of this sequence is kept to the minimum.

For example, the operation of summing the contents of the A register with constant 9 is executed by the `adc #9`[1] as a primary operation. However, to sum up the contents of the A register with constant 21 (15h), a secondary operation is needed. To form it, the instruction sequence `pfix #1; adc #5` needs to be executed. Adding constant -31 (FFFFFFE1h) can be done by either the instruction sequence `pfix #f; pfix #f; pfix #f; pfix #f; pfix #f; pfix #f; pfix #e; adc #1` or by using the `nfix` instruction: `nfix #1; adc #1`.

The `opr` instruction interprets the contents of the O register as a secondary operation code. In this case, operands are located in the processor's registers. Thus, the secondary operation code is specified as the operand of the primary operation `opr` and can have up to 2^{32} (4 G) different values. Contemporary transputers only use approximately 100 secondary operations.

For example, for a multiplication operation coded as f3h, the compiler will generate the `pfix #f; opr #3` instruction sequence.

4.2.5 Coprocessor Use

If a transputer has the on-chip floating-point coprocessor, instructions are decoded and issued for execution by hardware in the CPU. Calculation of the operand addresses and loading the operands into the FPU registers is also done by hardware in the CPU.

As shown in Figure 4.2, the FPU comprises two subunits: the exponent and the significand units, each with its own register set (two register stacks). All floating-point instructions are divided into two classes. To the first class belong fully

---

[1] The # character means that the following number is in hexadecimal format.

independent instructions that do not change the CPU state in any way. Instructions that pass operation execution results to the CPU belong to the second class.

After an instruction is issued for execution in the coprocessor, the CPU continues executing the instruction stream if the instructions are of the first class, or waits for the results if the instructions are of the second class.

4.2.6 Transputer Memory Distribution

The transputer can address up to 4 GB of memory. One of its distinctive features is that the addresses start in the negative value range. The low address in complement code is 80000000h.

Constructively, all operating memory is divided into on-chip and external. Depending on the transputer model, the on-chip memory can be 2 or 4 KB. Architecturally (that is, from the programmer's viewpoint), all memory, both the external and the on-chip, is equally accessible and has the same addressing mechanism; the on-chip memory is located at the bottom of the addressing range, the external memory at the top.

Because the on-chip memory has shorter access time (one processor cycle), it is normally used by programmers to place often-used data and subroutines in it. An example of memory distribution is shown in Figure 4.1.

Several low-memory words are used for special purposes: for status words of the hardware channels/links, event channel, timer registers, and as the memory dispatcher's work area.

4.2.7 Scheduling Processes

At any specific moment in time, each of the processes can be in one of the following states:

- Being executed
- Active (waiting in the corresponding priority queue to be executed)
- Waiting for a timer signal (in the queue to the corresponding priority timer)
- Waiting for input/output

For the processes being executed, the I register holds the address of the next instruction; the W register holds its descriptor.

To organize the active process queue, the scheduler uses register pairs Fptr0/Bptr0, Fptr0/ Bptr1 pointing to the beginning and the end of the high- and low-priority process queues, respectively. Process work areas are used as the queue elements. In addition to the process data, these work areas contain control information that is needed to save and restore process state during the scheduling stage (including the pointer to the work area of the next process in the queue).

Low-priority processes are executed for no more than 32 time slots of the low-priority timer; after this, the process (if it is still to be continued) is placed at the end of the active low-priority process queue and the scheduler schedules execution of the next process. First, the high-priority process queue is examined and then the low-priority process queue.

A high-priority process is not interrupted and is executed as long as possible (until completion or until switching into wait or input/output state or until a timer signal or an external event signal issued by the Event channel). A high-priority process may become active (receives a timer signal, for example) at the moment a low-priority process is being executed. When this happens, the latter is interrupted, its status is saved in the transputer's reserved address memory area, and the execution of the high-priority process, which has interrupted the low-priority process, is initiated. After the high-priority process completes execution, the interrupted low-priority process is resumed.

4.2.8 Input/Output

Process I/O in the transputer is organized in the same way, both when using hardware links (external I/O) and when using virtual links (exchange between processes of a same transputer). There always are two processes participating in an exchange: one of them inputs data, and the other outputs it.

At the moment a channel description instruction is being executed and after each exchange is completed, the channel status word (CSW) is initialized by the MinInt constant (80000000h), specifying the lowest integer: 0. The process that started executing the exchange instruction first stops and checks the contents of the channel. If it equals the MinInt, it means that the process has reached the rendezvous point first and must wait for the other process. The wait is organized as follows: the process descriptor is placed into the CSW, and the value of W along with the address of the data that are to be transferred from one process to the other are placed into the work area of the process; then the control is passed to the scheduler to schedule execution of other processes.

If the contents of the CSW are not the MinInt, it means that the given process has arrived to the rendezvous point second. In this case, the content of the CSW is the process descriptor. This descriptor is used to find the data for an input operation or data buffer for an output operation.

Exchange over the links is carried out in the same manner, with the difference being that the CSWs hold fixed memory addresses 8000000h–8000001Ch and during the exchange, process data are not passed from one memory area to another but are sent over the links in start/stop mode with handshaking for every byte.

When performing I/O, all the CPU has to do is to initialize the exchange. Upon the I/O instruction, all the information needed to conduct the exchange operation—the address and data length—are placed into the link's internal registers. The CPU is then freed from this task, and the transputer link controls the data exchange by itself.

4.2.9 Data Transfer Over Links

All transputer families T-2 (T-212, T-222, T-225), T-4 (T-414, T-400, T-425), and T-8 (T-800, T-801, T-805) use the same data transfer protocol to transfer data over the links. Transputers interact by exchanging byte sequence messages. Data is sent over one of the two wires forming the link. The other wire is used to send acknowledgement after receiving each byte. Transfers can be conducted concurrently in both directions, in which case data and acknowledgements in each of the links' wires alternate.

A transmitted byte is framed by service bits. The start bit comes first, followed by the control bit, followed by eight data bits, and finally the stop bit. Thus, there are three control bits supporting the transfer protocol for every eight information bits. An example of data transfer over a link is shown in Figure 4.3.

After a byte has been sent, the transmitter waits for the acknowledgement, which is composed of the start and control bits. Also, when the control bit of the data byte is 1, it is 0 in the acknowledgement. Moreover, data bytes transferred in one direction and the acknowledgements for data bytes transferred in the opposite directions are sent over the same wire. Acknowledgements have higher priority than data bytes. It is possible to set such transfer mode when acknowledgements are sent right after the start and control bits are received, thus transferring bytes without delays between them.

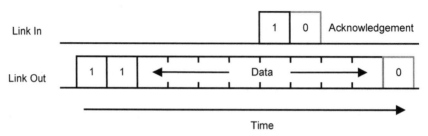

FIGURE 4.3 Data transfer over a transputer link.

If one of the processes that are conducting exchange over the link is not ready to receive data, bytes are accumulated in the link data register. When the register gets filled up, the next received byte is not acknowledged and the LinkOut suspends the transmission until the process receives the accumulated data and the link data register is unloaded.

At the beginning of the transmission of the first byte of a message, there is a delay to configure the DMA channels in the transmitting and receiving transputers, which is the reason for different transmission speeds of short (from several bytes to tens of bytes) and long (several thousands of bytes) messages. In the former case, the speed is 2–4 Mbps; in the latter, the maximum transmission speed ranges from 10 to 20 Mbps, depending on how the transmitting and receiving transputers are configured.

Transmission over the links is asynchronous, the receiving transputer being indifferent to the phase of the received signals. The only thing needed is an accurate 5 MHz quartz transputer clock generator.

The link operating mode unit allows the transmission speed over the transputer links to be set to 5, 10, or 20 Mbps; moreover, the transmission speed for the 0 link can be set independently of the other links. The speed is set by driving the `LinkSpecial`, `Link0Special`, and `Link123Special` inputs to corresponding levels.

Exchanges over the links are programmed individually in the transmitting and the receiving transputers. The transmitting transputer is programmed with an instruction to transmit a specified number of bytes into the link. The receiving transputer is programmed with an instruction to receive the specified number of bytes from the link. The exchange is carried out when the programs in both the transmitting and the receiving transputers arrive at the transmit and the receive instructions. If one of the transputers has arrived to its exchange instruction before the other one, it waits for the other transputer to arrive at its exchange instruction. An exchange that has been programmed incorrectly will result in an endless wait. Moreover, indefinite waits also are possible if different numbers of bytes to be transmitted and received are specified.

4.2.10 Waiting for Event Channel Input

The Event channel converts external logic levels at its input into a byte message with the value of 0 or 1, afterwards transmitted over the internal channel. Therefore, in software terms, the Event channel is perceived as a special channel, different from regular channels in that it can only be read. The CSW for the event channel has a fixed address of 80000020h. At any given moment, only one process can await for an input from the Event channel.

The Event channel is normally used to register external interrupts. The interrupt signal is placed onto the `EventReq` pin of the transputer chip. The reception of an interrupt signal (reading from the Event channel) is acknowledged by placing a high level on the `EventAck` pin of the transputer chip.

4.2.11 Waiting for Timer Input

The timer is perceived by software as a read-only channel, providing either the contents of the register of the timer with the corresponding priority or a signal indicating that the specified time has been reached.

All processes awaiting the specified time (lapsing of the time slice) are placed into the timer queue corresponding to their priority. The process queue is sorted in the arrival order of the specified times; it is organized by cross-referencing service words in the processes' work areas. The address of the first process in the queue of the corresponding priority is contained in the `TPtrLoc0` and `TPtrLoc1` service words in the low memory addresses. When the specified time is reached, the scheduler is passed the descriptor of the corresponding process for placing it at the end of the active process queue.

4.2.12 System Initialization Upon Power-Up

The transputer and its RAM are built on CMOS technology and do not preserve their states after the power is turned off. Therefore, for the transputer to begin functioning after the power is applied, some certain software minimum needs to be loaded. The transputer can be bootstrapped either from an external ROM or from any link. The `BootFromROM` pin output of the transputer chip is used to indicate the bootstrap mode. If its level is 1, the control is passed to the FFFFFFFFh address, at which an unconditional jump instruction to the bootstrap is usually located.

If the `BootFromROM` level is 0, the transputer is booted from a link. After the power is applied, the transputer switches into the state of waiting to receive data from the links. The first control byte received from any of the links controls the ensuing transputer operating mode. If this byte is greater than 1, it is interpreted as the length of the program code following it. The received data are written into the transputer's memory starting from the `MemStart` address; the program execution also starts from this address. As a rule, the bootstrap is loaded first, which then loads the rest of the needed software.

If the first received control byte is 0 or 1, the transputer switches into the memory control mode.

A control byte that equals 0 makes the transputer interpret the next four bytes as the memory address, to which the word coming in the next four bytes will be written. Thereafter, the transputer switches into the memory-control mode again, from which it can be taken out only by receiving a new control byte greater than 1.

If the control byte is 1, the next four bytes specify the memory address, from which four bytes will be read and transmitted over the same link, from which the previous four bytes came in the opposite direction. After this the transputer remains in the memory control mode.

The memory control mode is usually used for bootstrap loading and debugging.

4.2.13 Controlling the System

Applying a high level to the `Analyze` pin of the transputer running a program switches it into the memory-control mode. As was said earlier, a transputer's memory can be modified and read in this mode.

4.2.14 Error Handling

Software errors, such as arithmetic overflow, division by 0, and crossing array boundaries, set the `error` flag and drive high the `Error` pin of the transputer chip. The `HaltOnError` flag allows specifying a transputer's action in case of an error (the `error` flag getting set).

- If the `HaltOnError` flag is set to 1, when an error occurs, the `Error` pin is driven high and the transputer halts.
- If the `HaltOnError` flag is set to 0, when an error occurs, the error flag is set to 1 but the transputer continues operating.

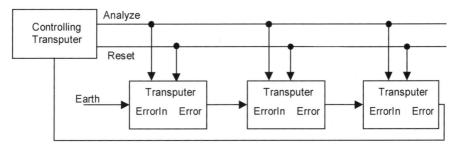

FIGURE 4.4 Diagram of connecting system service lines in a multitransputer system.

In multitransputer systems, the `Analyze`, `Reset`, `Error`, and `ErrorIn` pins of all transputers are usually connected as shown in the diagram in Figure 4.4. With these pins connected this way, an `Error` signal appearing on any transputer places the entire system into the memory-control mode. Appropriate software running on the control transputer determines the system state and the ways of handling the error situation.

4.3 T-9000 TRANSPUTER

4.3.1 Architectural and Structural Features

When they were first introduced, the T-8 family transputers were the fastest 32-bit microprocessors. Attempts by Inmos to preserve their lead over American microprocessor producers materialized in the development of the T-9000 transputer [25]. Its main technical characteristics follow:

- 200 MIPS, 25 MFLOPS performance
- 16 KB on-chip memory
- Four links
- 100 Mbps link bandwidth

The main distinctive architectural features of this transputer are the hardware support of the virtual link mechanism and the hardware instruction grouper, which increases loading of the processor's parallel units.

However, Inmos could not keep the promised supply deadlines for the T-9000; neither could it attain the announced performance because of not being able to reach the necessary clock rate with the VLSI manufacturing technological process used. Consequently, even though T-9000 units operating at lower-than-declared clock rate were produced, they were not commercially successful, because higher performance 32-bit microprocessors from American companies appeared by that time.

4.3.2 Virtual Links

The virtual links mechanism makes it possible to use one physical link to conduct exchanges between any number of process pairs taking place in different transputers.

The exchange is controlled by the on-chip virtual channel processor (VCP). The VCP divides the message transmitted from a transmitter process to a receiver process into packets. Each packet comprises 32 data bytes (the last packet can be from 1 to 32 bytes long), a packet header, and a packet terminator. The terminator of the last packet indicates the message end; in all other packets, it indicates the end of the packet. When the receiving transputer receives a VCP packet, it acknowledges the receipt by an empty packet, containing only the header and the terminator. The VCP uses the information, contained in the packet header, to route packets and to assemble the message from packets. Consequently, to the processes, data exchange appears the same as in the previous generation transputers, which makes for the continuity of the developed software.

Message routing in T-9000 transputer network being transparent to processes completely eliminates the difference between exchanges within confines of one transputer and exchanges in a transputer network. This feature makes developing programs for multitransputer systems substantially simpler and raises their performance levels, because it does not require additional expenditures on organizing routing and switching.

To increase the number of physical connections in the T-9000 transputer, the C104 programmable switch was developed. It routes messages from any of the 32 inputs to any of the 32 outputs according to their headers.

To allow use of the T-9000 transputer in systems containing previous-generation transputers, the C100 chip was developed. This chip matches signals' electric characteristics and converts the formats of data transmitted over the links.

4.3.3 Instruction Grouper

The instruction sets of the previous transputer generations are completely preserved in the T-9000. The performance level is raised by simultaneously executing a group of up to eight instructions.

The T-9000 has a hardware instruction grouper. The object of instruction grouping is to attain a high loading level of the processor units.

The following listing is given as an example of the instruction grouper operation [25]. Assume that the $a[i + 1] = b[j + 15] + c[k + 7]$ equation needs to be calculated. The compiler generates the code shown in Listing 4.1.

LISTING 4.1

```
ldl j              load local variable j

ldl b              load base address of array b

wsub               calculate address of b[j]

ldnl 15            load value of element b[j+15]
```

```
ldl k

ldl c

wsub

ldnl 7

add                 add two values on the top of stack

ldl i

ldl a

wsub

stnl 1              store into a[i+1]
```

The grouper converts this instruction sequence into three groups:

- `ldl, ldl, wsub, ldnl`
- `ldl, ldl, wsub, ldnl, add`
- `ldl, ldl, wsub, stnl`

The processor fetches four instructions from the memory in one cycle. Because some instructions require more than one cycle for execution, a number of instructions sufficient to form five groups of eight instructions each can be accumulated in the processor. This corresponds to the 100 percent loading of the processor units.

4.4 TRANSPUTER-LIKE MICROPROCESSORS OF KVANT SERIES

4.4.1 Architecture Basics

An example of transputer-like processor design is the Russian Kvant series of microprocessors [165, 166]. This is a family of 32-bit microprocessors with original architecture combining RISC approach with the VLIW method. The family is characterized by a high degree of the internal parallelism of processes, pipelined instruction execution, Harvard memory architecture, and the availability of communication channels/links.

- The Kvant-10 (clock frequency 10 MHz) was produced on 2.5-micron technology on three semi-custom 1537 XM2 matrix chips that were designed

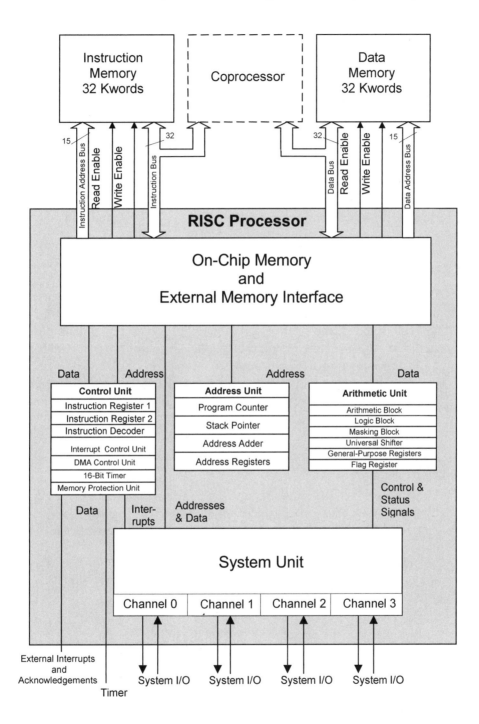

FIGURE 4.5 Block diagram of Kvant microprocessor.

at the Kvant Research & Development (R&D) Institute and produced at the Science & Research Institute of Exact Technologies at Zelenograd, Russia.

■ The Kvant-20 (clock frequency 20 MHz) was produced on 1.5-micron technology on a single U1700 chip, which was designed at the Kvant R&D Institute and produced by ZMD Company in Dresden, Germany.

The structure of the microprocessor is shown in Figure 4.5.

The microprocessor works with the external split data and instruction cache and can be equipped with a coprocessor. The data memory can store only data, whereas the program memory can hold both instructions and data (modified Harvard architecture).

4.4.2 Control Unit

The processor fetches 32-bit instructions from the program memory and places them first into the first instruction register and then into the second. Instructions are then decoded, and control signals for all processor functional units are generated. The interrupt control unit can handle 10 types of external and internal interrupts under the priority servicing mechanism. The DMA unit allows calculations to be performed concurrently with data exchange over four communications channels. The data protection unit protects areas of the program and data memories allocated to the operating system.

4.4.3 Address Unit

The processor is built using decoupled architecture concept [11]. Under this concept, a separate address unit calculates all addresses. This concept allows concurrent processing of data in the arithmetic unit and address calculation in the address unit.

The address unit holds in the program counter the address of the next instruction to be executed, supports memory stack of subroutine (interrupt handlers) return addresses, and performs all address calculations in the 16-bit address adder. The address register file of the Kvant-10 microprocessor has eight 16-bit address registers; the one of the Kvant-20 has four 16-bit ones.

4.4.4 Arithmetic Unit

The arithmetic block can perform 16 logic and 14 arithmetic operations, including byte operations, and incremental multiplication of a 32-bit multiplicand by two bits of the multiplier. The logic block can carry out logic operations in parallel with the arithmetic block operations. All operations are performed on 32-bit operands in a single cycle.

The universal shifter can perform logic, arithmetic, or circular left or right shifts of a 32-bit word, from 0 to 31 bits in a single cycle. It can also perform circular shifts within bytes, quads, and twains. The arithmetic unit's masking block allows results of any operation to be masked with the contents of one of the general-purpose registers. The general-purpose register file is used to store operands, results, masks, and data addresses. The flag register is used to store indicators of operation executions in the arithmetic unit.

4.4.5 System Unit

The system unit provides connection with four analogous processors over independent channels. Exchanges are carried out in word blocks bit-by-bit. At the beginning of each block, the number of the words to be transmitted and the memory address, into which the message is to be written, are specified. To ensure data integrity during transmission, even parity is used. A transmission error causes a corresponding interrupt to be generated.

The memory protection register is used to disable writes to any 4 Kwords memory block.

4.4.6 Processor Pipeline

The processor uses a three-stage instruction execution pipeline. Instructions are fetched from the program memory in the first stage. In the second stage, a data address for the future memory reference is formed and the address registers are modified. In the third stage, data is input/output from/to the memory using the address calculated in the second stage or arithmetic operations performed. Operations at each stage are executed in a single cycle, which on average allows instructions to be executed in a single cycle, providing there is a high level of the pipeline workload.

4.4.7 Instruction Set

Processor's instructions are divided into simple and complex. Each of the former performs one operation, whereas the latter specifies three-address arithmetic operations on data in registers concurrently with a memory data exchange and/or address register modification.

The following addressing modes are used to reference memory:

■ Based on the contents of an address register
■ Based on the contents of a general-purpose register
■ Autoincrement or autodecrement using any address register
■ Based indexed using two address registers

The Kvant-20 microprocessor has an additional addressing mode—based indexed with an 8-bit offset specified in an instruction field.

To make the control unit simpler and to allow for greater program flexibility, the microprocessor implements the following conditional and unconditional branch execution mechanism. For conditional branches, a special instruction checks whether the arithmetic operation result flag corresponds to one of the 16 possible branch conditions. If there is such a correspondence, then a NOP instruction is executed instead of the next instruction. The check for the condition is performed in the background, with execution of arithmetic instructions process not interrupted for this operation.

In order to not interrupt the pipeline operation (and avoid missing pipeline cycles), unconditional branches are executed under the postponed branch principle: the pipeline first executes the instruction following the branch command and then executes the branch. Subroutine calls are done the same way.

Pipelining internal processes and executing operations in different functional units simultaneously allow up to four instructions to be executed in a single cycle.

4.4.8 Microprocessor Performance

The microprocessor's flexible instruction set makes for its efficient use in both calculation and logic- and symbol-processing tasks. The processor's communication capabilities make it possible to build scalable MPP architecture systems on its base.

The unique architecture of the Kvant family processors provided better performance as compared with the Inmos transputers [166]. Performance of the Kvant-10 microprocessor running at 4 MHz is the same as that of the T-800 transputer operating at 20 MHz. Performance of the Kvant-10 and Kvant-20 microprocessors is 12–15 MIPS and 25–30 MIPS, respectively.

REVIEW QUESTIONS TO CHAPTER 4

1. Name the main architectural and structural features of the transputer family devices.

2. What properties make it possible to view the transputer as a basic building block of multiprocessor systems?

3. Name the main functional units of the transputer, and state their functions.

4. What is the process descriptor?

5. Explain the principle of forming transputer secondary operations.

6. What is the difference between scheduling high-priority and low-priority processes?

7. How is the transputer memory distributed?

8. Describe the data-exchange process between parallel processes over hardware and virtual links.

9. What control and diagnostics means are provided for in the transputer?

10. Describe the process of booting the transputer over a link.

11. Name the main functional units of the T-9000 transputer, and state their functions.

12. Describe the operating principle of the T-9000 transputer links.

13. State the purpose of the instructions grouper and its functions.

14. Name the microprocessors of the Kvant family, and describe their architectural features.

ENDNOTES

157. Goodman, S., Wolcott, P., Burkhart, G. Executive Briefing: An Examination of High-Performance Computing Export Control Policy in the 1990s. IEEE CS Press, Los Alamitos, California, 1996.

158. Wolcott, P. Soviet Advanced Technology: The Case of High-Performance Computing. Ph. D. Dissertation. University of Arizona. 1993.

159. Korneev, V. Architecture or Programmable Structure Computer Systems. Novosibirsk: *Science*, 1985, p. 168.

160. Levin, V. High Performance Multimicroprocessor Systems. *Information Technologies and Computer Systems*. 1995, Volume 1, #1, pp. 1–9.

161. InfiniBand™ Architecture Specification, Volume 1, Release 1.0. October 24, 2000 Final.

162. RapidIO: An Embedded System Component Network Architecture. Architecture and Systems Platforms Motorola Semiconductor Product Sector 7700 West Parmer Lane, MS: PL30 Austin, TX 78729.

163. Hull, M. Occam—A Programming System for Multiprocessor Systems. *Computer Languages*, 12 (1), pp. 27–37.

164. Hoar, C. Interacting Sequential Processes. Trans. Moscow: Mir, 1989, p. 264.

165. Viksne, P., Katalov, Yu., Korneev, V., Panfilov, A., Trubetskoy, A., Chernikov, V. Transputer-Like 32-Bit RISC Processor with Scalable Architecture. Issues in Radio Electronics. CT Series.1994. 2nd Issue. RESRI, pp. 49–59.

166. Viksne, P., Katalov, Yu., Konotoptsev, V., Korneev, V., and Yarmolinskiy, I. Comparative Performance Evaluation of the Kvant-10 and Kvant-20 Processors and the T-800 Transputer. Issues in Radio Electronics, CT Series, 1994. 2nd Issue, RESRI, pp. 60–65.

5 ▮ Neuroprocessors

5.1 GENERAL INFORMATION ABOUT NEURAL NETWORK COMPUTING

5.1.1 Problem-Oriented Nature of Neural Network Computing

The neural network approach to problem solving has proven to be effective in solving ill-formalized problems, such as recognition, clusterization, and associative search tasks, as well as in solving well-formalized but work-intensive multivariable function approximation and optimization problems.

Ill-formalized problems are characterized by the absence of worked out models that yield calculation formulas; neither can chains of simple actions be formed for them whose sequential, perhaps repeated, application would produce the result sought. A classic example of an ill-formalized task is the problem of recognizing images of a cat and dog. Man solves this problem without thinking twice; however, endeavors to design a recognition program based on rules "if A then B" run into the problem of the indeterminate number of parameters needed to tell the difference between these animals. Trying to increase the number of the considered parameters leads to complicating the problem and contradictory rules arising in it.

The problem of approximating multivariable functions using the traditional methods to receive the necessary degree of accuracy runs into snags when increasing function dimensions. This necessitates increasing the number of linear-combination members of fixed-basis functions, making these methods practically unacceptable for solving large-dimension problems [167]. Using the neural network method, the accuracy of approximation for any dimension of the function being approximated depends only on the number of linear members of the basis functions.

373

For optimization problems of the *NP*-complete class, there is no precise solution method other than searching through the entire $n!$ number of the possible solutions, where n is the dimension of the problem. Neural network algorithms provide ways of acceptable approximate solutions for this type of problems.

Practical importance of the just-enumerated tasks is indisputable. Following [168–170], these are the typical definitions of these problems:

Image recognition—The problem consists in associating an input set of data that represents the object to be recognized with one of the already known classes. Recognition of handwritten and printed characters for optical computer input, blood cell types, speech, and other patterns pertain to this type of problems.

Data clustering—The task is to group data by their intrinsic relations. The algorithm to determine these data relations (determining distances between vectors, calculating correlation coefficient, and other methods) is built into the neural network at its construction. A network clusters data into a number of clusters that is not known in advance. Most known uses of clustering have to do with data compression, analyses, and pattern search.

Function approximation—Consider the following problem: a set of experimental data $\{(X_1, Y_1), \ldots \ldots, (X_n, Y_n)\}$ is given that represents Y_i values of an unknown function of a multidimensional argument X_i, $i = 1, \ldots, n$. A function needs to be obtained that approximates the unknown function and that satisfies certain criteria. This is an important task when modeling complex systems and designing systems to control complex dynamic objects.

Predictions—A set $\{y(t_1), y(t_2), \ldots, y(t_n)\}$ of y values is given that represents a system's behavior at $t_1, t_2, \ldots, t_n$ time moments. The task is to predict the $y(t_{n+1})$ system behavior at the t_{n+1} time moment, based on the history of its past behavior. This type of problem is encountered in controlling warehouse stocks and in automated decision-making support systems.

Optimization—The goal of this type of tasks is to find a solution to an *NP*-complete problem satisfying a series of restrictions and optimizing the value of the target function. The traveling salesman problem pertains to this type of problems.

Content-addressed (associative) memory—Contents from this type of memory can be read using partial or distorted representation of the input data. Its main application area is multimedia databases.

Unlike programs based on "if A then B"-type rules, a neural network can extrapolate results. Another advantage of neural networks over rule-based programs is that to account for new facts, a network is simply trained using these facts, whereas a rule-based program needs to have its rules modified and itself be rewritten. Moreover, training a neural network on a large number of examples may not necessarily increase its processing time (when saving a network graph, for example), and adding new rules to a program slows it down.

Generally speaking, the more explored a problem is, the higher the probability of using algorithms based on equations and rules to solve it. However, when experimental data is limited, neural networks are the mechanism that allows the maximum use of the information available. A typical example [169] of a neural network use is optical character-recognition systems. The 10 best systems in this area include both those based on the neural network approach and those using rule-based programs.

5.1.2 Basic Principles of Organizing Neural Network Computations

The general idea of using neural network for ill-formulated problems is based, first, on training a network to remember examples presented to it at the network-creation stage and, second, on its producing results congruent with the learned examples during the problem-solving stage.

The first thing that a practical implementation of these principles presupposes is minimizing the memory area needed for storing the examples. The second thing is a rapid use of the memorized examples, which rules out using the traditional memory types.

The following posing of problems to be solved has been adopted in neural networks. Based on the problem, an n set of input parameters is determined, on which the researcher believes its solution depends. Afterwards, this set can be adjusted many times during the neural network training. Each input parameter x_i, $(i = 1, \ldots \ldots, n)$ of the problem is associated an i dimension of the multidimensional space whose number of dimensions equals to the n number of parameters. This downsizes the problem to determining properties of the $x_j = \{x_{j1}, x_{j2}, \ldots, x_{jn}\}$ points of the n-dimensional space, where x_{ji} is the value of the input parameter i of the point j, with the properties of the points belonging to the examples used in training already known.

Thus, assume this set of training examples is given:

$$<X_1, D_1> = <(x_{11}, \ldots\ldots, x_{1n}), D_1>;$$
$$<X_2, D_2> = <(x_{21}, \ldots\ldots, x_{2n}), D_2>;$$
$$\ldots\ldots\ldots\ldots\ldots\ldots\ldots\ldots\ldots\ldots$$
$$<X_m, D_m> = <(x_{m1}, \ldots, x_{mn}), D_m>.$$

$X_j = (x_{j1}, \ldots\ldots, x_{jn})$ are the input values of the jth example; D_j is the output value that has to be produced by inputting the jth example, $j = 1, \ldots \ldots, m$. A network is considered to be trained if the end-of-training criterion is met.

Usually, the following end-of-training criteria are applied, although other criteria can also be used:

- For all j: $\max|D_j - Y_j| \leq \delta$, where δ is the specified error magnitude; Y_j is the output value produced by the network when the jth ($j = 1, \ldots \ldots, m$) example is placed onto its inputs.

- $\sqrt{\sum_j (D_j - Y_j)^2} \leq \delta$.

In problems that can be effectively solved by neural networks, points in the multidimensional space in which the task is formulated form areas of points that have the same property (pertaining to the same class of objects), have the same values of some function specified over their range, and so on. Neural networks remember similar areas and not individual points that represent the examples used in the training.

Various methods of training neural networks to remember areas are employed. Currently, the most-often-used methods for this purpose are bounding areas by hyperplanes or covering areas with hyperspheres. Figure 5.1 shows examples of bounding areas in a two-dimensional space.

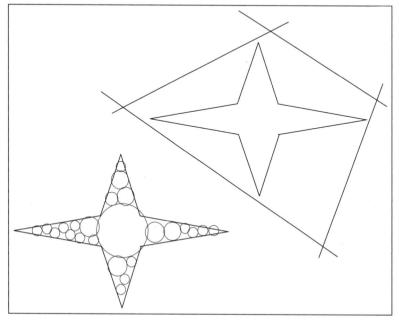

FIGURE 5.1 Bounding areas in a two-dimensional space.

To memorize one of the hyperplanes that bound the area, it is enough to save $n + 1$ values, where n is the number of dimensions. Accordingly, to memorize one hypersphere, $n + 1$ values are also needed: the coordinates of the center and the radius.

In neural networks, for memorizing each hyperplane or sphere an individual elementary computation unit, called a *neuron*, is used. Merging the component neurons into a parallel structure, a neural network, is used for memorizing all hyperplanes or spheres. It is exactly the coordinated parallel work of all the neurons that rapidly solves the problem of deciding whether a given point in the n-dimension space pertains to the area bounded when the neural network was created.

5.1.3 Neural Network Fundamental Concepts

A neuron j, $j \in \{1, 2, ..., N\}$ is defined by a collection of its inputs x_{ji}, $i \in \{1, 2, ..., n(j)\}$, weights w_{ji} of the inputs, the state function s_j, and the activation function f_j. The state function defines the neuron's state based on the values of its inputs, inputs' weights, and, perhaps, previous states. Most often, state functions independent of previous states are used. These are calculated either as the sum of the products of the input values and the corresponding input weights over all inputs ($\sum_{i=1}^{n(j)} x_{ji} \cdot w_{ji}$, where $n(j)$ is the number of inputs of the neuron j) or as the distance between the input vector $X_j = \{x_{ji}\}$ and the input weights vector $W_j = \{w_{ji}\}$, measured by some metric, $\sum_{i=1}^{n(j)} |w_{ji} - x_{ji}|$, for example.

The activation function $y = f(s)$ defines the neuron's output signal as a function of its state s. The most common activation functions are step (or binary) threshold, linear threshold, sigmoid, arctangent, linear, and gaussoid. Table 5.1 shows definitions for these functions.

TABLE 5.1 Neuron Activation Functions

Function	Definition
Step threshold	$y = f(s) = \begin{cases} 0 \text{ if } s < a \\ 1 \text{ if } s \geq a \end{cases}$
Linear threshold	$y = f(s) = \begin{cases} 0 \text{ if } s < a \\ ks + b \text{ if } a_1 \leq s \leq a_2 \\ 1 \text{ if } s \geq a_2 \end{cases}$ $a_2 = 1/k + a_1$

TABLE 5.1 Neuron Activation Functions *(Continued)*

Function	Definition
Sigmoid	$y = (1 + e^{-k(s-a)})^{-1}$
Hyperbolic tangent	$y = \text{th}(s) = (e^s - e^{-s})/(e^s + e^{-s})$
Arctangent	$y = 2\,\text{arctg}(s)/\pi$
Linear	$y = ks + b$
Gaussian	$y = e^{-k(s-a)^2}$

Linear neural networks use neurons with linear activation function, non-linear use non-linear activation functions, step or sigmoid, for example.

A neural network is created by linking outputs of some neurons to inputs of others using directed weighted connections. The graph of neuron connections can be acyclic or an arbitrary cyclic graph. The type of graph is one criterion by which neural networks are classified, dividing networks into feedforward and recurrent (feedback) types. Examples of these types of neural networks are shown in Figure 5.2 and Figure 5.3, respectively.

It is easy to see that having adopted some conventions as to the network clocking (timing of neurons' responses), an apparatus can be obtained for specifying algorithms by means of neural networks. Nothing limits the diversity of these algorithms, because neurons can be used with various activation and state functions and with binary, integer, real, and other values of weights and inputs. Therefore, neural network concepts can be used to describe solutions of both well-formalized problems, such as those of mathematical physics, and problems that don't yield easily to formalization, such as those involving recognition, classification, generalization, and associative memorization.

Neural networks can be constructed or trained. In a constructed network, the number and type of neurons, the neuron interconnection graph, and input weights are determined when the network is constructed based on the problem to be solved. For example, when constructing a Hopfield network [171] functioning as associative memory, each input vector of the previously defined set of remembered vectors takes part in determining weights of the network neurons' inputs. After the network has been constructed, it functions as follows: a certain time after a partial or distorted input vector is placed onto network inputs, the network switches into one of the stable states defined when it was constructed. In this situation, one of the remembered set of vectors appears on the network outputs that is recognized by the network as being the closest to the vector placed onto the network inputs.

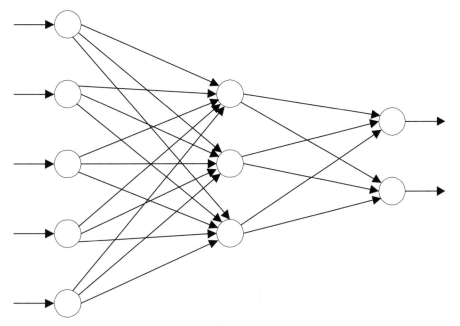

FIGURE 5.2 A feedforward neural network.

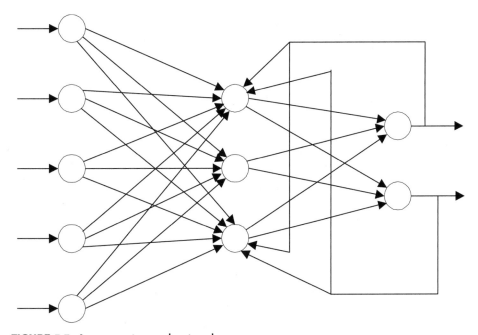

FIGURE 5.3 A recurrent neural network.

The number of the memorized input vectors M is related to the number of neurons in the network by the equation $M < N/4 \log N$, where N is the number of neurons.

In trained networks, their neuron interconnection graphs and input weights are modified in the process of executing the training algorithm. By the type of training algorithms, networks are divided into supervised, unsupervised, and hybrid types. A supervised network during training compares some already-known output with the output it obtains. An unsupervised network learns without knowing correct output values in advance but by grouping close input vectors in such a way so that they generate the same network output. In particular, unsupervised training is used in solving clustering problems. In hybrid training, some input weights are established by supervised training and some by unsupervised.

Supervised training is performed by presenting the network the examples that consist of collections of input data together with corresponding results. In unsupervised training, the corresponding output results are not used. The efficacy with which a neural network can solve problems depends on its neuroparadigm (the construction structure and the training algorithm used) and the completeness of the example database.

5.1.4 Making Network Function

When hyperplanes are used to bound areas, each threshold-activation function neuron j ($j \in \{1, \dots, N\}$, where N is the number of neurons in the network) defines a hyperplane by the weight values of its inputs:

$$a_j - \sum_{i=1}^{n(j)} w_{ji} \times x_{ji} = 0,$$

where $n(j)$ is the number of the jth neuron's inputs and a_j is the threshold value.

In this case, examples are learned by forming a neural network and setting input weights. Modifying input weights, the number of neurons, and the neuron interconnection graph changes the number and positions of the hyperplanes that subdivide the multidimensional space into areas.

Figure 5.4 shows a schematic depiction of the capabilities of two-input networks in bounding two-dimensional space areas. A one-layer network, also called a single-layer perceptron, cannot divide into two classes the points that correspond to the 0 and 1 values of the XOR function. Two- and multilayer networks can handle this type of problem. Using network with more than two layers and with n inputs, any arbitrary Boolean function of n variables can be specified [172].

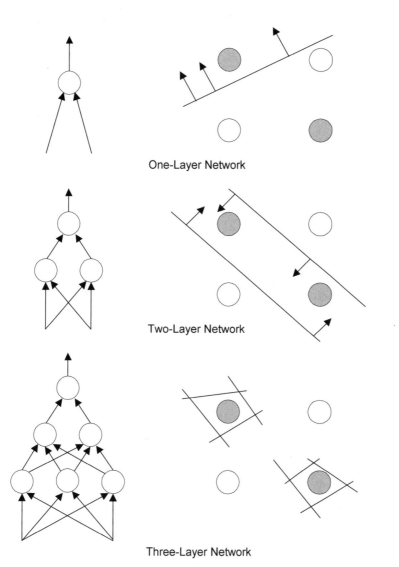

One-Layer Network

Two-Layer Network

Three-Layer Network

FIGURE 5.4 Bounding a two-dimension space by two-input neural networks.

As is demonstrated in [173–175], a two-layer neural network can approximate with any specified error level $\varepsilon > 0$ any continuous function $f(x_1, x_2, \ldots\ldots, x_n)$ defined over a bounded set:

$$f(x_1, x_2, \ldots, x_n) = \sum_{i=1}^{N} v_i \times \frac{1}{1 + e^{-(w_{i1}x_1 + w_{i2}x_2 + \cdots + w_{in}x_n)}},$$

where v_i are input weights of the second layer neurons with linear activation function; w_{ij} is the weight of the jth ($j = 1, ..., n$) input of the ith ($i = 1, ..., N$) neuron of the first layer with sigmoid activation function; N is the number of neurons in the first layer.

This type of network is called a multilayer perceptron network.

When hyperspheres are used to bound areas, each neuron defines the coordinates of the center of the hypersphere by the values of its weights and also remembers the radius of this hypersphere.

These networks are called radial basis function networks.

As can be seen, in both cases collective distributed memorizing by the neurons of the examples presented to the network takes place during training. Naturally, the diversity of neural networks is not limited to these two cases, because, for example, second- or higher order hyperplanes can be used instead of hyperplanes for bounding areas.

During functional operation, a network assigns a set of values placed onto its inputs to one or the other area, which is the result sought. Note that a set of input values presented to the network does not necessarily have to be one of those used in training the network. However, because there are collections of areas formed by other sets of input values, this set is placed into one of them. If the result is correct, the network is functioning properly; otherwise, errors were made in its construction or training. Therefore, the essence of the training or construction consists in bounding collections of points, that is each area, without either including extraneous points or losing those that belong to an area.

5.1.5 Training Algorithms of Multilayer Perceptron Networks

Most training algorithms use heuristic methods of forming networks graphs and connection weights. Training of a multilayer perceptron network starts with setting up the initial network (a heuristically selected graph) that has the number of inputs and outputs specified when the problem the network is to solve was posed. In [170], for example, it is recommended to start with a three-layer network with the number of neurons in the hidden (internal) layer equal to half the sum of the network inputs and outputs. Each neuron in the hidden layer must be connected with the outputs of the neurons of the network's input layer. Each output neuron must be connected by its inputs to the outputs of the hidden layer neurons.

Then the network is trained to adjust the weights of the input neurons to such values, at which the network solves the posed problem. If this cannot be achieved with the given network graph, then, using some heuristic methods (those based on genetic algorithms [176]), an alternative network graph is constructed and trained. This process continues until an acceptable result is obtained.

The best algorithm for supervised network training, for which process convergence has been proven, is the back propagation algorithm [168–170]. During training, the error signal propagates backward over the network. Neuron input weight values are corrected, which prevents this error from appearing again.

5.2 HARDWARE FOR INTERPRETING ALGORITHMS SPECIFIED BY NEURAL NETWORKS

5.2.1 Approaches to Implementing Neural Networks in Hardware

Currently, computation systems that interpret neural network algorithms are built using the traditional elementary building base. However, the possibility of implementing the basis operation (calculating scalar product) within the physical medium of the signal carrier looks quite promising. Primarily, this concerns performing addition in an electromagnetic field, but in living organisms, operations in other media can also be performed, addition on the biochemical level, in particular. Implementing scalar multiplication by electrical field summation (including the optical band) allows the response time of the scalar multiplier element to be exceptionally short—comparable to the time it takes light to pass through the element's physical dimension.

Algorithms specified by neural networks can be interpreted by conventional universal computers or by some specialized devices [177, 178].

Comparing hardware used to interpret neural network algorithms is difficult because of a great variety of parameters. Some of these parameters follow: the number of the neurons and connections interpreted, precision of the input/output values and input weights, and precision of the multiplier and addition circuitry (especially in analog implementation). For example, if 8-bit inputs and weights are used with a 16-bit adder, problems with the result precision arise when interpreting networks with multi-input neurons.

In the neurocomputer world, the adopted performance unit is the number of connections per second (CPS). A connection is the multiplication of the input value by its weight and adding the product to the accumulated sum.

Another index, which measures the training speed, is the connection weights update per second (CUPS).

These indexes are quite relative, because they do not take into account the widths of the processed inputs and outputs, the precision of the intermediate results, and other details.

When evaluating traditional computers using these indexes, operations to fetch operands from the memory and to organize computations must be taken into account in addition to the computation operations proper.

Analyses of conversions performed when interpreting neural network algorithms allow the following conclusions to be made:

- Some of the neural network algorithms application areas are solving ill-formalized problems for which, in the course of training or at construction, multidimensional space areas are created that are comprised of points to which always correspond the same network output values. Some of these are modeling, prediction, and recognition problems. When implementing algorithms for such types of networks, input values and weights can be represented by short numbers, and multiplication and addition operations done on fixed-point numbers. This is made possible by normalizing input values (reducing them to the [0, 1] or [−1, 1] interval) and a small, as a rule, number of values for each input. At the same time, fixed-point operations, being executed significantly faster and with lesser equipment expenditures as compared to floating-point operations, provide acceptable precision in computing locations of the bounding hyperplanes or hyperspheres.

- When solving well-formalized problems that are formulated in neural network terms, computation precision is of a substantial importance. This circumstance necessitates the use of floating-point operations on wide numbers. The self-evident goal of developing neural network algorithms for solving well-formalized tasks is to understand how to build neurocomputers and how to develop enough software for neurocomputers that are built on new physical principles and calculate scalar products rapidly. More important, however, creation of a class of parallel algorithms, constructed with new approach, is sought. In most cases, these algorithms are obtained not as a result of formal conversions of the already-existing serial algorithms, but as a result of creative discoveries. The great degree of parallelism intrinsic to these algorithms, which includes the ability to search through all possible solutions, from which only one is known to be needed, gives freedom for researching into hardware and time expenditures needed to implement calculations. It is possible that such parallel programs will execute faster in massively parallel systems implemented, for example, on a silicon wafer than programs with fewer operations but with a structure that does not provide complete loading of all computing units.

Consequently, one aspect of specializing computing equipment for execution of neural algorithms consists in raising efficiency of the exchange between memory and functioning in parallel operational units; another aspect involves reducing the execution time of multiplication with accumulation operations by shortening operands and using fixed-point operations. The former undoubtedly produces performance gains in execution of any types of algorithms; the latter, however, is specific to neurocomputers and allows building embedded systems with record values of performance/cost and performance/dimension-weight indexes. This results from an important property of neural algorithms' being able to solve problems with varying precision degree of the result.

For example, when using multilayer perceptrons, the same area can be bounded by different numbers of hyperplanes with area points being included with a different degree of accuracy or points not belonging to the area being included. Neural networks of different complexities correspond to these different sets of hyperplanes. This results in different problem-solution complexities and, consequently, different performance requirements for solving the same problem within the specified time. Therefore, such a version of implementing the neural network can be chosen that will solve the problem sufficiently well and meet the execution time parameters on the hardware available.

5.2.2 Neurochips

Neuro chips come in digital and analog varieties and hybrids of the first two. They can either have circuitry for adjusting connection weights during training or not have such circuitry and provide for loading weights from outside. The biggest problem in creating neurochips is multiplication circuitry, because it is precisely this circuitry that limits calculation speeds. Following the [178] review, first-generation neurochips are presented.

5.2.3 Digital Neurochips

One of the first commercially available neurochips was the MD1220 from Micro Devices® [179]. This chip has eight neurons and eight connections with 16-bit weights stored in the on-chip memory and one-bit inputs. The inputs have one-bit serial multipliers. The cycle is 7.2 mcs long, which provides 8.9 MCPS performance. The chip's adders also are 16-bit ones. By cascading these neurochips, neurocomputers can be built that in a single cycle interpret by hardware the combined number of neurons in the interconnected neurochips.

The NLX-420 chip [180] was designed by Neuralogix®. It has 16 processor elements (PE), each of which has a 32-bit adder. Weights and input data are loaded as 16-bit words but can be used as 16 one-bit words or as four 4-bit words, or as two 8-bit words, or as one 16-bit word. Weights are stored off-chip. All PEs have one common input, which allows up to 16 multiplication operations to be executed concurrently. After all inputs have been computed, the 16 results are multiplexed to the user-defined threshold function to calculate outputs. The neurochip has a 300 MCPS performance. Chips can be cascaded.

The Lneuro [181] chip from Philips has 16 PEs with 16-bit registers. Each PE can function as 16 one-bit, eight 2-bit, four 4-bit, two 8-bit, or one 16-bit PE. The chip has 1 KB of weight memory, which makes it possible to use 1,024 8-bit or 512 16-bit weight coefficients. The activation function is implemented off-chip, which allows large networks to be interpreted when cascading chips by assembling 32-bit sums from different chips' data. The outputs of the 16 PEs are multiplexed. The chip is oriented toward using a transputer as the master processor. The performance is 100 MCPS in the 256 one-bit PE mode and 26 MCPS in the 64 eight-bit PE mode. During training, respective performances of 160 and 32 MCUPS are achieved.

The Wafer Scale Integration chips [182] are a Hitachi® product. These semiconductor wafers house a 576-neuron Hopfield network, with each of the neurons having 64 eight-bit weight coefficients.

5.2.4 Digital Chips for Systolic and Single Instruction Stream Systems

These chips have a lower degree of specialization for neurocomputing and are mostly 16- or 32-bit microprocessors close to the conventional RISC processors.

The N64000 [183] chip from Inova® is used in the Connected Network of Adaptive Processors (CNAPS) systems. It has 80 PE, from which 64 are main and 16 reserve. Each PE has nine 16-bit multipliers and a 32-bit adder. The chip holds 4 KB of 8- or 16-bit weight memory and 32 register. All chips synchronously execute one external instruction stream.

The 100 NAP chip from Hecht-Nielson Computers® has four PEs processing 32-bit floating-point data. A PE can address 512 KB of off-chip memory and has a 160 MFLOPS performance.

The MA-16 [185] chip from Siemens® is used to build systolic networks oriented toward neural network algorithms. The MA performs matrix operations on 4 x 4 matrices with 16-bit elements. Its multipliers and adders are 48 bits wide. Weights are stored off chip and so is implemented the activation function.

The MT19003-Neural Instruction Set Processor [186] from Micro Circuit Engineering® is a RISC processor with seven instructions oriented toward neural computations. Because the instruction set is problem-oriented, programs for interpreting neural networks have a small size, which allows loading them into the neurochip. Inputs and weights are represented by 13-bit operands. The chip has a 16-bit multiplier and a 35-bit adder. Weights are stored off chip. The processing speed is 40 MCPS.

5.2.5 Radial Basis Function Neurochips

Radial basis function neural networks manipulate distances between the input vector and the memorized prototype weight vectors of the neurons' inputs.

If the distance to a prototype vector does not exceed the threshold value, then the input vector is considered to pertain to the given prototype. If the distance between the input vector and any of the prototype vectors is greater than the threshold distance, the input vector is remembered as a prototype vector. If the distance from the input vector to several prototype vectors does not exceed the threshold distance, the threshold value of these prototype vectors is lowered. Consequently, the multidimensional input space is segmented into a collection of areas defined by the prototype vectors. These networks are easy to train and simple to implement in hardware.

The ZISC036 (Zero Instruction Set Computer) chip was produced by IBM [187]. The first in a series of similar chips, this chip allows working with 36 prototypes. The fact that only the prototype values have to be specified for chip's work determined its name. The chip is easily cascaded for increasing the number of prototypes. Its vectors have 64 eight-bit elements. The threshold distance can be set to any value. The input vector is loaded serially in 3.5 mcs with the result produced in 0.5 mcs.

The Ni1000 [188] chip was designed by Intel and Nestor®. This chip is similar to the ZICS but is more powerful. It contains 1,024 256-dimension prototype vectors of 5-bit elements. At 40 MHz, 40,000 vectors per second are processed.

5.2.6 Analog Neurochips

Analog neural chips use basic physical phenomena for implementing neural network transformations. Analog elements are usually smaller and simpler than digital ones. On the other hand, the chip must be designed and manufactured carefully to provide the needed degree of precision.

The 80170NW ETANN [189] chip from Intel has 64 neurons and two banks of 64×80 weights. Several network configurations are possible. The chip has 64 analog inputs (0–3 V) and 16 internal offsets. The chip can be used to implement a two-layer network with 64 inputs, 64 hidden, and 64 output neurons with the performance of 2 GCPS. Other possible configurations include three or more layered networks or one-layer 128-input networks.

The ETANN's precision is 5–6 bits for weights and outputs. The BrainMaker® software package from California Scientific Software® is used to train the chip.

5.2.7 Hybrid Neurochips

Hybrid neurochips combine the analog and digital approaches. For example, their inputs can be analog, with the weights and outputs digital.

The CLNN-32 [190] chip from Bellcore® has 32 neurons. The inputs, outputs, and internal processing are analog, whereas the 5-bit weights are digital.

The ANNA chip [191] from AT&T® is basically a digital chip but uses internal capacitors to store weights. The chip has 4,096 weights, with the number of neurons varying from 16 to 256. Neurons have from 256 to 16 inputs for 16- and 256-input chips, respectively. One-layer networks with 64 inputs and 64 neurons achieve speeds of 2.1 GCPS.

There are neurochips, in which data are represented by the frequency or width of pulses.

5.2.8 Signal Microprocessors and Microprocessors with Multimedia Instruction-Set Extensions

Signal microprocessors are oriented toward processing data vectors, which can be used in interpreting neural network algorithms [177] for multiplying with accumulation weight and input vectors of the network's neuron inputs. The same factor is responsible for efficient interpretation of networks by microprocessors with MMX or other multimedia instruction-set extensions.

It can be noted that whereas neural network algorithms have been developing rapidly, no new digital neurochips have been introduced for the past several years. Apparently, this can be explained by the fact that all problem-orientation specifics of the first-generation neurochips have been adequately taken into account by the multimedia instruction set extensions of such microprocessors as, for example, Pentium III and Motorola's Power PC AltiVec. The high clock frequency and the capability to execute batches of integer instructions in a single cycle make these processors an almost ideal means for neural network interpretations. An important aspect of using microprocessors for this purpose is their capability to pre-process data and process the results of neural network operations. However, this is true only for digital neurochips; problem-oriented analog neurochips continue to be introduced.

5.3 NEUROMATRIX NM6403 NEUROPROCESSOR

5.3.1 Main Architectural Characteristics

The NM6403 [193–195] neuroprocessor was designed at the Russian science and technology Modul [191] center. It uses a scalar processor (a scalar RISC core) for executing logic, integer arithmetic, and shift operations and for generating memory access addresses; it also has a vector processor for processing binary arbitrary width vectors within 1 to 64 bit limits. In this respect, in a single cycle the vector processor can process up to 64 vectors, whose combined length does not exceed 64 bits. The scalar processor does all the work preparing data for the vector processor. As a whole, the NM6403 neuroprocessor is classified as a decoupled architecture processor.

The neuroprocessor has two identical interfaces to work with the memory: one interfaces the global memory and the other interfaces the local memory. Each interface can address 2 GB of 32-bit words. Memory exchanges are carried over a 64-bit data bus with two adjacent words fetched in one memory access.

The NM6403 neuroprocessor has two on-chip links that, in terms of the logical and physical protocols, are compatible with the links of the TMS320C4X signal microprocessor. Moreover, the local and global memory interfaces of the neuroprocessor have on-chip arbitration circuitry, which allows interfaces of two NM6403s to be connected to one shared memory block. The availability of two links and two shared memory interfaces makes it possible to compose a wide range of parallel structures, some of which are shown in Figure 5.5.

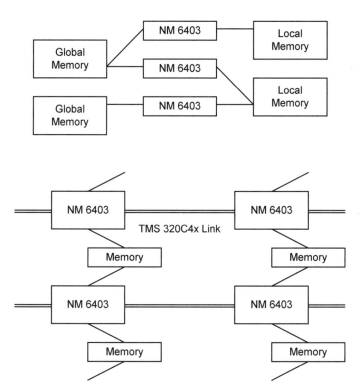

FIGURE 5.5 Examples of implementing parallel structures using the NM6403 neurochip.

5.3.2 NM6403 Instruction Set

Neuroprocessor's instructions are of two main types: scalar and vector processor instructions. The scalar instructions are executed in a single cycle. The vector processor instructions take from one to 32 cycles to execute.

The scalar instructions have a 32- or 64-bit format. In the latter case, the second 32-bit word specifies a constant, a branch address, or a branch address offset. The scalar instructions have three fields—OPER, MOVE, and P—and specify execution of two instructions. The functions of the instruction fields follow:

- The 16-bit OPER [0:15] field specifies one of the arithmetic, logic, or shift operations on the contents of the general-purpose registers.
- The 15-bit MOVE [16:30] field specifies the following:
  - Modifications of the address registers' contents
  - Register/register, register/memory, memory/register, and constant/register type moves of constants and register contents
  - Modifications of the program counter to control branches when executing conditional and unconditional branches, subroutine calls, and returns from subroutines and interrupts
- The one-bit P [31] field enables execution of the given instruction against the background of the incomplete previous vector instruction.

The scalar processor has the following 32-bit registers:

- Eight address registers AR0–AR7 that are used in memory accesses. AR7 is used as the stack pointer
- Eight general-purpose register GR0–GR7
- Program counter
- Program Status Word Register containing information about the processor state, flags set by the executed instructions, processor resources, and the current values of the interrupt masks

The vector processor instructions have 32-bit width, specify four operations, and consist of the following six fields:

- The one-bit L [0] field specifies move of WBUF to WOPER
- The 12-bit VOPER [1:12] field specifies arithmetic or logic operations on packed data vectors
- The 5-bit COUNT [13:17] field specifies the number of times the instruction is to be executed; this field is not used in the WFIFO to WBUF move, which always executes in 32 cycles; moreover, it must be noted that if a WBUF

to WOPER move is specified for a repeated instruction execution, this move is performed only once: when the instruction is executed the last time

■ The one-bit W [18] specifies a move from WFIFO to WBUF

■ The 12-bit VMOVE [19:30] field specifies external memory exchange through a read/write operation on a packed data vector

■ The one-bit P [31] field enables execution of the given instruction against the background of incomplete previous vector processor instruction

How the vector processor operates will not be considered here. Those readers who are interested in this subject can consult site [192] and works [193, 194]. Some basic explanations need to be provided, however. The WBUF and WOPER are 32 × 64 bit memory cell arrays that store weight coefficients in 64-bit words. Moving WBUF to WOPER takes one cycle.

The WFIFO is a dual-ported FIFO buffer 32 × 64 bits large. It is used for swapping 64-bit packed data words from the memory. A move from WFIFO to WBUF takes 32 cycles and is performed against the background of other operations performed by the processor using the WOPER.

Consequently, vector processor idle time for data waits is eliminated by the following:

■ The WFIFO buffer, filled at the rate allowed by the memory

■ Shadow memory WBUF

■ The WOPER memory, which is used directly by the processing unit of the vector processor in performing operations

5.3.3 NM6403 Neuroprocessor Performance

The NM6403 neuroprocessor was manufactured on 0.5 micron CMOS technology and has 50 MHz operating frequency. The architectural relatedness of the MN6403 scalar processor and Kvant series microprocessors, considered in Chapter 4, allows a conclusion to be made that the performance of the former is 60–75 MOPS (five times that of the Kvant-10, as prorated to the NM6403 higher clock frequency).

In Viksne [194], an equation is given for determining NM6403 performance when executing neural network algorithms with varying width of neuron weights and inputs. According to this equation, NM6403 performance is given as follows:

■ 50 MCPS with 32-bit weights and inputs

■ 51.2 GCPS with one-bit weights and inputs

REVIEW QUESTIONS TO CHAPTER 5

1. What class of problems does neural network algorithms efficiently deal with?

2. How is the memorization of training examples by neural networks organized?

3. How must a task be formulated for solving using neural network algorithms?

4. Give definitions of the concepts *neuron* and *neural network.*

5. What are neuron state and activation functions? Give examples of state functions for perceptron networks and radial basis function networks.

6. Of what does the neural network training process consist?

7. Express the main ideas of training multilayer perceptron networks using the back propagation algorithm.

8. State the main principles of training radial basis function networks.

9. What indexes are used to measure neural network performance?

10. How can performances of the traditional von Neumann computers and neural networks be compared?

11. What main specialization directions of computing means are used to raise the performance level of interpreting neural network algorithms?

12. What modifications of neural network algorithms make it possible to achieve record values of such indexes as the performance/cost, performance/equipment amount, and so on?

13. Enumerate the main types of neural microchips.

14. State the signal processor features that raise the performance levels of interpreting neural network algorithms.

ENDNOTES

167. Barron, A. Universal Approximation Bounds for Superposition of Sigmoid Functions, IEEE Transactions Information Theory. 1993, Vol. 39.

168. Wasserman, F. Neurocomputer Hardware. Moscow: *Mir,* 1992.

169. Jain, A., Mohiuddin, J. Artificial Neural Networks: A Tutorial, *Computer.* 1996. No. 3, pp. 31–44.

170. Lawrence, J. *Introduction in Neural Networks: Design, Theory and Applications.* California Scientific Software, 1994, p. 423.

171. Hopfield, J. Neural Networks and Physical Systems with Emergent Collective Computational Abilities. Proc. Nat. Acad. Sci. USA, Vol. 79, 1982, pp. 2554–58.

172. Gavrilkevich, M. *Introduction into Neural Mathematics. Review of Applied and Industrial Mathematics.* Moscow: TVP, 1994, pp. 377–88.

173. Hornick, K., Stinchcombe, M., White, H. Multilayer Feedforward Networks are Universal Approximators, *Neural Networks.* 1989, Vol. 2, No. 5, pp. 359–66.

174. Cybenko, G. Approximation by Superpositions of a Sigmoidal Function, *Mathematics of Control, Signals and Systems.* 1989, No. 2, pp. 303–14.

175. Funahashi, K. On the Approximate Realization of Continuous Mappings by Neural Networks, *Neural Networks,* 1989, Vol. 2, No. 3, pp. 183–92.

176. Evolution Computing and Genetic Algorithms. *Applied and Industrial Mathematics Review.* Moscow: TVP, Volume 3, Issue 5, 1996.

177. Marguerat, C. Artificial Neural Network Algorithms on a Parallel DSP System. In: Transputers'94 Advanced Research and Industrial Applications. Proc. of the International conf. 21–23 Sept. 1994. IOS Press 1994, pp. 278–87.

178. Lindsey, C., Lindblad, T. Survey of Neural Networks Hardware, *SPIE,* Vol. 2492, pp. 1194–1205.

179. *MD1220 Data Sheet.* March 1990, Micro Devices, 30 Skyline Dr., Lake Mary, FL 32746-6201, USA.

180. *NLX420 Data Sheet.* June 1992, Neurologix, Inc., 800 Charcot Ave., Suite 112, San Jose, CA. USA.

181. Mauduit, N., Duranton, M., Gobert, J. Lneuro 1.0: A Piece of Hardware LEGO for Building Neural Network Systems, IEEE Trans. on Neural Networks. 1992, Vol. 3, pp. 414–22.

182. Yasunga, M., Msuda, N., Yagyu, M., Asai, M., Yamada, M., Masaki, A. Design, Fabrication and Evaluation of a 5-Inch Wafer Scale Neural Network LSI Composed of 576 Digital Neurons, Proceedings International Joint Conference on Neural Networks. IJCNN'90, June 1990.

183. Hammerstrom, D. A VLSI Architecture for High-Performance, Low-Cost, On-chip Learning, Proceedings International Joint Conference on Neural Networks. IJCNN'90, June 1990.

184. Means, R., Lissenbee, L. Extensible Linear Floating Point SIMD Neurocomputer Array Processor, Proceedings International Joint Conference on Neural Networks. IJCNN'91, July 1991.

185. Beichter, J., Bruels, N., Meister, E., Ramacher, U., Klar, H. Design of General-purpose Neural Signal Processor. Proceedings of the 2nd International Conferance on Microelectronics for Neural Networks, Munich, Germany, Oct. 1991.

186. *MT19003 Data Sheet.* May 1994, Micro Circuit Engineering, Alexander Way, Tewkesbury, Gloucestershire GL20 GTB.

187. LeBouquin, J-P. IBM Microelectronics ZISC, Zero Instruction Set Computer, Proceedings of the Word Congress on Neural Networks, Supplement, San Diego, 1994.

188. Holler, M., Park, C., Diamond, J., Santoni, U., The, S., Glier, M., Scofield, C., Nunez, L. A High Performance Adaptive Classifier Using Radial Basis Function. Procedings of Government Microcircuit Application Conference, Las Vegas, Nevada, USA, Nov. 1999.

189. *80170NX Electrically Trainable Analog Neural Network*, Data Sheet, Intel Corp., Santa Clara, CA, 1991.

190. Alspector, J., Jayakumar, T., Luna, S. Experimental Evaluation of Learning in a Neural Microsystem, Proceedings of NIPS'91 in Advances in Neural Information Processing Systems-4, pp. 871–78, Morgan-Kaufmann Pub., San Mateo, CA, 1992.

191. Boser, B., Sackinger, E., Bromley, J., LeCun, Y., Jackel, L. Hardware Requirements for Neural Network Pattern Classifiers, *IEEE Micro.* 1992, No. 2, pp. 32–40.

192. *www.module.ru.*

193. Viksne, P., Fomin, D., Chernikov, V. A Variable Operand Length Single-Chip Digital Neuroprocessor. College Proceedings. *Instrument-Making.* 1996, Volume 366, Issue 7, pp. 13–21.

194. Viksne, P., Fomin, D., Chernikov, V., Shevchenko, P. Architectural Features of NM6403 Neuroprocessor. Report Collection of the 5th All-Russia Conference Neurocomputers and their Applications. Moscow, 17–19 February 1999.

195. Viksne, P., Fomin, D., Chernikov, V., Shevchenko, P. Using NM6403 Microprocessor for Neural Network Emulation. Report Collection of the 5th All-Russia Summit Neurocomputers and Their Applications. Moscow, 17–19 February 1999.

Index